T0315356

Chirality in Transition Metal Chemistry

Inorganic Chemistry

A Wiley Series of Advanced Textbooks
ISSN: 1939-5175

Editorial Board

Derek Woollins, University of St. Andrews, UK
Bob Crabtree, Yale University, USA
David Atwood, University of Kentucky, USA
Gerd Meyer, University of Hannover, Germany

Previously Published Books in this Series

Bioinorganic Vanadium Chemistry
Dieter Rehder
ISBN: 978-0-470-06516-7

Inorganic Structural Chemistry, Second Edition
Ulrich Müller
ISBN: 978-0-470-01865-1

Lanthanide and Actinide Chemistry
Simon Cotton
ISBN: 978-0-470-01006-8

Mass Spectrometry of Inorganic and Organometallic Compounds: Tools – Techniques – Tips
William Henderson & J. Scott McIndoe
ISBN: 978-0-470-85016-9

Main Group Chemistry, Second Edition
A. G. Massey
ISBN: 978-0-471-49039-5

Synthesis of Organometallic Compounds: A Practical Guide
Sanshiro Komiya
ISBN: 978-0-471-97195-5

Chemical Bonds: A Dialog
Jeremy Burdett
ISBN: 978-0-471-97130-6

Molecular Chemistry of the Transition Elements: An Introductory Course
François Mathey & Alain Sevin
ISBN: 978-0-471-95687-7

Stereochemistry of Coordination Chemistry
Alexander Von Zelewsky
ISBN: 978-0-471-95599-3

Bioinorganic Chemistry: Inorganic Elements in the Chemistry of Life – An Introduction and Guide
Wolfgang Kaim
ISBN: 978-0-471-94369-3

For more information on this series see: http://eu.wiley.com/WileyCDA/Section/id-302900.html

Chirality in Transition Metal Chemistry: Molecules, Supramolecular Assemblies and Materials

Hani Amouri and Michel Gruselle

Institut Parisien de Chimie Moléculaire (IPCM), Université Pierre et Marie Curie, Paris-6, France

A John Wiley and Sons, Ltd, Publication

This edition first published 2008.
© 2008 John Wiley & Sons Ltd

Registered office
John Wiley & Sons Ltd, The Atrium, Southern Gate, Chichester, West Sussex, PO19 8SQ, United Kingdom

For details of our global editorial offices, for customer services and for information about how to apply for permission to reuse the copyright material in this book please see our website at www.wiley.com.

The right of the author to be identified as the author of this work has been asserted in accordance with the Copyright, Designs and Patents Act 1988.

All rights reserved. No part of this publication may be reproduced, stored in a retrieval system, or transmitted, in any form or by any means, electronic, mechanical, photocopying, recording or otherwise, except as permitted by the UK Copyright, Designs and Patents Act 1988, without the prior permission of the publisher.

Wiley also publishes its books in a variety of electronic formats. Some content that appears in print may not be available in electronic books.

Designations used by companies to distinguish their products are often claimed as trademarks. All brand names and product names used in this book are trade names, service marks, trademarks or registered trademarks of their respective owners. The publisher is not associated with any product or vendor mentioned in this book. This publication is designed to provide accurate and authoritative information in regard to the subject matter covered. It is sold on the understanding that the publisher is not engaged in rendering professional services. If professional advice or other expert assistance is required, the services of a competent professional should be sought.

The publisher and the author make no representations or warranties with respect to the accuracy or completeness of the contents of this work and specifically disclaim all warranties, including without limitation any implied warranties of fitness for a particular purpose. This work is sold with the understanding that the publisher is not engaged in rendering professional services. The advice and strategies contained herein may not be suitable for every situation. In view of ongoing research, equipment modifications, changes in governmental regulations, and the constant flow of information relating to the use of experimental reagents, equipment, and devices, the reader is urged to review and evaluate the information provided in the package insert or instructions for each chemical, piece of equipment, reagent, or device for, among other things, any changes in the instructions or indication of usage and for added warnings and precautions. The fact that an organization or Website is referred to in this work as a citation and/or a potential source of further information does not mean that the author or the publisher endorses the information the organization or Website may provide or recommendations it may make. Further, readers should be aware that Internet Websites listed in this work may have changed or disappeared between when this work was written and when it is read. No warranty may be created or extended by any promotional statements for this work. Neither the publisher nor the author shall be liable for any damages arising herefrom.

The cover picture represents a chiral two-bladed propeller cobalt complex $[(Co_2(CO)_4\mu,\eta^2,\eta^2\text{-}(-H_2CC \equiv CCH_2-)(\text{-}dppm)_2][BF_4]_2$ (drawn by H. Amouri). This complex was prepared and resolved by the authors using a chiral auxiliary anion. (Note: the carbonyl groups on the cobalt were removed for clarity.)

Library of Congress Cataloging-in-Publication Data

Amouri, Hani.
  Chirality in transition metal chemistry : molecules, supramolecular assemblies and materials /
Hani Amouri, Michel Gruselle.
    p. cm.
  Includes bibliographical references and index.
  ISBN 978-0-470-06053-7 (cloth) – ISBN 978-0-470-06054-4 (pbk.) 1. Transition metal
compounds. 2. Chirality. 3. Chemistry, Inorganic. I. Gruselle, Michel. II. Title.
  QD172.T6A44 2008
  546'.6–dc22   2008031731

A catalogue record for this book is available from the British Library.

ISBN 9780470060537 (Cloth) 9780470060544 (Paper)

Typeset in 10/12pt Times by Thomson Digital Noida, India.

Contents

Preface

This book deals with chirality in transition metal chemistry. Chirality is an ever-fascinating topic and is a field which occurs in various subjects of modern chemistry. Surprisingly, a check of the literature for textbooks on the topic yielded a series of titles that only partially covered the subject. To our knowledge *Stereochemistry of Coordination Compounds,* by Professor Alex von Zelewsky, is the only recent book which gives a specialized treatment of this subject and was published by John Wiley & Sons, Ltd, in 1996. In the last decade a great number of papers have been reported in the literature describing chirality in different fields of chemistry: organometallics, catalysis, coordination chemistry, supramolecular assemblies and nanomaterials. Therefore a new book devoted to this topic will no doubt provide a very interesting tool for students and researchers who are interested to this increasingly important field. We feel that a complementary book, *Chirality in Transition Metal Chemistry: Molecules, Supramolecular Assemblies and Materials,* which covers new discoveries in the field since 1996, and establishes the interesting connectivity between the various aspects of chirality in different fields of chemistry, will be a valuable work to advance research in this area. These different subjects are treated in Chapters 2 to 6 of this book. Chapter 1 represents the introduction and summary of the book. We have attempted to cite examples and key references which cover the full breadth of the field in all its diversity but, bearing in mind that the selection of references necessary for a textbook is always arbitrary to a certain degree, we apologize in advance to the many researchers whose work could not be included. No book is free from errors and this one will be no exception, so we would appreciate it if colleagues around the world would point out errors and mistakes and suggest any improvements. We hope this book will serve as a guide for students and researchers working in this area.

This work could not have been accomplished without the help, comments and advice of many people who agreed to read the chapters and contributed to improving the quality of the book and to whom we are very grateful. Thus we thank Professors Peter J. Stang (University of Utah) and Gérard Coquerel (University of Rouen) for reviewing Chapter 2. We also thank Professors Henri Brunner 'Emeritus' (University of Regensburg), John A. Gladysz (University AM-Texas) and Antonio Togni (University of Zurich-ETH) for Chapter 3. We are grateful to Professor Jacqueline K. Barton (California Institute of Technology) for Chapter 4. We also thank Professors Kenneth N. Raymond (UC-Berkeley) and Rolf W. Saalfrank (University of Erlangen) for reviewing Chapter 5. We are grateful to Professors Vincent L. Pecoraro (University of Michigan) and Gérard Coquerel (University of Rouen) for Chapter 6. We thank Debra and Simon Greenfield for translating Chapters 2 and 6, which were originally written in French by Dr Michel Gruselle, into English. Chapters 3, 4 and 5 were written in English by Dr Hani Amouri, while the introduction, (Chapter 1) was written by both authors.

We also thank the Wiley team at Chichester, West Sussex: Paul Deards, Richard Davies, Rebecca Ralf, and their colleagues for improving the quality and the presentation of the book and for the production, as well as Lyn Hicks for the copy-editing process, especially when the authors were writing in English, which is not our first language.

We are indebted to Professor Alex von Zelewsky who agreed to read the whole document and to write the Foreword to this book; these few words are not enough to express our gratitude and our immense respect to him. Finally, we would like to thank our life partners Jeannine and Anita for their support and patience throughout this period of writing.

H. Amouri
M. Gruselle
Paris, July 2008

Foreword

One hundred years before the present book was written by two colleagues in Paris, a giant of modern science Alfred Werner, who originated from Mulhouse (now in France), was heading a group of students from several countries at Zurich University. He was obsessed by the idea that coordination compounds should be able to exhibit optical activity, that is rotate the plane of polarized light in a way similar to purely organic compounds, as had been known for decades. Until 1893 stereochemistry was completely dominated by the ideas of Le Bel and van't Hoff already published in 1874. That meant limitation to a tetrahedral arrangement of neighbours around an atomic centre in a molecule. Werner's coordination theory of 1893, which generalized the arrangements of atoms in space, notably to octahedral but also to square planar geometries, opened up new possibilities for isomerism. The proof for the correctness of these ideas was consequently largely based on stereochemical arguments. In 1899 Werner realized that the ultimate confirmation for the hypothesis of octahedral coordination would be the experimental demonstration of optical activity in a compound which did not contain any 'asymmetric' carbon atoms. It was not until 1911 that L. King in Werner's laboratory showed that, indeed, this could be achieved with a certain cobalt complex, as described in Chapter 2. It is known that Werner lived his relatively short life according to the principle of 'hard work followed by great feasts' and the story tells that an exuberant party was celebrated in the laboratory when King reported the result to his boss.

In the twentieth century organic stereochemistry continued to develop at a rapid pace reaching a high degree of sophistication, especially in view of the synthesis of natural compounds, which often comprise a large number of stereogenic centres. Highly stereoselective reactions were designed and correct absolute configurations were often obtained starting from simple 'chiral pool' natural products. In the past almost 40 years enantioselective catalysis has been added to the arsenal of methods yielding many enantiopure compounds, especially those for pharmaceutical applications. The field of chiral drugs has become a multibillion euro business since the importance of the correct absolute configuration of compounds that interact with the chiral molecular environment in our body was realized some decades ago. It is not an exaggeration to say that 'chirality' has become one of the most central concepts in molecular sciences – a Google search yields 3.5 million hits for this entry.

Despite these achievements in organic chemistry, stereochemistry of coordination compounds found much less attention. Textbooks of organic stereochemistry still do not generally cover coordination numbers higher than four, as realized in most carbon compounds. The stereochemistry of coordination compounds is basically a much more complex subject, due to the high number of possible arrangements of atoms in molecules containing one or several centres of higher coordination numbers. However, from a general point of view, all stereochemistry is based on the same principle: the geometrical properties of three-dimensional space. Today stereochemistry should no longer be strictly divided into the traditional branches of chemistry. In many cases stereochemical problems require considerations that combine principles from purely organic chemistry with those from coordination chemistry. This is especially true in enantioselective catalysis, where the active centre is often a metal, but also in organometallic chemistry.

It is the merit of the present book to treat chirality in a unified manner and to contribute in this way to bridging the gap between organic and coordination chemistry. One hundred years after Alfred Werner's fundamental work, *Chirality in Transition Metal Chemistry: Molecules, Supramolecular Assemblies and Material* is the definitive text about this important subject and it will serve as an introduction for students and as a reference for researchers for a long time to come.

Alex von Zelewsky

Hani, Haniel Amouri

Hani, Haniel Amouri, was born in Anapolis Goias (Brazil) and obtained his Ph.D. degree (1987) in chemistry from Université Louis Pasteur Strasbourg (France), with Professor John A. Osborn, on the subject of homogeneous catalysis (hydrogenation). In 1988 he spent one year at Gif-sur-Yvette (France) as a post-doctoral fellow with Dr Hugh Felkin where he studied C-H activation of saturated hydrocarbon with transition metal polyhydrides. In 1992–1993 he spent one year at UC-Berkeley (USA) with Professor K. Peter C. Vollhardt and was working on the synthesis of oligocyclopentadienyl metal complexes and their behaviour as electron transfer reagents. He is a Research Director in CNRS and is currently the director of the 'ARC' group (*Auto-assemblage, Reconnaissance et Chiralité*) of the IPCM at Université Pierre et Marie Curie Paris-6. His main research interests are chirality, organometallic and coordination chemistry, and he has had over 90 research papers and reviews published in international scientific journals.

Michel Gruselle

Michel Gruselle was born in Decazeville (France) and obtained his Ph.D. degree (doctorat d'Etat) in the CNRS laboratory of Thiais, a suburb of Paris, in 1975 with Dr Daniel Lefort where he worked on stereochemical problems in radical chemistry. In 1974 he joined Bianca Tchoubar's group and started working on nitrogen activation by organometallic complexes, and he spent some time collaborating with Prof. A. E. Shilov in Moscow. He is a Research Director in CNRS at Université Pierre et Marie Curie Paris-6 and was the director of the ARC group (*Auto-assemblage, Reconnaissance et Chiralité*) at the IPCM from 1996–2000. His main research interests are enantioselective synthesis in coordination chemistry and in material science and he has had over 110 research papers and reviews published in international scientific journals.

1 Introduction

Pasteur in his laboratory in Arbois (1864) by Callot, reproduced with the permission of the 'Société des Amis de Pasteur' (Dole France), from J. Jacques, La Molécule et son double, Hachette, Paris, 1992

'C'est se tromper entièrement que de croire qu'on fait de la dissymétrie quand on produit des racémiques'

'One is completely mistaken in believing that dissymmetry is created when racemates are produced'

Louis Pasteur[1]

This statement of Pasteur is free of any ambiguity, it is fundamental. Its interpretation simply implies that in order to create dissymmetry it is therefore necessary to separate physically the object in question from its mirror image; an object and image which, in the case of a chiral molecule, are by definition nonsuperimposable enantiomers and whose mutual association forms a racemate. Thus the goal of the chemist is not simply to record the geometric property of a molecular structure, but physically to obtain the possible enantiomers that result from this geometry. This clarification is necessary due to the

Chirality in Transition Metal Chemistry: Molecules, Supramolecular Assemblies and Materials   H. Amouri and M. Gruselle
© 2008 John Wiley & Sons, Ltd

frequent confusion caused by the incorrect use of the concept of chirality, in particular the tendency to use it in place of the physical reality of the molecule. It is in this way that the absolute configuration, which unambiguously describes the geometry of a chiral molecule, is confused with chirality itself. The definition of the latter, which we will clarify in Chapter 2, concerns only the molecular symmetry and says nothing about the physical nature of the molecule under consideration. Is it a racemic derivative or an enantiomer? This question is essential, since the properties of a racemic or enantiomeric chiral molecule differ when placed in an asymmetric environment, whether this environment is produced by another chiral molecule or by polarized electromagnetic radiation.

One of the distinguishing features between a racemic compound and its enantiomers lies in their respective crystal organization, which are different for a racemate and an enantiomer, and which lead to specific properties for each.

The creation of dissymmetry at a molecular level, that is the preparation of enantiopure compounds, inevitably requires either the resolution of racemic compounds or asymmetric synthesis, so we shall pay particular attention to the problems associated with these two methods. The formation of enantiopure crystals by spontaneous resolution during crystallization, whether the crystals are separable or not, is of undeniable theoretical and practical importance to the technique of resolution by entrainment. In general, however, we will not consider this in itself as a technique for resolution.[2]

The importance of the topic of chirality is reflected by the large number of books, reviews and articles devoted to it. However, most of these books treat the stereochemistry in organic compounds.[3–8] Surprisingly the only recent book devoted to this topic in coordination compounds was published in 1996 by A. von Zelewsky[9] and therefore a new book is needed to update the recent evolution of this field. Our book is a continuation of von Zelewsky's book and is devoted to chirality in organometallic and coordination complexes as well as an introduction to chirality in molecular material science. In addition to the theoretical problems[10] raised by molecular asymmetry, in particular those relating to the origins of chirality in nature, the production of enantiopure molecules is an important issue in the fields of pharmaceutics and agriculture. Thus, the US Food and Drug Administration (FDA's policy statement for the development of new diastereomeric drugs, www.fda.gov./cder/guidance/stereo.htm) recommends that in the case of chiral molecules, the properties – particularly toxicity – of both enantiomers should be studied, which implies the total resolution of the racemates.

However, from the point of view of chirality, what is the specific role of organometallic and coordination chemistry? Is it merely a question of showing that the same general rules apply, or are there indeed concrete applications significant enough to stimulate research in this field?

In organic and organometallic chemistry, asymmetric catalysts play an important role in enantioselective synthesis, and the majority of these are mononuclear organometallic compounds. It was for their work on asymmetric catalysis that W. S. Knowles, R. Noyori and K. B. Sharpless won the Nobel Prize for Chemistry in 2001,[11–13] and hence the study of chirality for this class of compounds is of the utmost importance. For this reason, following on from Chapter 2, which lays the necessary groundwork in organometallic and coordination chemistry, in Chapter 3 we will develop the study of the chemistry of chiral mononuclear organometallic complexes and asymmetric catalysis in brief.

We will focus in particular on those complexes where the chiral element is the metallic centre itself and also on those bearing a chiral ligand. Many of these compounds are asymmetric catalysts, thus enabling the chemist to make use of them in order to produce a large amount of enantiopure active molecules. In these reactions we encounter the phenomenon of 'chiral recognition', which is just as essential for synthesis as for the separation of enantiomers, but what exactly does this concept mean? It can perhaps be defined as follows: there is chiral recognition when a chiral compound of defined stereochemistry interacts differently with the stereoisomers of another chiral compound. Here it is, in fact, a matter of there being a distinction between diastereomeric forms that are more or less strongly associated via hydrogen bonds, or from electrostatic or van der Waals interactions between chiral molecules, but also from steric and/or electronic repulsion. These interactions may also result from an association between a cationic complex and a chiral anion to form an ion pair. The latter has been highlighted as the 'chiral anion strategy' and has a profound effect in the area of enantioselective catalysis and also in chiral discrimination in coordination chemistry and is evolving rapidly. In particular the interaction between cationic transition metal complexes and the optically pure chiral anions such as Δ-Trisphat and (*R*-)- or (*S*)-BNP anions (Figure 1.1).[14,15] Finally a brief introduction on chiral recognition between such cationic octahedral complexes and biological molecules, particularly DNA[16] will be discussed in Chapter 4.

While mononuclear complexes and those of low nuclearity represent an important part of organometallic and coordination chemistry, those of higher nuclearity obtained through self-assembly currently form an extremely active area, defined as metallosupra-molecular chemistry. We will explore the self-assembly of building blocks of metal ions and bridging ligands, leading to various types of chiral supramolecular structures – an important field in which J. M. Lehn won the Nobel Prize for Chemistry in 1987.[17,18] We will attempt to unravel the forces that guide and control the chiral information in these self-assemblies leading to enantiopure structures. The transfer of the chiral information that is contained in the constituent bricks of the resulting supramolecular structure is, in general, under thermodynamic control. As Lehn has already remarked, this leads to great regio- and stereoselectivity for these reactions. We will see that the same is also true for the formation of enantiopure supramolecular compounds. While many of the parameters relating to the transfer of chiral information remain poorly defined, thus occasionally leading to unexpected results, rational synthesis is nevertheless progressively replacing

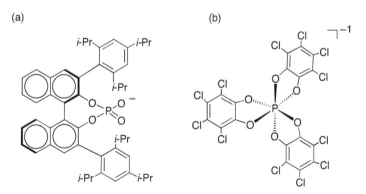

Figure 1.1
Gold asymmetric catalysis (Chapter 3) (a) (*R*)-BNP. Chiral recognition (Chapter 4) (a) (*R*)-BNP and (b) Δ-Trisphat.

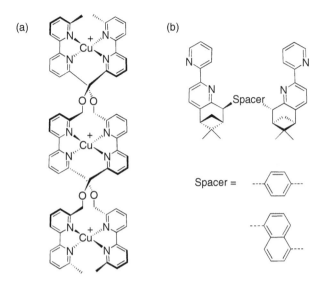

Figure 1.2
Chirality in supramolecular coordination chemistry (Chapter 5). (a) Double-stranded helicates;[19] (b) chiragen ligands.[20]

the role of chance in these matters. This theme will be addressed and developed in Chapter 5 (Figure 1.2).

The process of self-assembly not only occurring in solution but also on metallic surfaces leads to supramolecular structures with specific properties which lie at the root of what we term 'molecular materials' or nanometer scale materials. Those that will be described here are enantiopure, and we will explore their principal features. This chemistry is in its infancy, and while concrete results in terms of properties are still modest, the rapid growth in these chiral materials is such that their applications are becoming promising. An appraisal of the chemistry of chiral molecular materials is the subject of Chapter 6 (Figure 1.3).

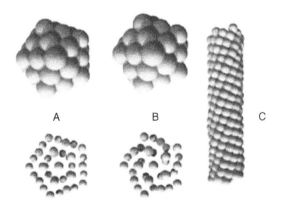

Figure 1.3
Achiral and chiral structural motifs for nanoscale metal structures: A represents an achiral 39-atom structure; B represents the chiral 39-atom structure of D5 symmetry; C illustrates a compact helical nanowire (Chapter 6) (reproduced with permission from reference[21], copyright American Chemical Society).

One of the important aspects of this chemistry is the ability of the materials to be multifunctional. From this, multifunctionality properties can emerge that are more than the simple addition of the individual effects.

To close this brief introduction, we hope this book will be useful to readers not currently working in chirality, as well as informative to those who are more familiar with this subject. We would also expect to stimulate further interest and experiments in order to push the horizon of the chirality field further since Pasteur's and Werner's first discoveries up to the present day and beyond.

References

[1] J. Jacques, *La Molécule et son Double*, Hachette, Paris, **1992**.
[2] J. Jacques, A. Collet and S. H. Wilen, *Enantiomers, Racemates and Resolution*, John Wiley & Sons, Inc., New York, **1981**.
[3] K. Mislow, *Introduction to Stereochemistry*, **1965**, W. A. Benjamin, New York.
[4] K. Mislow, *Introduction to Stereochemistry*, 2nd edn, **1978**, W. A. Benjamin, New York.
[5] H. B. Kagan, *La Stéréochimie Organique*, Presses Universitaires de France, Paris, **1975**.
[6] E. L. Eliel, S. H. Wilen and M. P. Doyle, *Basic Organic Stereochemistry*, John Wiley & Sons, Ltd, Chichester, **2001**.
[7] E. L. Eliel, S. H. Wilen and L. N. Mander, *Stereochemistry of Organic Compounds*, John Wiley & Sons, Inc., New York, **1994**.
[8] A. Collet, J. Crassous, J.-P. Dutasta and L. Guy, *Molécules Chirales: Stéréochimie et Propriétés*, EDP Sciences/CNRS Editions, Les Ulis, Paris, **2006**.
[9] A. von Zelewsky, *Stereochemistry of Coordination Compounds*, John Wiley and Sons, Ltd, Chichester, **1996**.
[10] V. I. Sokolov, *Introduction to Theoretical Stereochemistry*, **1991**, Gordon and Breach, New York.
[11] W. S. Knowles, *Angew. Chem. Int. Ed.* **2002**, *41*, 1998.
[12] R. Noyori, *Angew. Chem. Int. Ed.* **2002**, *41*, 2008.
[13] K. B. Sharpless, *Angew. Chem. Int. Ed.* **2002**, *41*, 2024.
[14] J. Lacour and R. Frantz, *Org. Biomol. Chem.* **2005**, *3*, 15.
[15] G. L. Hamilton, E. J. Kang, M. Mba and F. D. Toste, *Science* **2007**, *317*, 496.
[16] K. E. Erkkila, D. T. Odom and J. K. Barton, *Chem. Rev.* **1999**, *99*, 2777.
[17] J. M. Lehn, *Angew. Chem. Int. Ed.* **1988**, *100*, 91.
[18] J. M. Lehn, *Supramolecular Chemistry: Concepts and Perspectives*, **1995**, Wiley-VCH Verlag GmbH, Weinheim, Germany.
[19] J. M. Lehn, A. Rigault, J. Siegel, J. Harrowfield, B. Chevrier and D. Moras, *Proc. Natl. Acad. Sci. USA* **1987**, *84*, 2565.
[20] A. von Zelewsky, *Coord. Chem. Rev.* **1999**, *190–192*, 811.
[21] T. G. Schaaff and R. L. Whetten, *J. Phys. Chem. B* **2000**, *104*, 2630.

2 Chirality and Enantiomers

2.1 Chirality

2.1.1 Brief Historical Review

2.1.1.1 Alfred Werner and the First Resolved Cobalt (III) Coordination Complex

It is fitting to begin this brief historical overview of chirality in organometallic and coordination chemistry with the name of Alfred Werner (1866–1919) who, as far back as 1893, applied van't Hoff and Le Bel's stereochemical ideas of the tetrahedral nature of the carbon atom to the structure of hexacoordinated metal complexes.[1] He established their octahedral structure and predicted that some could exist in an enantiomeric form with the power of optical rotation. This prediction was followed in 1911 by the resolution of the two enantiomers of the complexes $[Co^{III}(en)_2(NH_3)X]X_2$ (X = Cl, Br) **(2.1)-X$_2$** (en = ethylene diamine) (Figure 2.1).[2,3] This overall work won him the Nobel Prize for Chemistry in 1913, following which he then went on to resolve the inorganic complex $\{Co^{III}(OH)_6[Co^{III}(NH_3)_4]_3\}Br_6$ **(2.2)-Br$_6$** (Figure 2.2).[4]

In 1897 Werner began the search for a definitive proof of his hypothesis of the octahedral structure of chiral cationic Co^{III} complexes, and to do so he required an efficient anionic resolving agent. Whether or not he himself was familiar with the work of Pope[5–7] on the resolving power of the (+)-3-bromo-camphor-9-sulphonate anion **(2.3)** (Figure 2.3), it was his student L. King who chose to use this anionic agent to resolve the complexes $[Co^{III}(en)_2(NH_3)X]X_2$ (X = Cl, Br) **(2.1)-X$_2$**.[8]

To give an idea of the difficulty of his task, King recalls that, prior to achieving this resolution by crystallization of one of the diastereomeric salts resulting from the combination of **(2.1)** with the (+)-3-bromo-camphor-9-sulfonate, he had attempted no fewer than 2000 unsuccessfully experiments.[9] His eventual choice of the anion **(2.3)** was due to its large power of optical rotation, which he believed made it the most capable of increasing the dissymmetry of the resulting diastereomers. Werner also observed that the solubility of the oxalato(bis)ethylenediamine cobalt bromides **(2.4)-Br** (Figure 2.4) differed between the racemic and optically active forms.[10] He therefore successfully attempted their resolution by starting with a supersaturated aqueous solution, slightly enriched in one of the enantiomers, and initiating crystallization by the addition of methanol (Scheme 2.1).

Addition of methanol results in the crystallization of the (+)-**(2.4)-Br** salt.

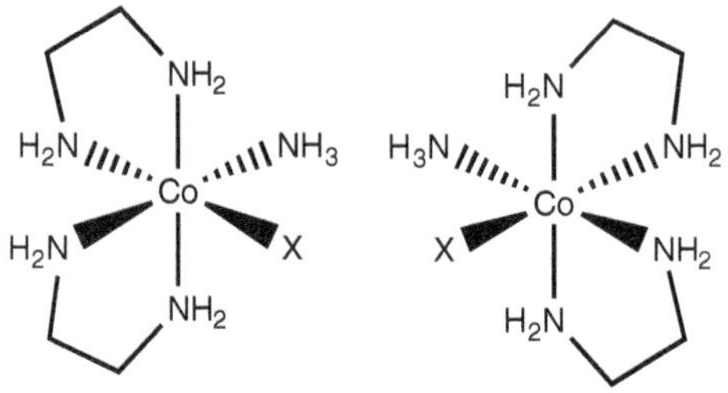

Figure 2.1
View of the cationic part of the two enantiomers of **(2.1)-X$_2$**.

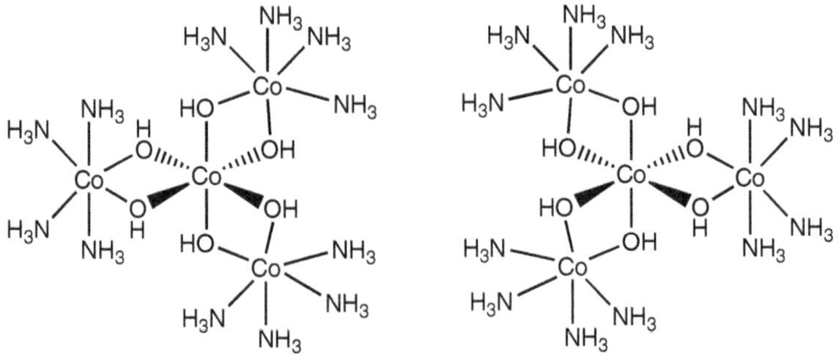

Figure 2.2
View of the cationic part of the two enantiomers of **(2.2)-Br$_6$**.

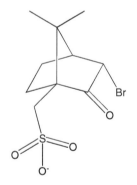

Figure 2.3
View of the (+)-**(2.3)** anion.

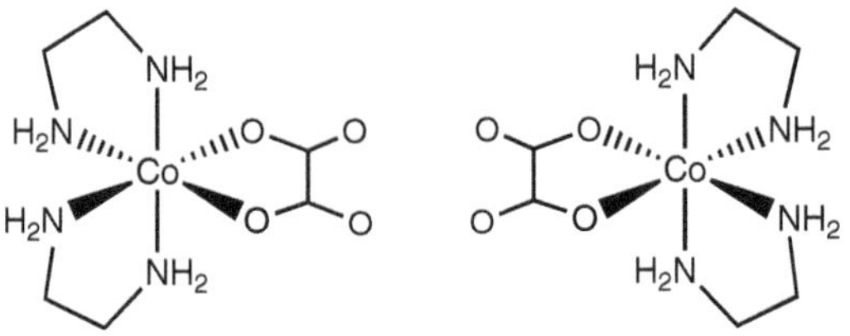

Figure 2.4
View of the cationic part of the two enantiomers of **(2.4)-Br**.

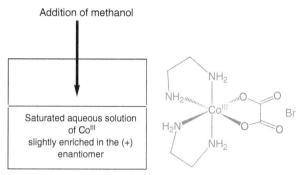

Scheme 2.1
Enantioselective crystallization of the (+)-**(2.4)**-**Br** salt starting from a slightly enriched (+)-**(2.4)**-**Br** solution by addition of methanol.

This experiment was in fact a rediscovery, several years later, of the work of Gernez (a student of Pasteur) on the crystallization of one of the enantiomeric tartrates from a supersaturated solution by seeding with a pure enantiomeric crystal.[11] This method is known as resolution by entrainment.

We will return later to the first resolution of octahedral complexes as it embodies numerous concepts, which we will need to explain the general principles of chirality in the field of coordination chemistry. Meanwhile, we will quickly retrace the long route, which led to the concept of molecular asymmetry and chirality.

These concepts, which seem so familiar today, only became clear with the atomic theory of Le Bel and van't Hoff[12,13] which provided a solid theoretical basis for the understanding of the three-dimensional structure of molecular objects.

2.1.1.2 Polarized Light and Rotation of the Polarization Plane

At the origin of the concept of dissymmetry lies the work of the physicists who studied the interaction of polarized light with crystalline minerals. The work of Bartholin (1625–1698), who discovered the phenomenon of double refraction in Iceland spar ($CaCO_3$), forms the basis of the wave theory of light that was developed by Huygens (1629–1698). The discovery of polarized light by Malus (1775–1812) rapidly led to intense research by physicists in this area. In 1811, Arago (1786–1853) introduced the concept of optical rotation by showing that pieces of quartz placed in the path of polarized light could turn the plane of polarization in one sense or the other depending on the crystal. Biot (1774–1862), in 1812, went further, as he was able to prove that the rotation of polarized light is not exclusive to certain crystals. It is also shown, for example, by solutions of natural camphor and oil of turpentine, as well as tartaric acid. In 1817, he proved that the vapour from oil of turpentine was able to rotate the plane of polarized light. This was a decisive step since, whether in solution or the gas phase, the observed phenomenon could not be due to the molecular arrangement in the crystal, but rather to an intrinsic molecular property.

2.1.1.3 Hemihedrism and Enantiomorphism

At this time, whereas the physicists had developed a theory of light to account for its interactions with highly organized crystalline matter, the chemists did not yet have a theory of molecular structure, but were themselves also interested in crystals. It was to be the studies of the latter group that would play a decisive role in the understanding of the phenomenon of structural asymmetry, whether this originated from the actual organization of the crystal or from the molecular structure itself.

In this context it was the discovery of hemihedrism that proved to be critical. The most important works in this area are those of Haüy (1743–1822) and Mitscherlich (1794–1863), and from their studies emerged the idea that the morphology of a crystal reflects the underlying form of its constituents. Haüy hypothesized that the exterior form was just a consequence of a periodic arrangement of matter in the solid. This idea was of great importance, even if it appeared to contradict Mitscherlich's discovery of isomorphism, that is the possibility of different substances having identical crystalline forms.

At this point we should recall that there are seven crystal systems. All crystals of the same crystal system have lattices with the same orientation symmetry. In each of the seven crystal systems, the group which defines it is the one with the highest order, and is termed holohedral. Within the same system, all the groups which are not holohedral are termed merihedral and, in particular, hemihedral for the case where the group order is half that of the holohedral. Those which possess a centre of symmetry are called centrosymmetric hemihedral, whilst those which lack an inversion centre are enantiomorphic hemihedral.

To summarize, there is a hemihedral form which, while having a reduced level of symmetry, nevertheless conserves the planes of symmetry in the crystal. There is also a 'nonsuperimposable' hemihedral form for which truncation of the crystal faces or edges gives rise to objects that are nonsuperimposable with their mirror image. Thus, certain crystals with the same chemical composition are mirror images of each other and are nonsuperimposable. Herschel (1792–1871) showed that Haüy's right-handed and left-handed quartz *plagièdre* were able to rotate the plane of polarized light in opposite senses.

2.1.1.4 Pasteur and Tartaric Acid

It was the observation of the hemihedral crystals of sodium ammonium tartrate tetrahydrate that enabled Pasteur (1822–1895) to make a decisive step forward in stereochemistry. The problem he encountered was the contamination of the potassium salt of tartaric acid with that of another acid (which Gay-Lussac (1778–1850) called the racemic acid) that made it unsuitable for commercial use. The two acids had the same chemical composition, and Biot showed that whereas tartaric acid and its salts could rotate the plane of polarized light, the racemic acid itself was inactive. In 1848, Pasteur found the solution to this problem.[14] He noticed that crystals of tartaric acid, like its salts, have hemihedral faces, but that the racemic sodium ammonium tartrate exists as two distinct crystals in which the hemihedral faces are mirror images of each other. One of these crystalline forms is identical to the optically active tartrate. In solution, it rotates the plane of polarized light in a dextrorotatory manner, while the other form (a mirror image of the first) is levorotatory, that is in solution it rotates the plane of polarization towards the left (Figure 2.5).

The sense of rotation is termed right if an observer placed behind the object through which the light passes sees the plane of polarization turn in a clockwise sense; rotation in the inverse sense is termed left.

Pasteur noted that the crystal forms were enantiomorphic, and he supposed that the molecules that comprised them would be also and called them dissymmetric. It was

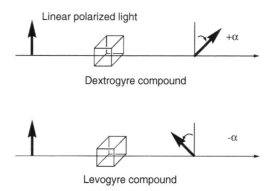

Figure 2.5
Right and left rotation of polarized light through a dextrogyre or levogyre compound.

widely agreed that Pasteur was lucky to observe what, in reality, was a spontaneous resolution of the enantiomorphic crystalline forms of racemic sodium ammonium tartrate **(2.5)-NaNH$_4$.4H$_2$O** (Figure 2.6).

However, this fundamental discovery had nothing to do with chance, but rather with Pasteur's insight and knowledge of both chemistry and crystallography.[15] Following Pasteur, the work of his student Gernez revealed the properties of conglomerates and

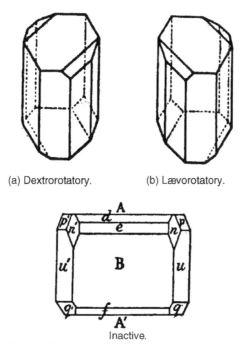

Figure 2.6
Enantiomorphous crystals of (+)-**(2.5)NaNH$_4$.4H$_2$O** and (−)-**(2.5)NaNH$_4$.4H$_2$O**. Crystal of the (*rac*)-**(2.5)NaNH$_4$.4H$_2$O** (reproduced with permission from reference[15]).

opened the way to resolution by external entrainment through seeding a supersaturated solution of a racemic compound, slightly enriched in one of the enantiomers, with an enantiomerically pure crystal of this configuration.[11]

2.1.1.5   Le Bel and van't Hoff and the Discovery of the Asymmetric Tetrahedral Carbon

While the work of Pasteur paved the way to a concept of molecular structure, it was the contributions of Le Bel (1847–1930) and van't Hoff (1852–1911), which gave theoretical support with their idea, developed simultaneously, of the asymmetric carbon atom. In his remarkable paper presented in the *Bulletin de la Société Chimique de Paris*,[12] Le Bel, starting from the point of Pasteur's discovery, affirmed two fundamental structural principles:

(i) *First General Principle: 'Let us consider a molecule of a chemical compound having the formula MA_4; M is a simple or complex radical combined with four monoatomic atoms A, capable of being replaced by substitution; let us replace three of them by monoatomic, simple or complex radicals which are different from each other and nonidentical to M; the body thus obtained will be dissymmetric. In fact, the presence of radicals R, R', R'' and A, which are all different, produce a structure that is nonsuperimposable with its image and it is not possible for the residue M to regain symmetry. Therefore, in general, a body derived from our original molecule MA_4 by substitution of A with three distinct atoms or radicals will be dissymmetric and have the power of optical rotation.*

There a two distinct exceptions: 1. if the base molecule possesses a plane of symmetry containing the four atoms A . . . . . .2. if the last radical substituted at A is composed of the same atoms as the whole group into which it is entering, whereby the effect these two equal groups have on polarized light may either compensate or add to each other.'

(ii) *Second General Principle: 'If in our fundamental structure we only substitute two radicals R and R', either symmetry or dissymmetry may occur depending on the constitution of the base molecule MA_4. If this molecule originally had a plane of symmetry passing through the two atoms which were replaced by R and R', this plane will remain a plane of symmetry after the substitution; the structures obtained will therefore be (optically) inactive. Our knowledge of the composition of certain simple structures allows us to confirm that compounds derived from them by two substitutions are inactive. In particular, if it happens that not only one substitution results in a single derivative, but also that two or three substitutions give only one and the same chemical isomer, then we have to admit that the four atoms A occupy the vertices of a regular tetrahedron, the symmetry planes of which will be identical to the whole molecule MA_4, and in this case no bisubstituted structure will have the power of rotation.'*

These two principles are summarized in Figure 2.7.

These principles having been established, Le Bel then reviewed the organic compounds corresponding to the lactate, tartrate and amyl groups, as well as the sugars. In the light of his second principle he examined derivatives of ethylene, but clearly lacked the data to advance his stereochemical ideas in this area. In his article Le Bel proposed a theory which in fact examines the origin of chirality and forms the basis of asymmetric synthesis: *'When a dissymmetric body is formed in a reaction where only symmetric bodies have been introduced in each others presence, the two isomers of inverse*

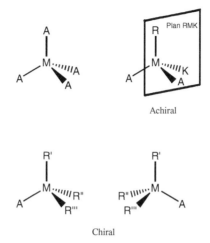

Figure 2.7

According to Le Bel a compound derived from the original molecule MA_4 by substitution of A with three distinct atoms or radicals is dissymmetric.

symmetry will be formed in the same proportions' and he adds: *'This is not necessarily the same for dissymmetric compounds that are formed in the presence of active bodies themselves, or through which circularly polarized light is passed, or even subjected to whatever factor favours the formation of one of the dissymmetric isomers.'*

So, after a long gestation, a new branch of chemistry emerged with a solid theoretical foundation: stereochemistry.

After this brief historical overview, we will now follow stepwise the path that opened the way to a rational understanding of chirality in the field of organometallic and coordination chemistry.

2.1.2 Definition of Chirality

The most general definition of chirality is the one given by Lord Kelvin (1824–1907):[16] *I call any geometrical figure, or group of points, chiral, and say that it has chirality, if its image in a plane mirror, ideally realized, cannot be brought to coincide with itself.* Today, chirality is defined by IUPAC as:[17] *The geometric property of a rigid object (or spatial arrangement of points or atoms) of being nonsuperimposable on its mirror image; such an object has no symmetry elements of the second kind (a mirror plane, $\sigma = S_1$, a centre of inversion, $i = S_2$, a rotation reflection axis, S_{2n}). If the object is superimposable on its mirror image the object is described as being achiral.* A chiral object, whatever its nature, exists in the form of two enantiomeric mirror images, as shown in Figure 2.8. The word chirality comes from the Greek $\chi\varepsilon\iota\rho$ = Kheir (hand).

In the application of this general definition, it is possible to include an examination of the symmetry group to which a molecule belongs. Knowledge of its symmetry group allows us to say whether a molecule is chiral or not. The condition for a molecule to be chiral is that it has no element of inverse symmetry, that is it does not have a centre, a plane or an improper axis of symmetry (Figure 2.9).

Before defining the symmetry point groups for achiral and chiral molecules, we will hypothesize that the positions of their atoms may vary in a continuous manner. In this

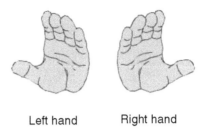

Left hand   Right hand

Figure 2.8
The left and right hands are not superimposable objects.

case, none of the possible conformations can be distinguished from any other in terms of its energy. We will therefore choose the conformation with the highest possible symmetry as being characteristic of the molecular structure. Of course, this is an ideal situation and we will see that, in practice, chiral structures can be attained as a result of a particular conformation.

We will now examine the point groups of achiral and chiral molecules.

(i) *Achiral molecules:*
   -The group C_s (a plane of symmetry and the identity E)
   -The group C_i or S_2 (a centre of symmetry and the identity E)
   -The group S_n (an improper axis and the identity E, S_2 is equivalent to C_i)
   -The group C_{nv} (an axis of symmetry C_n and n vertical planes of symmetry s_v that contain the axis C_n and intersect each other at an angle of p/n, and the identity E)
   -The group C_{nh} (an axis of symmetry C_n, a plane of symmetry s_h perpendicular to the axis C_n and the identity E)

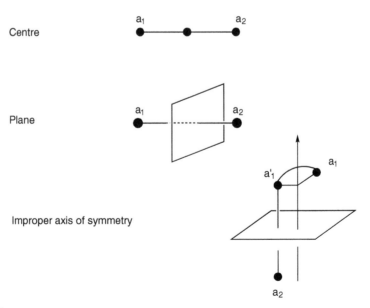

Figure 2.9
A chiral object has no inverse symmetry element.

-The group D_{nd} (a principal axis C_n, n vertical planes of symmetry s_d and n C_2 axes perpendicular to C_n bisecting each angle between two adjacent s_d planes and the identity E)
-The group D_{nh} (a principal axis C_n, n C_2 axes perpendicular to C_n, n planes of symmetry s_v and a plane of symmetry s_h and the identity E)
-The groups T_d, O_h and I_h (more than one axis of symmetry and the identity E). These groups correspond to tetrahedral, octahedral and icosahedral structures.
(ii) Chiral molecules
-The group C_1 (the identity E)
-The group C_n (an axis C_n and the identity E)
-The group D_n (an axis C_n, n C_2 axes perpendicular to C_n and the identity E)
-The groups T, O, I. These groups correspond to tetrahedral, octahedral and icosahedral structures.

Figure 2.10 shows a classical diagram that allows determination of the symmetry point group and distinguishes between achiral and chiral molecules.[18]

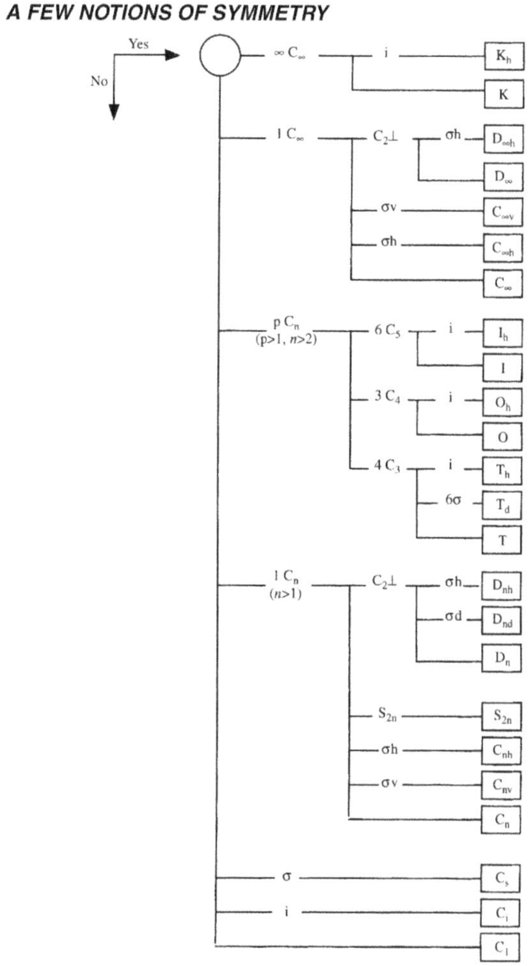

Figure 2.10
Procedure for finding the point symmetry group of a molecular unit (reproduced with permission from reference[18]).

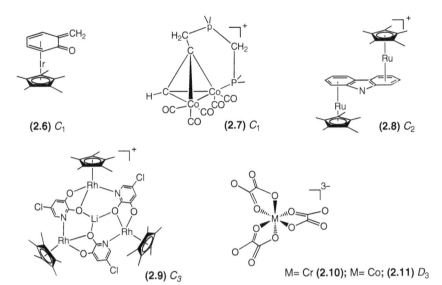

Figure 2.11
Some examples of chiral organometallic compounds **2.6-2.11** with different symmetry point groups.

Some examples of chiral molecules are given in Figure 2.11.

We hypothesized that all the possible conformations are iso-energetic, but this situation is not general. In certain cases some particular conformations are more stable than others. These conformations, which become configurational isomers, can sometimes be separated by energy barriers that are sufficient for them to be isolated at a given temperature. This type of isomerism is known as *atropisomerism*.

This is the case for 2,2'-Bis(diphenylphosphino)-1,1'-binaphthalene, known as BINAP, which exists in the form of two enantiomers (Figure 2.12).

BINAP is a bidentate ligand, very highly chelating towards metal centres, particularly palladium (II). Obtaining such chiral complexes is of great interest in the field of enantioselective catalysis which will be considered in Chapter 3.

It is possible for an achiral conformation to be transformed into a chiral one, for example in bis- and tris(chelate) complexes. Starting from a bis(chelate) complex with D_{2h} symmetry (the torsion angle between the two (MAA) planes being $0°$), if the torsion angle is increased in the sense right to left, for angles up to but less than $90°$, the symmetry of the complex becomes D_2. The complex thereby acquires a chiral structure. For an angle equal to $90°$ the symmetry becomes D_{2d}, signifying that the structure is once again achiral (Figure 2.13).

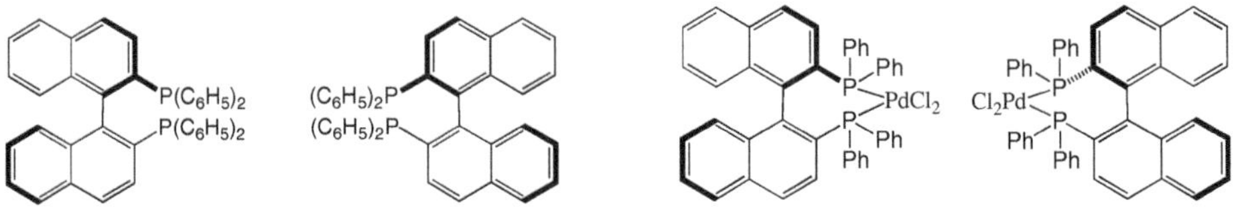

Figure 2.12
The two enantiomers of the BINAP ligand (left) and of the PdCl$_2$BINAP complex.

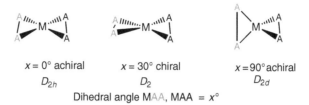

$x = 0°$ achiral $x = 30°$ chiral $x = 90°$ achiral

D_{2h} D_2 D_{2d}

Dihedral angle MAA, MAA $= x°$

Figure 2.13
The chiral character of an MA_4 tetracoordinated complex depends on the dihedral angle MAA, MAA.

Considering now the case of tris(chelate) complexes, starting from an achiral trigonal prismatic conformation of D_{3h} symmetry, a rotation of an angle of 60° around the C_3 axis leads to a chiral structure with D_3 symmetry (Figure 2.14).

The quantitative study of the transformation of achiral to chiral structures has led to the concept of 'continuous chirality measures in transition metal chemistry' developed by Alvarez and Avnir.[19,20]

2.1.3 Definition of a Prochiral Object

We have just seen that a chiral object is defined in a three-dimensional space. However, it is possible to reduce the space to two dimensions and to define chiral objects in a plane. To say that a molecule is two-dimensional appears at first sight to be an abuse of language. In effect, no molecule, even benzene, can be reduced to a plane. We will nevertheless treat molecules such as benzene and ethylene as two-dimensional molecules. These molecules are achiral in three-dimensional space since they possess at least the plane of symmetry in which they lie (Figure 2.15).

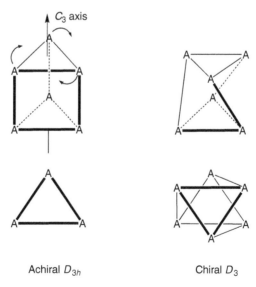

Achiral D_{3h} Chiral D_3

Figure 2.14
A 60° rotation around the C_3 axis leads to an achiral structure.

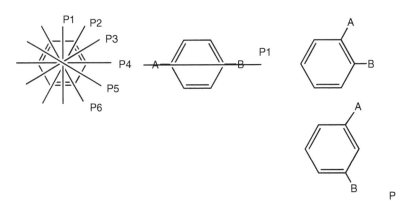

Figure 2.15
Changes in symmetry from a nonsubstituted to a disubstituted benzene ring.

However, taking the case of benzene, if we introduce two different, achiral substituent groups A and B then, depending on the relative position of the substituents (i.e. 1,2 ; 1,3 ; 1,4), there are two different possible outcomes for the symmetry of the system. While the non-substituted benzene possesses, in addition to its 'natural plane', six other planes of symmetry perpendicular to the first, the 1,4 substitution leaves only one of these passing through the two substituted carbon atoms (Figure 2.15), whereas 1,2 and 1,3 substitution retain only the 'natural plane'. In two-dimensional space these substituted derivatives are chiral. In effect, it is impossible to superimpose the two 'isomers' by only using movements within the plane (Figure 2.16).

Of course this is merely an abstract operation, which has no physical sense, but it allows us to show that the two faces of such a molecule are not equivalent. These two faces (a) and (a′) (Figure 2.17) are termed *enantiotopic*.

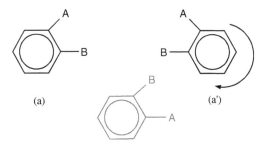

Figure 2.16
(a) and (a′) are not superimposable in the plane, there are enantiomers in a two-dimensional space.

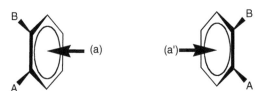

Figure 2.17
Enantiotopic (a) and (a′) face of a disubstituted benzene ring.

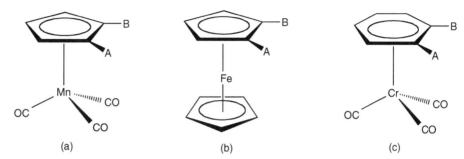

Figure 2.18
(a) Chiral cymanthrene, (b) ferrocene and (c) benchrotrene.

This concept of the enantiotopic faces of a prochiral molecule is very important as it allows us to show that the introduction of an atom or group of atoms in the third dimension leads to a chiral molecule. This is the case for all such complexes of benchrotrene, ferrocene, cymanthrene, and so on (Figure 2.18).

2.1.4 Definition of Elements of Chirality

What are the stereogenic elements of a molecule? First of all, it should be said that the terms 'chirality element' or 'stereogenic element' of a molecule are preferable to 'chiral centre' or 'stereogenic centre'. This choice arises from the fact that by definition a centre cannot be stereogenic. Thus the term 'asymmetric centre' or 'chiral centre' for a tetrahedral carbon atom substituted by four different groups does not take account of the fact that the chirality of the molecule depends exclusively on the asymmetric environment of the carbon atom. This remark also applies to chiral complexes. Thus, in a chiral ferrocene derivative where one of the cyclopentadienyl rings is substituted in the 1,2 or 1,3 positions by two different groups, it is impossible to define a centre which is the origin of the chirality, no more so than in a complex hexacoordinated by three bis(chelate) ligands (Figure 2.19).

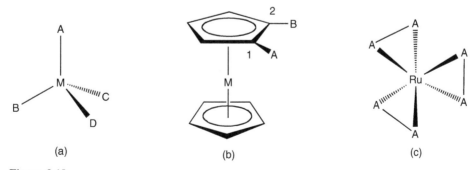

Figure 2.19
Main type of chirality encountered in organometallic compounds: (a) centred, (b) planar and (c) helical.

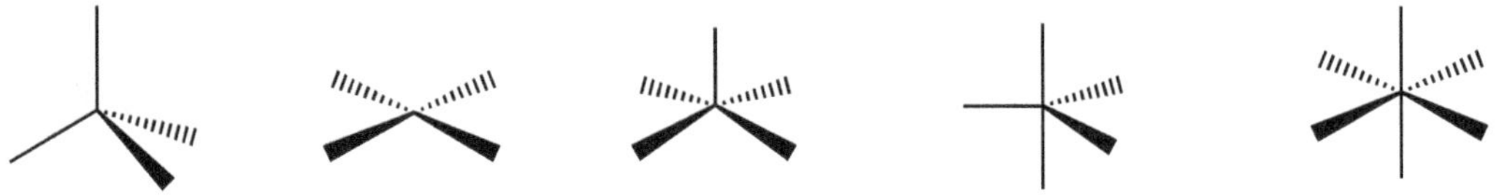

(a) d^3s Tetrahedral   (b) dsp^2 Square planar   (c) dsp^3 Square pyramid   (d) dsp^3 Trigonal bipyramid   (e) d^2sp^3 Octahedral

Figure 2.20
Coordinated metal centres: when *d* orbitals as well as *s* and *p* orbitals are available an important set of hybrids can arise, therefore, a large variety of bricks can be generated by the highly directional formation of coordination bonds.

2.1.5   Principal Elements of Chirality Encountered in Organometallic and Coordination Chemistry

(i) Metal coordinated to different types of ligands
In organic chemistry the tetrahedral carbon bonded to four different groups is the archetypal chiral element. However, while carbon through its sp^3 hybridization is capable of forming triangular based pyramids, the metals, and in particular the transition metals, can, due to the greater possible number of valence electrons, have coordination numbers higher than four. As a result, the geometries of possible structures are more numerous than in the case of sp^3 carbon. Figure 2.20 gives several possible spatial arrangements. In these different cases the metal and its ligands form a stereogenic element with the metal at the centre.

The structure of type (a) is well known in organometallic chemistry, in particular among the complexes of manganese[21] and rhenium[22,23], for example the resolved enantiomeric complexes of [Mn(C$_5$H$_5$)(NO)COP(C$_6$H$_5$)$_3$]X **(2.12)-X** and [Re(C$_5$H$_5$)(NO)CH$_3$P(C$_6$H$_5$)$_3$]X **(2.13)-X** (Figure 2.21).

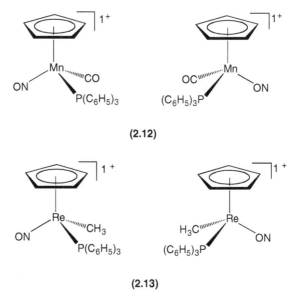

(2.12)

(2.13)

Figure 2.21
View of the cationic parts of the **(2.12)** and **(2.13)** enantiomers.

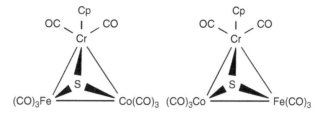

Figure 2.22
View of the two enantiomers of a chiral trinuclear cluster.

Note that in Figure 2.20 (e) M(a,b,c,d,e,f) (OC-6), if all the ligands (a,b,c,d,e,f) are different then there are 15 possible isomers.

(ii) Tetrahedral complexes
Many clusters with low nuclearity (from two to four), which do not have an atom at the centre of the pyramid, exist in a tetrahedral structure and are chiral if the four vertices, metallic or not, are different. This is the case, for example, of the tetranuclear cluster $[FeCoCr(Cp)(CO)_8][\mu_3S]$ **(2.14)** (Figure 2.22).[24]

(iii) Planar chirality
This group includes an important part of organometallic chemistry, that of arene, ferrocene and cymanthrene complexes and certain complexes of rhodium and iridium. In this case the chiral element is composed of the metal and the prochiral ligand. Figure 2.23 represents the two enantiomers of the 1,2-disubstituted ferrocene complex**(2.15)**.[25]

Note that in the case of ferrocene derivatives, 1,2- or 1,3-substitution leads to a chiral structure, whereas in the case of the arenes, only 1,2- and 1,3-substitution leads to chiral compounds.[26]

Where there are two equivalent coordinated rings that are both substituted, two different situations can arise. In the first, the two rings are substituted by identical achiral ligands in the same position, that is either 1,2 or 1,3, and the complex is achiral as it contains a plane of symmetry. In the second case, each prochiral plane has the same configuration, the molecule has C_1 symmetry and it is chiral, possessing two chirality elements (Figure 2.24).

(iv) Metal with achiral chelating ligands leading to helical geometries
This type of chiral element is often found in coordination chemistry, in particular for all the hexacoordinated octahedral complexes of type $M(L^2)_2A_2$, in which at least two ligands in a *cis* position are bis(chelate).

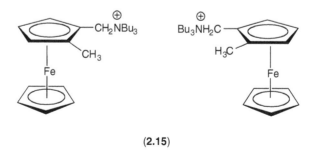

(2.15)

Figure 2.23
View of the two enantiomers of **(2.15)**.

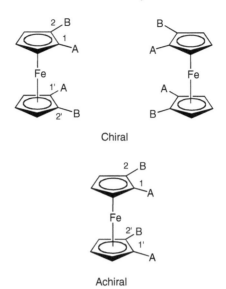

Chiral

Achiral

Figure 2.24
Chiral and achiral 1,2-disubstituted ferrocenes.

In the case of $M(L^2)_2A_2$ the complex has C_2 symmetry, whereas for $M(L^2)_3$ the symmetry is D_3. This is shown in Figure 2.25 for the cationic part of the complexes $[Ru(bpy)_2py_2]X_2$ (**2.16**) (py = pyridine) and $[Ru(bpy)_3]X_2$ (**2.17**).[27,28]

The two or three bis(chelate) ligands that are coordinated to the metal centre may in fact be equivalent, it is the actual geometry of the whole structure, resembling a three-bladed propeller, which determines the chiral character of the molecule. In the area of helical chirality, other types of chiral complexes have been described, in particular the two-bladed dinuclear acetylenic complexes (**2.18**)(**BF$_4$**)$_2$ which represent a rare example in organometallic chemistry (Figure 2.26).[29]

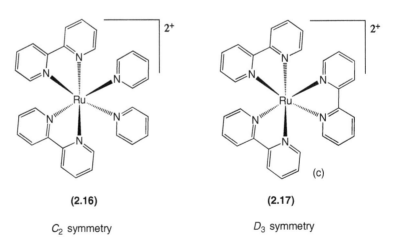

(2.16)

(2.17)

C_2 symmetry

D_3 symmetry

Figure 2.25
Hexacoordinated Ru(II) complexes of C_2 symmetry (left) and D_3 symmetry (right).

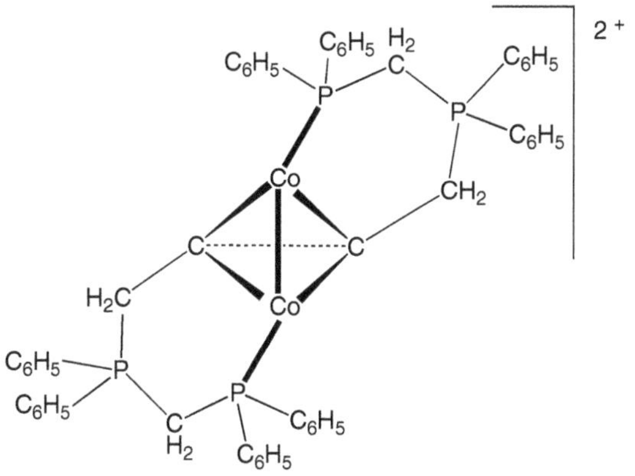

Figure 2.26
Two-bladed helical dication of C_2 symmetry.

(v) *Metal surrounded by four ligands or two bis(chelate) ligands leading to a complex with an axis of chirality or axial chirality*

A square planar complex of the type ML_4 can be chiral even if the ligands surrounding the metal are achiral. This atropisomerism can result from a particular conformation related to the position of the ligands situated along the long axis of the molecule. Axial chirality may also originate from two *trans* coordinated monodentate ligands formed of planar rings having substituents large enough to prevent free rotation about the axis that joins them. Figure 2.27 shows an achiral complex of Pd (**2.19**) in which a plane of symmetry passes through the two pyridine rings, while in the complex (**2.20**), the two planes containing the pyridine rings form an angle of 30°.[30]

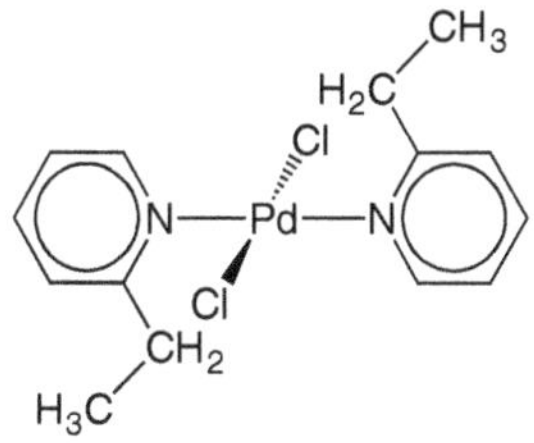

Achiral (**2.19**)

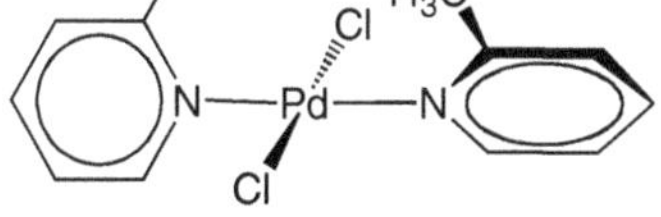

Chiral C_2 symmetry (**2.20**)

Figure 2.27
Achiral and chiral tetracoordinated palladium (II) complexes.

2.2 Enantiomers and Racemic Compounds

2.2.1 Enantiomers

We have seen that any object whose mirror image is nonsuperimposable with it, or in other terms does not have any element of inverse symmetry, is chiral. It exists in the form of two isomers which have the same chemical composition and the same connectivity of the constituent atoms. These two isomers are named enantiomers. Enantiomers have the same physical properties, only by their vectorial properties are inverted and in particular their ability to rotate the plane of polarization of light. Rotation to the right corresponds to the dextrorotatory (+) isomer, rotation to the left to the levorotatory (−) one. Molecules, whether in the gas, liquid or solid phase, are not unique, isolated objects. In the chemist's flask there are a great many molecules. If they comprise just one single enantiomer the product is termed enantiopure or homochiral. In the case where both enantiomers are present but with a majority of one of them, it is referred to as an enantio-enriched or heterochiral mixture (Figure 2.28).

Note that the terms 'homochiral' and 'heterochiral' are ambiguous and we will avoid using them to designate mixtures enriched in one enantiomer. In effect, they are often used in the chemistry of chiral materials to refer to the situation where chiral elements of the same type are present in a molecule and have either the same or opposite configurations.

2.2.2 Racemic Compounds

When two enantiomers are present in equal quantities, that is in a 1:1 stoichiometry, it is termed a racemic compound or racemic mixture. Such a mixture has no effect on polarized light since, through compensation, each enantiomer rotates the plane of polarization by the same amplitude and in the opposite sense. A racemic mixture can occur, particularly in the crystalline state, in three different forms. This is an important point since the possibility, or not, of directly separating the two enantiomers depends on the nature of the racemic mixture in the solid state. This separation of enantiomers, or resolution, is an essential task for the chemist who is interested in chirality and obtaining enantiopure compounds.

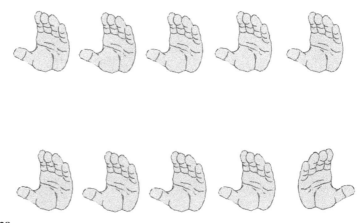

Figure 2.28
Homochiral collection of left hand (top) and heterochiral collection enriched in left hand (bottom).

[Ru(bpy)$_3$](PF$_6$)$_2$ **(2.17)**(PF$_6$)$_2$ *P3c1*

P3c1[Ru(bpy)$_2$ppy]PF$_6$ **(2.21)**PF$_6$ *C2/c*

[Ru(bpy)$_2$Quo]PF$_6$ **(2.22)**PF$_6$ *C2/c*

[Ni(bpy)$_3$](PF$_6$)$_2$ **(2.23)**(PF$_6$)$_2$ *P3c1*

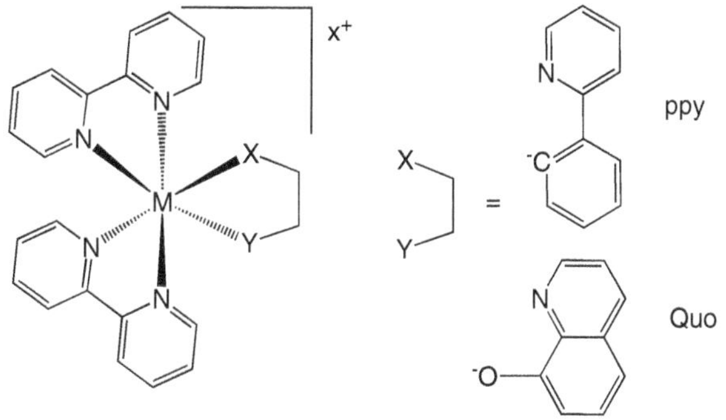

Figure 2.29
Space groups of some coordination compounds crystallizing as racemic compounds.

(i) Racemic compounds

This is the most common type of crystalline racemic compound. The two enantiomers, in equal quantities, are present in a well-defined arrangement within the crystal structure. This type of racemic compound is sometimes called a 'true racemic compound'. In this arrangement the unit cell contains at least the two enantiomers, and the space group is centrosymmetric. According to J. Jacques,[31] 60 to 80% of racemic crystals come from the groups $P2_1/c$, $C2/c$ and P-1. Note that racemic compounds can crystallize in any space group, mostly in a centrosymmetric, rarely in a noncentrosymmetric and exceptionally in a chiral space group. Some examples are given in Figure 2.29, corresponding to complexes of the type $[M(bpy)_2L^1L^2]^{n+}X_n^{n-}$ **(2.17)**, **(2.21–2.23)**.[32,33]

(ii) Conglomerates

A conglomerate is an equimolecular, mechanical mixture of crystals of the two enantiomers. Thus, one single enantiomer is present in the unit cell. The principal space groups (up to 90%) in which conglomerates are found are: $P2_12_12_1$, $P2_1$, C_2 and $P1$.

To put it simply, it is possible to say that the two enantiomers present in the solution separate and form distinct crystals. This phenomenon, which is called spontaneous resolution, is what allowed Pasteur to separate the levo- and dextrorotatory cystals of sodium ammonium tartrate tetrahydrate mechanically. In this case, the crystals of each enantiomer can be physically distinguished by their geometry as a result of their hemihedral character. Later, Werner made the same observation for the crystals of [Rh^{III}(C$_2$O$_4$)$_3$]K$_3$ **(2.24)** where the levo- and dextrorotatory forms are mirror images of each other[34] (Figure 2.30).

Note that, while spontaneous resolution resulting in a conglomerate is a rare phenomenon in organic chemistry (in less than 5% of cases according to J. Jacques[31]) it is more common in coordination chemistry. Thus, for example,

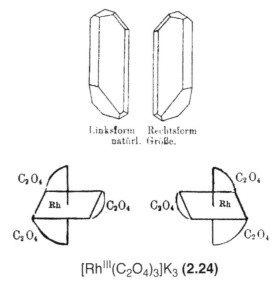

Linksform Rechtsform
natürl. Größe.

$[Rh^{III}(C_2O_4)_3]K_3$ **(2.24)**

Figure 2.30
View of the levorotatory form (left) and dextrorotatory form (right) of the $[Rh^{III}(C_2O_4)_3]K_3$ **(2.24)** crystals. Reprinted from [4].

according to Kaufmann[8] most of the complexes resolved by Werner crystallize in enantiomorphic space groups. However, it is difficult to establish the reason for this tendency. Breu[32] showed, for example, that, although very similar, complexes of the type $[M(bpy)_3](PF_6)_2$ with (M = Ni **(2.25)**, Zn **(2.26)**, Ru **(2.17)**, bpy = bipyridine) crystallize for no apparent reason into achiral or chiral space groups $P3c1$ and/or $P3_1$, $P3_1$ and $P3c1$ respectively.

The fact that the crystallization of chiral coordination compounds leads to conglomerates by spontaneous resolution, can give rise to two types of confusion. The first is thinking that spontaneous resolution leads to crystals that are visually distinguishable, which is generally not the case. This problem returns us to that of the isomorphism raised by Mitscherlich.[35] Conditions for apparition of hemiedrism are detailed by Coquerel.[36,37] The second confusion is related to the fact that conglomerates, far from forming distinct levorotatory and dextrorotatory crystals, can exist as unique twin crystals, the space group of which is certainly enantiomorphic.

For example, for the three-dimensional, chiral, helical lattices of the type $C^{n+}[M^1M^2(C_2O_4)_3]^{n-}$ (where C^{n+} is a monovalent or divalent cation, M^1 is a divalent metal ion and M^2 is a di- or trivalent metal ion) which have cubic symmetry, it is impossible to distinguish the levorotatory and dextrorotatory enantiomers by observation of their single crystals. In general, the right and left modification of cubic crystals results in twin crystals (agglomerations of two forms of single crystal). In this case, it is impossible to determine the absolute configuration, that is to distinguish between the space groups $P4_132$ and $P4_332$ for monometallic lattices such as $[(Fe(bpy)_3)(Fe_2^{II}(C_2O_4)_3)]$ **(2.27)** (phen = phenantroline)[38,39] or $P2_13$ (right or left) for bimetallic lattices such as $[(Ru(bpy)_2ppy)(Mn^{II}Cr^{III}(C_2O_4)_3)]$ **(2.28)**.[40] Such conglomerates, composed of single crystals, are not mechanically separable and do not have any optical activity.

(iii) Solid solutions or pseudoracemic compounds

There is a third form of solid state organization of racemic compounds, called the solid solution or pseudoracemic compound. This is a rare form in which the arrangement of the enantiomers in the crystal is variable and not defined. The significance of this is that, while the stoechiometry of the two enantiomers in the whole crystal may well be 1:1, it is possible to find homochiral unit cell structures, one enantiomer being substituted for the other. Such disorder arises from the strong similarity of the two enantiomers, particularly in the case of quasispherical molecules. This arrangement is also called a racemic solid solution of the two enantiomers.

2.2.3 Diastereomers

Up to now we have limited ourselves to compounds with a single chiral element, but we will now consider the case where several chiral elements are present in the same compound. If the latter has n distinct chiral elements there will be a maximum of n^2 isomers that carry the name of diasteromer. We will see, however, that taking into account elements of inverse symmetry can reduce this number.

Let us take the simplest example of a compound where two distinct chiral elements (the foot and the hand) are present. The four possible combinations (diasteromers) are therefore:

right foot–right hand,
left foot–left hand,
right foot–left hand, and
left foot–right hand.

Figure 2.31 shows that these combinations can be classified into two distinct groups.

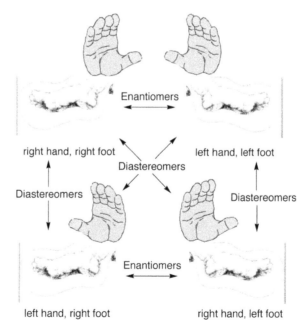

Figure 2.31
Possible combinations for an object possessing two stereogenic centres.

(i)   *'Right foot–right hand, left foot–left hand' and 'right foot–left hand, left foot–right hand'*
      In each group the diasteromeric combinations are mirror images of each other and so
      they are enantiomers. For each pair of enantiomers there is a corresponding racemic
      compound '*rac*-(right foot–right hand, left foot–left hand)' and '*rac*-(right foot–left
      hand, left foot–right hand)'.

(ii) *'Right foot–right hand, right foot–left hand' and 'left foot–right hand, left foot–left hand'*
      Here, in each group we have diastereomers. Overall there are two pairs of
      diastereomers, each comprising two enantiomers.

Although the enantiomers of a compound have the same physical properties (except
the sense of rotation of plane polarized light), this is not true for diastereomers, which
have different properties and, apart from the chemical formula and the connectivity of
their atoms, must be considered as different compounds. This is reflected in differing
solubilities and melting or boiling points, which can be exploited in their separation.

We have just considered the case where the two chiral elements are different. Now we
will look at what happens when they are identical (two hands for example). There are
four possible combinations:

right hand–right hand,
left hand–left hand,
right hand–left hand, and
left hand–right hand.

While 'right hand–right hand' and 'left hand–left hand' are enantiomers, 'right hand–
left hand' and 'left hand–right hand' are one and the same achiral product which has a
plane of symmetry (Figure 2.32).

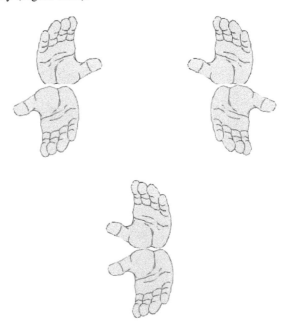

Figure 2.32
Two right and left hands are enantiomeric forms (top), while the combination of a right and left
hand is achiral (bottom).

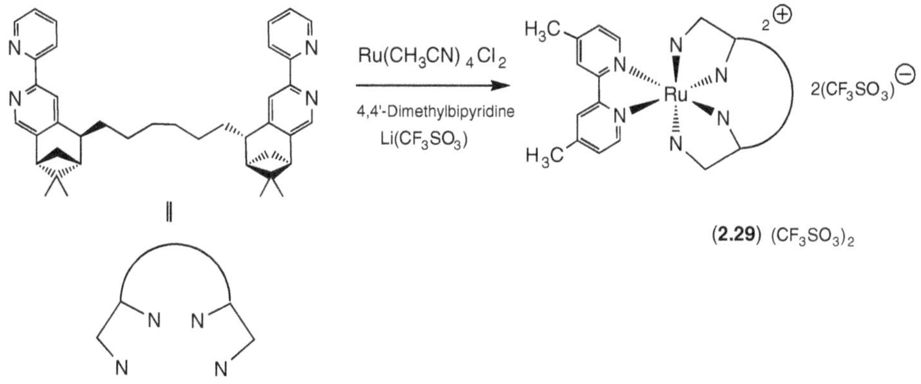

Figure 2.33
Synthesis of the (Δ)-[Ru((+)-chiragen[6](4,4'-dimethylbipyridine](CF$_3$SO$_3$)$_2$ **(2.29)**-(CF$_3$SO$_3$)$_2$.

This achiral material is termed a *meso* compound. This situation had been predicted by Le Bel.[12] He considered that two identical chiral groups could compensate for each other and so cancel out the effect of rotation of plane polarized light.

In organometallic chemistry, and even more so in coordination chemistry, the existence of diastereomers is governed by the ionic nature of the compounds (chiral cations and anions). We can distinguish two types of diastereomers. First, the class where the chiral elements are joined by covalent or dative bonds, well represented by complexes of ruthenium (II) hexacoordinated to three chiral bis(chelate) ligands. In these complexes, the metal surrounded by its ligands constitutes an element of helical chirality, while the ligands themselves contain one or more asymmetric carbon atoms. Here the counter ion (the anion) is achiral. This chemistry has been explored in particular by von Zelewsky[41] who used chiral bipyridine ligands in an enantiomerically pure form. Figure 2.33 shows the complex of (Δ)-[Ru((+)-chiragen[6](4,4'-dimethylbipyridine](CF$_3$SO$_3$)$_2$ **(2.29)** obtained from Ru(CH$_3$CN)$_4$Cl$_2$, (+)-chiragen[6] and 4,4'-dimethylbipyridine.

The second type of diastereomer comprises those salts where both the anionic and cationic parts are chiral. The nature of the chiral elements involved here can be variable: for example, a cation **(2.16)** with C_2 symmetry and an anion containing asymmetric carbon atoms such as dibenzoyl tartrate **(2.30)**,[27] a cation relevant to planar chirality **(2.31)** with an anion **(2.32)** of D_3 symmetry,[25] or finally an anion **(2.32)** and a cation **(2.17)** both of D_3 symmetry.[42] These three examples are shown in Figure 2.34.

2.2.4   Enantiomeric and Diastereomeric Excesses

We have seen that a mixture of two enantiomers can exist in either an enantiopure (one single enantiomer present) or enriched form (two enantiomers present, but with one of them in excess over the other). The same is also true for diastereomers. In order to characterize such mixtures quantitatively, we will use the important terms of enantiomeric and diastereomeric excess, (*ee*) and (*de*) respectively.

(i) Enantiomeric excess

The enantiomeric purity (*p*), or the enantiomeric excess (*ee*), of a mixture (+)-**a**, (−)-**a** is defined by the formula: (*p*) or (*ee*) = {[(+)-**a**] − [(−)-**a**]}/}[(+)-**a**] + [(−)-**a**]},

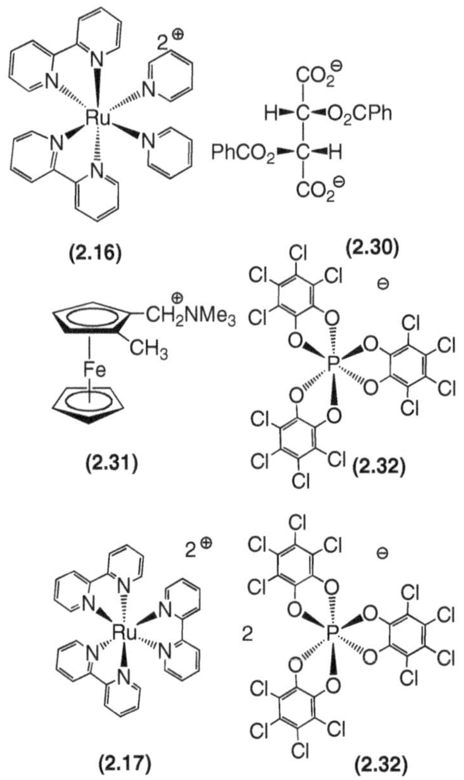

Figure 2.34
Diastereomeric ion pairs relevant to different types of chirality.

where $[(+)\text{-}\mathbf{a}] > [(-)\text{-}\mathbf{a}]$, and where $[(+)\text{-}\mathbf{a}]$ and $[(-)\text{-}\mathbf{a}]$ represent the concentrations of the two enantiomers.

(ii) Diastereomeric excess

The diastereomeric excess *(de)* of two diastereomers **a** and **b** is defined by the formula: $(de) = \{[\mathbf{a}] - [\mathbf{b}]\}/\{[\mathbf{a}] + [\mathbf{b}]\}$, where $[\mathbf{a}] > [\mathbf{b}]$. Note that the diastereomeric excess tells us nothing about the enantiomeric purity of the diastereomers **a** and **b**. It is sometimes seen in the literature that *(ee)* and *(de)* are considered as being equivalent, in particular when the *(ee)* is determined by NMR with the use of a chiral shift reagent. This only makes sense if the *(ee)* of the shift reagent itself is equal to one.

2.2.5   Racemization and Configurational Stability

Until now we have considered that the form of chiral objects is invariant with time. However, we know that molecules are prone to geometric changes, which are more or less rapid, due to the nature of the bonds (more or less fragile) that join their atoms to each other, while certain atoms, such as nitrogen or phosphorus, can undergo a rapid inversion of their environment (Figure 2.35).

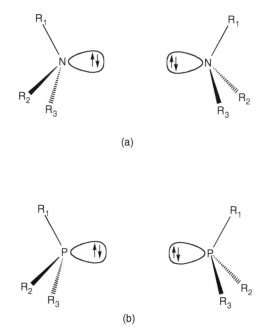

(a)

(b)

Figure 2.35
Inversion of configuration at the (a) nitrogen and (b) phosphorous centres.

The study of configurational stability is of prime importance since, for example, it allows us to know the timescale during which a reaction must occur in order for the configuration of a chiral reactant to be maintained. Taking, for example, the coordination compound $[Cr^{III}(C_2O_4)_3]K_3$ **(2.10)** K_3, obtained as an enantiomer with a negative sense of rotation $([\alpha]_D^{20°} = -1800$ $(c = 0.02,$ $H_2O))$,[43,44] the value of the rotation diminishes to zero over a period of several hours.[45] This phenomenon, which converts an enantiomer to a racemic compound, is called a process of racemization. Taking into account the reciprocity of this phenomenon for the two enantiomers we can write:

$$(-)\text{-}[Cr(C_2O_4)_3]K_3 \rightleftarrows (+)\text{-}[Cr(C_2O_4)_3]K_3.$$

The thermodynamic constants have been measured and they show that it is a complex process, including the solvent (in this case water) and possibly ammonium salts (i.e. the counter ion plays a part in the racemization). In organometallic and coordination chemistry, because of the numerous dative bonds involved, the question of racemization must be studied to ensure that the product obtained as an enantiomer does not racemize over time. This can happen in solution, where the solvent can act as a ligand and thereby participate in the process, but it is also true of the solid state in the presence of an external agent, such as temperature or exposure to light. We say that an enantiomer has configurational stability if it does not racemize under given conditions.

We will see later that, while in most cases racemization is an inconvenience that destroys what the chemist has spent much time in achieving, it can be put to use in the resolution of a racemic compound to obtain one enantiomer in a yield close to one.

2.3 Absolute Configurations and System Descriptors

2.3.1 Definition of the Absolute and Relative Configuration of a Molecule

2.3.1.1 Absolute Configuration

The configuration of a molecule is nothing more than a description of its general geometric appearance or, in other words, the unequivocal description of the spatial position of the atoms it contains. The absolute configuration of an enantiomer makes it possible to designate in an unequivocal manner the spatial arrangement that is the cause of the chirality. Two enantiomers will have opposite absolute configurations. We will see that, by using descriptors that are universal, it is possible to designate the absolute configuration of the chiral elements of a chemical compound.

Returning to the history of chirality, it must be noted that the question of absolute configuration is not a simple problem to resolve. In the beginning, it was the property of enantiomers to rotate the plane of polarized light that was important to chemists. Molecules with this property are designated (+) associated with the symbol (d), or (−) with the symbol (l), but the sign of rotation cannot be directly related to the absolute configuration of the molecule. There is no simple relationship between the sense of rotation of polarized light and the absolute configuration. Moreover, the sense of rotation is a function of the wavelength and for a given enantiomer may be inverted as the wavelength changes. Before X-ray diffraction became available as a tool for determining absolute configuration – a possibility from 1952 following the work of Bijvoet[45] – chemists had attempted to relate the optical rotation of a molecule to its absolute configuration. If we look once again at the case of tartaric acid and its salts, it is clear that the *meso* compound, which has no effect on polarized light, can only be the one where the two chiral elements are mirror images of each other in an internal symmetry plane of the molecule. On the other hand, for the optically active derivatives, only an arbitrary decision associated the dextrorotatory enantiomer with an absolute configuration (Figure 2.36).

Fischer (1852–1919) who received the Nobel Prize for Chemistry in 1902, proposed just such an association in sugar chemistry.[46,47] He had a 50:50 chance of being wrong, but luck was with him, and the work of Bijvoet showed definitively that his choice was in fact correct.

2.3.1.2 Relative Configuration

Once the configuration of one chiral element in a molecule is fixed, arbitrarily or not, the configuration of the other chiral elements in the molecule can be related to it. This is

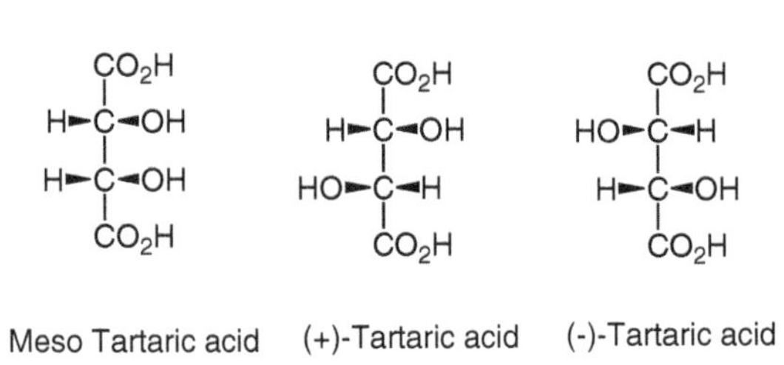

Meso Tartaric acid   (+)-Tartaric acid   (-)-Tartaric acid

Figure 2.36
View of the (meso)-, (+)- and (−)-tartaric acid.

possible, in particular, by chemical relationship but also by physicochemical methods since the presence of several chiral elements results in diastereomers with different properties. Such a configurational relationship, based on an arbitrarily chosen configuration, is called the *relative configuration* of the chiral elements.

However, we must still bear in mind that, without having arrived at the absolute configuration, the chemist's work is not finished. In order to determine the absolute configuration there are two requirements:

   (i) *a universal system that can unequivocally describe the geometry of chiral elements;*
  (ii) *one or more physical methods which are able to relate molecular properties to an absolute configuration.*

2.3.2   Absolute Configuration and Universal Descriptors

In their article:[48] 'Specification of configuration about quadrivalent asymmetric atoms', Cahn and Ingold recall that before their work it was the concept of a '*correlation rule*' associated with a '*standard substance*' which prevailed in designating the absolute configuration of substances containing stereogenic elements bound to a quadrivalent carbon atom. Straightaway they asserted the superiority of their proposed descriptor system: '*The time will come when absolute configuration can be determined with certainty, and the "standard substance" will be a redundant concept.*'

However, once again showing that theories live, develop and die, they do not completely reject the previous system: '*Anticipating this, we shall express the "correlation rule", calling it a "sequence rule", so that it can be used without change after the concept of "standard substance" has been discarded. But for the time being, we retain this concept.*'

Thus, from 1951, Cahn and Ingold, later associated with Prelog,[49–51] defined a method for attributing absolute configuration based on a simple and universal principle, and we will now detail the application of this principle to organometallic and coordination compounds. In order to describe these compounds two systems are used:

   (i) *the steering-wheel system, and*
  (ii) *the skew-line system.*

Before reaching the stage of defining absolute configuration, it is appropriate to describe coordination geometry precisely (square planar, tetrahedral, octahedral, etc.) and to define the elements of chirality.

For example, the square planar arrangement of M(a,b,c,d) is achiral while the tetrahedral form of the same chemical formula is chiral (Figure 2.37).

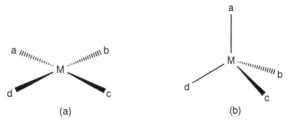

(a)   (b)

Figure 2.37
(a) Achiral and (b) chiral Mabcd structures.

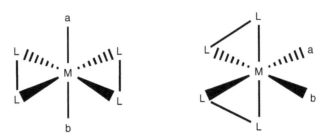

Figure 2.38
(a) Achiral and (b) chiral Mab(L-L)$_2$ structures.

An octahedral complex $M(a,b,L_2^2)$ is achiral if the ligands L^2 are in the *trans-* position but chiral if they are in the *cis-* position (Figure 2.38).

2.3.2.1 The Steering-wheel System

This is the generalization of the principle put forward by Cahn, Ingold and Prelog,[52] consisting of an oriented plane (or a circle) traversed by a vector oriented perpendicularly to the plane. The orientation in the plane or around the circle, either to the right or to the left, is unambiguous and will be called (*R*) and (*S*) or (*C*) and (*A*) (for rectus, sinister or clockwise, anticlockwise) (Figure 2.39).

The priority rules (sequence rule) allow the oriented plane and axis to be defined, as well as the (*R*) or (*S*) character of the configuration. The priority rules are the same in coordination chemistry as in organic chemistry.[53–55]

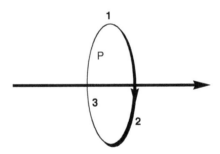

Priority order 1 2 3, the oriented plane P is (*R*) or (*C*)

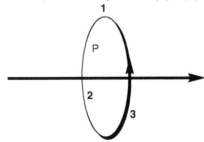

Priority order 1 2 3, the oriented plane P is (*S*) or (*A*)

Figure 2.39
The steering-wheel system.

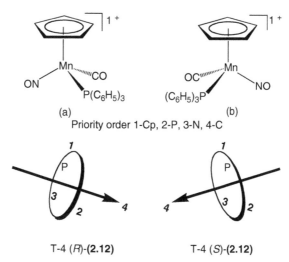

(a)   (b)

Priority order 1-Cp, 2-P, 3-N, 4-C

T-4 (*R*)-(**2.12**)   T-4 (*S*)-(**2.12**)

Figure 2.40
Determination of the absolute configuration using the steering-wheel system. Enantiomer (a) is designated as T-4 (*R*)-(**2.12**) and (b) as T-4 (*S*)-(**2.12**).

(a) The case of T-4 tetrahedral complexes. Taking, for example, the T-4 manganese complex [Mn(Cp)(CO)(NO)P(C_5H_5)$_3$]X (C_5H_5) = Cp (**2.12**)-X, the manganese atom is located in the centre of a tetrahedron in which all the vertices are different. In order to define the configuration of each enantiomer, an oriented plane and axis must first be defined. The order of the substituents is determined by the priority rules: 1 Cp, 2-P(C_5H_5)$_2$, 3-NO, 4-CO. Consequently the axis will be oriented from the manganese to the CO, and in order to go from 1 to 2 and then 3, it is necessary to turn to the right (*R*) configuration for enantiomer (a) and to the left (*S*) configuration for enantiomer (b) (Figure 2.40).

This description is also valid for complexes of the type TPY-3 (trigonal pyramidal) by imagining a coordination site with a dummy atom, which is assigned the lowest priority (4) (Figure 2.41).

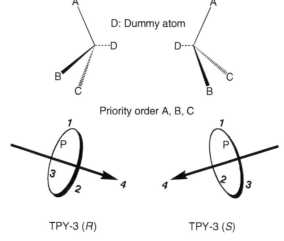

D: Dummy atom

Priority order A, B, C

TPY-3 (*R*)   TPY-3 (*S*)

Figure 2.41
Determination of the absolute configuration of a TPY-3 type complex.

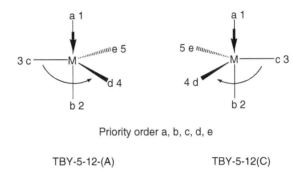

Priority order a, b, c, d, e

TBY-5-12-(A)   TBY-5-12(C)

Figure 2.42
Determination of the absolute configuration for a TBY-5 type complex.

However, in organometallic and coordination chemistry, many other geometric shapes are found besides the tetrahedron. It is possible to extend the rules used above to other polyhedrons. In order to avoid confusion with the tetrahedral systems, the absolute configurations will be designated *'clockwise'* (*C*) and *'anticlockwise'* (*A*).

(*b*) *Trigonal bipyramidal with the metal in the centre TBY-5*. By convention, the oriented vector goes from the substituent with the highest order of priority (1) towards the metal occupying the centre of the polyhedron. The three other substituents that are located in the plane perpendicular to the oriented vector are numbered according to the priority order, and the direction of rotation is observed to go from the one with the highest priority (3) towards the lower priority (4) (Figure 2.42).

(*c*) *Square-based pyramid SPY-5*. In the case of SPY-5, the oriented axis will be the formal C_4 axis, without taking account of the ligands situated in the plane. The axis is oriented from the axial ligand (1) towards the metal (Figure 2.43).

(*d*) *Octahedron OC-6*. Octahedrons with bis(chelate) ligands will be dealt with later (see skew-line convention). For the other types, the oriented axis goes from the ligand atom with the highest priority (1) towards the *trans* ligand with the lowest possible priority. The ligands situated in the perpendicular plane are numbered according to the priority rules. The sense of rotation is then that which goes from the ligand with the highest priority towards its immediate neighbour with the lowest priority. Some examples are shown in Figure 2.44.

(*e*) *Trigonal prisms TPR-6*. In this case, the ligands are first numbered according to their priority order. The oriented axis is defined as going from the trigonal face with the highest priority (that with the lowest overall sum of the ligand indices) towards the other trigonal face. The sense of rotation is determined by the ligand order (from the one with highest priority to its neighbour with lower priority) on the lower priority face (Figure 2.45).

This completes our description of polyhedrons, but a more detailed examination of this problem can be found in the IUPAC nomenclature rules.[54,55]

(*f*) *Metal coordinated to a prochiral plane or planar chirality*. For this type of chirality it is possible to apply the same rules that we have developed for centred tetrahedrons by considering the substituent in the complexed ring with the highest priority as being the chiral element. This substituent is thought of as being substituted, not only by the atoms to which it is directly bonded, but also by the metal that is complexed to the prochiral ring. In order to establish the priority order for the substituents adjacent

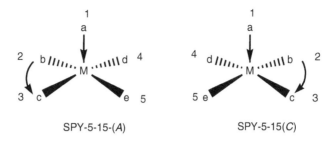

SPY-5-15-(*A*)   SPY-5-15(*C*)

Priority order a,b,c,d,e

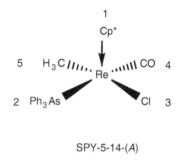

SPY-5-14-(*A*)

Figure 2.43
Determination of the absolute configuration for SPY-5 type complexes.

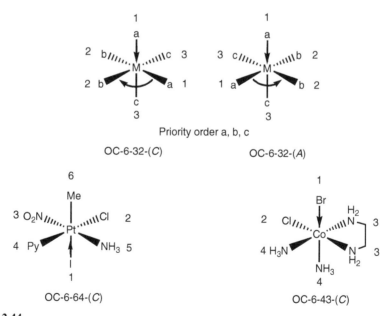

Priority order a, b, c

OC-6-32-(*C*)   OC-6-32-(*A*)

OC-6-64-(*C*)   OC-6-43-(*C*)

Figure 2.44
Determination of the absolute configuration for OC-6 type complex. Note that for the Pt complex on the left, the viewer should look from the iodide side before applying the priority rules.

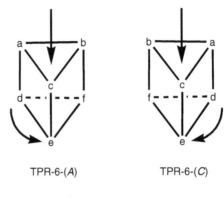

TPR-6-(*A*)   TPR-6-(*C*)

Priority order a, b, c, d, e, f
with a + b + c < d + e + f

Figure 2.45
Determination of the absolute configuration for TR-6 type complex.

to the central carbon atom, the adjacent carbon atoms are themselves also considered to be bonded to the metal through a formal covalent bond. Figure 2.46 shows the application of this treatment to a ferrocene derivative with 1,2-substitution in one ring, and for a 1,2-disubstituted arene chromium tricarbonyl complex, with in each case the substituent (a) being of higher priority than substituent (b) (Figure 2.46).

For ferrocene compounds the notation of Schlögl[56] is often used. The priority order of the substituents is determined by the Cahn–Ingold–Prelog rules. An axis passing through the substituted cyclopentadienyl ring is oriented towards the metal, and the direction of rotation about this axis is in the sense of decreasing priority, with *pR* for a rotation to the right and *pS* to the left (Figure 2.47). In the case where both cyclopentadienyls are substituted, a configuration is given to both of the 'chiral planes', specifying the plane. Note that if the substituents (a) and (b), with (a) being higher priority than (b), are the same for the two rings and in the same relative position (1,2) or (1,3), then the configuration (*pR*, *pS'*) corresponds to a *meso* compound, due to the fact that there is a plane of symmetry passing through the metal atom and parallel to the plane of the cyclopentadienyls.

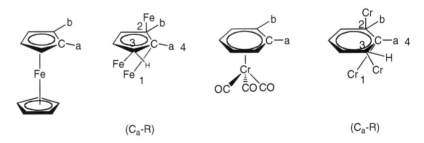

(C_a-R)   (C_a-R)

Figure 2.46
Determination of the absolute configuration for a 1,2-disubstituted ferrocenic and 1,2-disubstituted arene chromium tricarbonyl compounds. Carbon C bearing the priority substituent (a) is considered as the central atom, the priority order is determined following the Cahn–Ingold–Prelog rules.

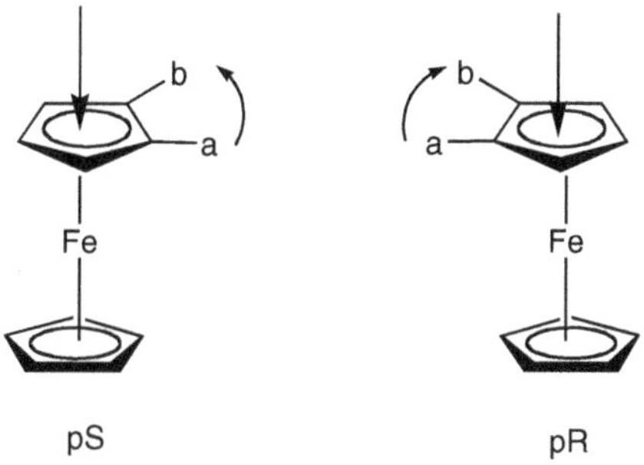

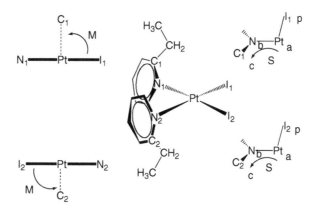

Priority order a, b

Figure 2.47
The Schlögl notation for the absolute configuration description of a 1,2-disubstituted ferrocenic compound.

Another type of 'planar chirality' can be found in tetra-coordinated complexes such as, for example, those of platinum, when two aromatic ring ligands in a *cis* position are substituted by bulky groups, and the symmetry is reduced to a C_2 axis. This is an example of conformational chirality depending on planar chirality. This is the situation in, for example, compound (**2.33**) (Figure 2.48) which has two 'chiral planes' with configuration (*pS, pS*) or (M, M) if the torsion planes are used as the descriptor.[30]

(g) *Axial chirality arising from metal coordinated to four ligands.* This is the case for tetra-coordinated square planar compounds in which two axial aromatic ligands are '*ortho/ortho*' di-substituted with (a)/(a') and (b)/(b'), the former having the higher priority. The resulting conformational chirality arises from the hindered rotation about the axis joining the ligands to the metal. (Note that if (a) = (a') and (b) = (b') the complex is also chiral) (Figure 2.49).

Figure 2.48
The absolute configuration (pS, pS)-(**2.33**) is attributed using iodide 1 and 2 as pilot atoms (p). Using the torsional angles $\tau(I_1PtN_1C_1)$ and $\tau(I_2PtN_2C_2)$ and according to the Cahn–Ingold–Prelog priority rules the absolute configuration of the two planar chiral elements can be named as (M, M).

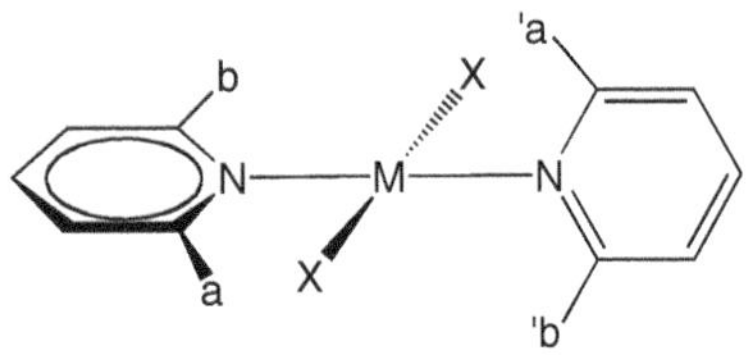

Figure 2.49
Example of conformational chirality arising from the hindered rotation around the N-M-N axis.

There are two possible methods for assigning the configuration in this case. According to the Cahn–Ingold–Prelog rules the axis is defined as joining the metal centre to the atoms of the substituted axial ligands. The priority order is determined by taking into account the fact that, starting from the highest priority atom, the number (2) is the *nearest substituted atom*. The sense of rotation is defined as going from (1) to (2) and then (3). Note that the direction of the axis may be inverted without changing the configuration (*aR*) or (*aS*), and the symbol '*a*' preceding the configuration signifies that we are dealing with axial chirality (Figure 2.50).

Another way of designating the configuration of this type of complex is based on the torsion axis. In this case as well the dihedral torsion angle is defined according to the priority rules, including the one where the nearer has precedence over the further away. In the example shown in Figure 2.49, the configurations are (*aR*) and (*aS*), or (P) and (M) if based on the torsion angle. Note that in axial chirality the (*aR*) configuration corresponds to the torsion angle (M).[57]

2.3.2.2 The Skew-line Convention

In coordination chemistry helicoidal chirality is frequently encountered. It is the chirality of OC-6 complexes possessing two (*cis* geometry) or three bis(chelate) substituents, and also of helicates in supramolecular chemistry.

The principle of the 'skew-line convention'[58] arises from the geometry of a helix, which is by nature chiral, and is generated by simultaneous movements of translation and rotation. The sense of a helix is unequivocally determined by the relative sense of the translation and rotation vectors. We take the translation movement as a nonoriented axis

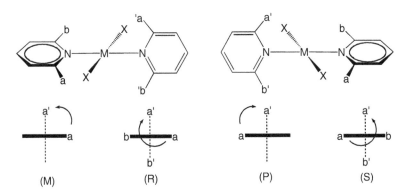

Figure 2.50
Axial chirality: assignment of the absolute configurations (M) or (P) using the torsional axis; (R) or (S) using the steering-wheel system.

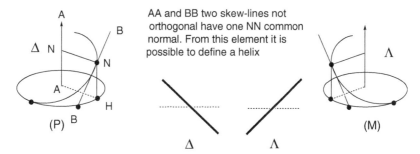

Figure 2.51
The skew-line convention for the determination of the absolute (Δ) or (Λ) configuration of a propeller.

(AA), and define a tangent (BB) to the helix at a point N (these two straight lines having a common perpendicular NN). With the tangent (BB) placed at the front, the sense of the helix is then defined as a function of the relative position of these two straight lines. If the sense is clockwise the helix is right (Δ) or (P), while in the opposite case it is (Λ) or (M) (Figure 2.51). (Δ) and (Λ) are used for configurations, and (δ) and (λ) where it is a question of conformations, as is the case, for example, for chelating ligands that form penta-coordinated rings with the metal centre.[53]

(Δ) and (Λ) are used for mononuclear complexes, while (P) and (M) are used in the case of polynuclear complexes that form a helix. In the following example of the three-dimensional anionic network (3D) $[Li^ICr^{III}(C_2O_4)_3]^{2-}$ (**2.34**)$^{2-}$ with a (Δ) configuration, the octahedrons surrounding each metal produce an (M) supramolecular helix (Figure 2.52).[59]

If we now take a compound of OC-6 geometry with a metal centre surrounded by three identical bis(chelate) ligands, we can represent this structure, which has D_3 symmetry, by (a) and following a rotation of 90° about the axis XX' by (b) (Figure 2.53).

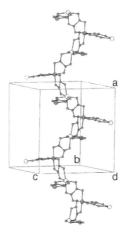

Figure 2.52
View of the (M)-handed helix of the 3D $[Li^ICr^{III}(C_2O_4)_3]^{2-}$ anionic network, in which the hexacoordinated metallic Li(I) and Cr(III) centres are of (Δ) absolute configurations (reproduced with permission from reference[59], copyright 1999, American Chemical Society).

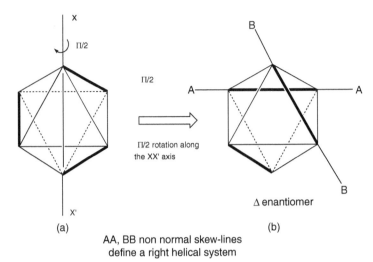

AA, BB non normal skew-lines
define a right helical system

Figure 2.53
Determination of the absolute configuration for a OC-6 tris chelated metal complex using the skew-line convention.

In this type of complex the helicity is determined by the relative position of the lines (AA) and (BB). In the case shown in Figure 2.53, the absolute configuration of the complex will therefore be OC-6-(Δ). This rule that we have just defined can also be applied to complexes of the type $M(L^2_2)a_2$ (symmetry C_2) and $M(L^2_2)ab$ (symmetry C_1).

Note that in simple cases such as [Ru^{II}(bpy)$_3$](PF$_6$)$_2$ **(2.17)-(PF$_6$)$_2$**, the determination of the absolute configuration can easily be accomplished by locating three metal–ligand bonds in front of the plane of the paper and three behind. If one turns to the right when going from the front to the rear the complex is (Δ), while it is (Λ) in the opposite case (Figure 2.54).

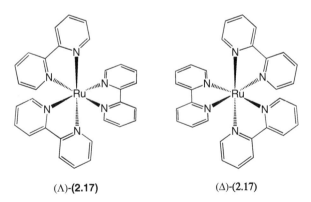

(Λ)-**(2.17)**   (Δ)-**(2.17)**

Figure 2.54
View of the (Δ) and (Λ) absolute configuration of the cationic part of [Ru^{II}(bpy)$_3$](PF$_6$)$_2$**(2.17)**-(PF$_6$)$_2$.

Physical Properties of Enantiomers and Racemics

2.4.1 Optical Properties

2.4.1.1 Rotation of Plane Polarized Light, Optical Dispersion and Circular Dichroism

(a) *Optical rotation*. The fundamental property which distinguishes the two enantiomers of a chiral molecule from the racemic compound is the rotation of plane polarized light, the sense of the rotation being opposite $(+)/(-)$ for each of the enantiomers. What is the origin of this phenomenon, which is the oldest manifestation of enantiomeric molecules? How can it be measured and what is the relationship between its direction and magnitude and the absolute configuration?

The propagation of an electromagnetic wave in a given medium is determined by the interactions between the wave and the material. Important information on these interactions can be obtained from the material's optical properties by spectroscopy. When chirality, in the form of a given enantiomer, is present at the molecular level there is a breakdown in the spatial symmetry of the medium through which the wave passes. It is this breakdown in symmetry that is responsible for the phenomenon of the rotation of the plane of polarization of light.

Without going into the physical details of this phenomenon (more detailed information may be found in the works cited in the Reference Section[60,61]), we can say that the two circularly polarized components of linearly polarized light, having the opposite sense but the same phase and amplitude, do not propagate through a chiral medium at the same speed. The median plane between the right and left components of the light is no longer the initial plane of polarization, but a plane which has turned through an angle (α).

Fresnel's law expresses this rotation as a function of:

   (i) *the difference in the indices of refraction for right- and left-polarized light* $(n_L - n_R)$,
   (ii) *the path length (l) of the sample, and*
   (iii) *the wavelength (λ) of the incident light,*

$$\alpha = [(n_L - n_R)\pi l]/\lambda.$$

In practice, the specific rotation of a substance is defined by the relationship:

$$[\alpha]_{\text{specific}} = \alpha_{\text{measured}}/lc$$

where $[\alpha]_{\text{measured}}$ is the observed rotation expressed in degrees, l is the path length measured in dm and c is the concentration in g cm^{-3} (for a pure material c is the density expressed in g cm^{-3}).

At this point two important aspects of the specific rotation must be emphasized.

   (i) *The measured value of α can lie between 0 and 360°. It is important to check this point carefully by diluting the solution or changing the path length.*
   (ii) *The specific rotation, which is the physical magnitude associated with an enantiomer, depends on the concentration, the solvent, the temperature and the wavelength of the incident light. As such, the specific rotation only makes sense if these values are stated, and should be expressed as:*

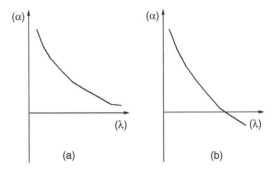

Figure 2.55
The two types of optical rotatory dispersion $[\alpha] = f(\lambda)$ with (a) positive values, and (b) positive and negative values.

$$[\alpha]^{T}_{\lambda} = x \, (C, \, solvent) \, with \, T \, in \, ^{\circ}C, \, \lambda \, in \, nm, \, C \, in \, g \, 100 \, cm^{-3} \, and \, [\alpha]^{T}_{l} \, in \, 10^{-1} \, deg \, cm^{2} \, g^{-1}.$$

For example, for (+)-$[Cr^{III}(C_2O_4)_3]K_3$ **(2.10)** K_3 we write: $[\alpha]^{20}_{589} = +1900$ $(C = 0.04, \, H_2O)$.[62] In general, the optical rotation is measured at the wavelength of the sodium D ray (589 nm) or of the mercury rays at (578, 546, 436 and 365 nm).

(b) *Optical dispersion (OD).* The specific rotation of a compound, either pure or in solution, varies as a function of the wavelength. This phenomenon is called the 'optical rotatory dispersion'. Two types of variation can be distinguished in the curves of $[\alpha] = f(\lambda)$.

(i) *Monotonic variations*

In this case, the function $[\alpha] = f(\lambda)$ varies according to $1/\lambda^2$. The observed curve can show positive or negative values of $[\alpha]$ (Figure 2.55 (a) and (b)).

A positive or negative monotonic curve can only be observed when the substance through which the light passes has no absorption at the wavelengths under consideration.

(ii) *Cotton effect curves*

In the case where there is absorption, the difference in refractive indices $\delta(n) = (n_G - n_D)$ passes through a maximum and a minimum with a point of inflection. This is a positive or negative Cotton effect (Figure 2.56 (a) and (b)).

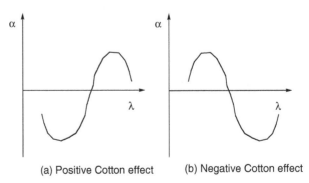

(a) Positive Cotton effect   (b) Negative Cotton effect

Figure 2.56
$[\alpha] = f(\lambda)$ curves showing (a) positive and (b) negative cotton effect.

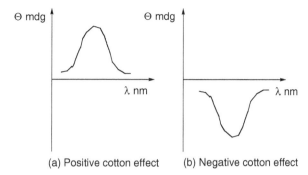

(a) Positive cotton effect   **(b)** Negative cotton effect

Figure 2.57
(a) CD positive and (b) CD negative curve for $\Theta = f(\lambda)$.

For two enantiomers, these curves are strictly symmetrical with respect to the *x*-axis. It is possible that a single substance can give multiple Cotton effects of opposite sense.

(c) *Circular dichroism (CD)*. This is a phenomenon of the same nature as optical dispersion. It arises from the fact that the extinction coefficients for two waves with left- and right-circular polarization, εL and εR respectively, are different when the waves pass through an optically active substance possessing a chromophore at the wavelength under consideration, in this case in the ultraviolet-visible region. The right- and left-circularly polarized light becomes transformed into elliptically polarized light, and the Cotton effect can be observed in the absorption region of the compound. The ellipticity Θ is defined by: $\Theta = K(\varepsilon R - \varepsilon L)$, where Θ is measured in millidegrees (mdg). The curves $\Theta = f(\lambda)$ are symmetrical for the two enantiomers (Figure 2.57).

In order to be able to compare curves, the value Θ is converted into the molar ellipticity $[\Theta] = \Theta/10cl$, where *c* is the molar concentration, *l* is measured in cm and the molar ellipticity has units of deg cm^2 dmol^{-1}. If the phenomenon is expressed as molar circular dichroism, where $\delta\varepsilon = (\varepsilon L - \varepsilon R)$ and $\delta\varepsilon$ has units of M^{-1}cm^{-1}, then $\delta\varepsilon$ and $[\Theta]$ are related by: $\Delta\varepsilon = [\Theta]/3298$.

In Figure 2.58, the CD curves $\delta\varepsilon = f(\lambda)$ are shown for the (Δ) and (Λ) enantiomers of the compound [Ru(bpy)$_2$Hcmbpy](PF$_6$)$_2$ **(2.35)** where Hcmbpy = 4-carboxy-4'-methyl-2,2'-bipyridine.[63]

It can be seen that for the two enantiomers the curves are opposite and symmetrical with respect to the *x*-axis. Each enantiomer shows positive and negative Cotton effects of differing intensities. In the case of ruthenium (II) complexes with two 2,2'-bipyridine ligands, Bosnich[64,65] demonstrated that the Cotton effect observed around 249 nm is strongly negative at low energy and positive at high energy if the absolute configuration is (Δ). This is attributed to metal–ligand charge transfer (MLCT).[60,66,67]

(d) *Induced circular dichroism*. Induced circular dichroism is observable for electronic transitions of an achiral molecule when, for example, it is associated in the form of a coordination complex with a chiral molecule. In the example given by Stang[68] of a tetranuclear platinum complex where two achiral fragments are linked by two (*S,S*)-tartrate anions **(2.36)** (Figure 2.59), a Cotton effect is observed for the transitions of the achiral anthracene unit.

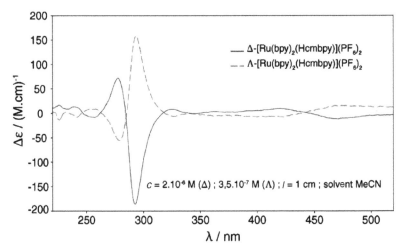

Figure 2.58
CD curves ($\Delta\varepsilon = f(\lambda)$) for the ($\Delta$) and ($\Lambda$) enantiomers of [Ru(bpy)$_2$Hcmbpy)](PF$_6$)$_2$ (**2.35**))-(PF$_6$)$_2$.

(e) Vibrational circular dichroism (VCD). Vibrational circular dichroism is the equivalent of CD in the infrared region,[69,70] a technique that is now in almost routine use and which has both pros and cons. On the one hand, one of the most important advantages is the ability to identify in the spectrum the contribution of possible optically active impurities. On the other hand must be noted the restriction on

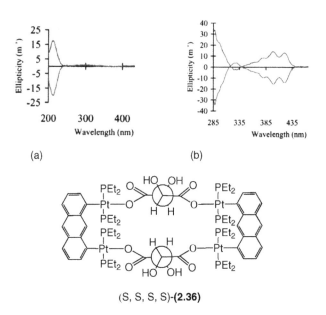

(S, S, S, S)-(**2.36**)

Figure 2.59
Induced circular dichroism of (R, R)-tartaric acid (red a) and (S, S)-tartaric acid (blue a) ($c = 2.25 \times 10^{-4}$ M, H$_2$O) and of (R, R, R, R)-(**2.36**) (red b) and of (S, S, S, S)-(**2.36**) (blue b) (reproduced with the permission from reference[68], copyright 2006, American Chemical Society).

molecular size, fewer than one hundred atoms, and the complexity of the simulation calculations which must take into account all the possible conformers.

(f) Raman optical activity (ROA). This method, developed by Barron[71–73], permits the determination of enantiomeric purity of chiral molecules in solution with high accuracy. However, examples of studies of chiral transition metal complexes are rare since traditional methods have insufficient accuracy. In our opinion, the most likely reason for this is the specificity of ROA using a laser light source with wavelength in the visible or ultraviolet range, as most transition metal ions have absorption bands in that range. The resulting strong absorption of the incoming radiation does not allow the study of inelastic coherent scattering (i.e. the Raman effect) or lattice vibrations (phonons).

(g) Magnetochiral dichroism (MChD) or magnetochiral anisotropy (MChA). Magneto-optical effects, like the Faraday effect, are intimately related to the breaking of time-reversal symmetry by the magnetic field. The breaking of space-inversion symmetry in chiral systems leads to optical activity. In both cases, the polarization of light is rotated upon propagation through the medium. Interpreting the Faraday effect as a sign of magnetically induced chirality, Pasteur was the first to investigate (followed by many others) whether magnetic fields can discriminate between left- and right-handed chiral systems, but without success.

The recently discovered magnetochiral anisotropy (MChA) effect[74–77] can be regarded as an optical cross effect between chirality and magnetism or, alternatively, as the consequence of simultaneously breaking time- and space-symmetry. It is described by an extra term in the dielectric constant for right/left $(+/-)$ circularly polarized light and right/left (d/l) media which is proportional to the dot product $(\mathbf{k} \cdot \mathbf{B})$ where $\mathbf{k}$ is the wave vector of the light and $\mathbf{B}$ is the magnetic field.

The essential features of MChA are:

(i) the dependence on the relative orientation of $\mathbf{B}$ and $\mathbf{k}$,

(ii) the dependence on the handedness of the chiral medium, and

(iii) the independence of the polarization state of the light.

The relative strength of MChA in optical absorption with unpolarized light is given by:

$$g_{MChA} \equiv 2 \frac{A\left(_B^l \uparrow\uparrow {}_k^l\right) - A\left(_B^l \downarrow\uparrow {}_k^l\right)}{A\left(_B^r \uparrow\uparrow {}_k^r\right) + A\left(_B^r \downarrow\uparrow {}_k^r\right)} \approx \frac{c''}{\varepsilon''} \left(_B^r \cdot {}_k^r\right), {}^{[78]}$$

where the A's are the optical extinction coefficients for the respective configurations.

2.4.1.2 Solid State Organization and X-ray Diffraction

We have seen in Section 2.2 that racemic compounds can exist in three crystalline forms: racemic compounds, conglomerates and pseudo-racemic compounds, each of which has particular properties. Thus, racemic compounds and conglomerates crystallize in forms whose space groups are centro- and noncentrosymmetric respectively. The crystallization of enantiomers always leads to noncentrosymmetric structures.

Particular properties result from the type of organization in the solid state. Thus, while the infrared spectra of a conglomerate and the pure enantiomers that make it up are identical, this is not the case for a racemic compound and the corresponding pure enantiomers.

2.4.2   Determination of Absolute Configuration

The determination of the absolute configuration of a compound is a decisive step in the characterization of an enantiomer. A range of physical and chemical methods is available for this, and often it is a combination of the two that leads to a solution. The problem of determining the absolute configuration can be posed as follows: by using a descriptor (*R*), (*S*); (Δ), (Λ); (P), (M); (*pR*), (*pS*); (*aR*), (*aS*) we know how to represent the two enantiomers in space. We have measured the physico-chemical properties of each enantiomer. How do we then relate these properties to one of the spatial representations?

2.4.2.1   X-ray Diffraction

We have seen that the enantiomers of a molecule have the same properties (melting and boiling points, refractive index, etc.) and that spectroscopy that does not involve polarized light is incapable of distinguishing between them, so that their infrared, NMR, ultraviolet-visible and Raman spectra are identical. On the other hand, the optical rotation, optical dispersion and circular dichroism give results that are opposite in sign for the two enantiomers. We have also seen that in certain cases it is possible to determine the absolute configuration of a molecule in the crystal by X-ray diffraction.

It should be added that, unlike Pasteur and Werner, we are not starting from zero as the absolute configuration of many enantiomers is already known and indexed. This knowledge is a very powerful tool. In effect, if we take the case of a compound with several chiral elements, the knowledge of their relative configuration, coupled with the absolute configuration of just one of them, allows us to define the absolute configuration of all the chiral elements. In coordination chemistry, where many compounds are salts, it is possible (if of course the anion and cation are chiral) from the relative configuration determined by X-ray diffraction on a monocrystal, and knowing the absolute configuration of one of the ions, to determine the absolute configuration of the other.

In the following example of the hexacoordinated phosphate [P(C$_6$Cl$_4$O$_2$)$_3$]-Cinchonidinium[79] Trisphat Cinchonidinium (**2.32**)-Cinchonidinium, the configuration of the cinchonidinium cation (**2.37**) is known, as it results from the protonation of(−)-cinchonidine (**2.38**) (a naturally occurring enantiomerically pure amine). In these circumstances, the relative configuration determined by X-ray diffraction leads us directly to the absolute configuration (Δ) of the anion (Figure 2.60).

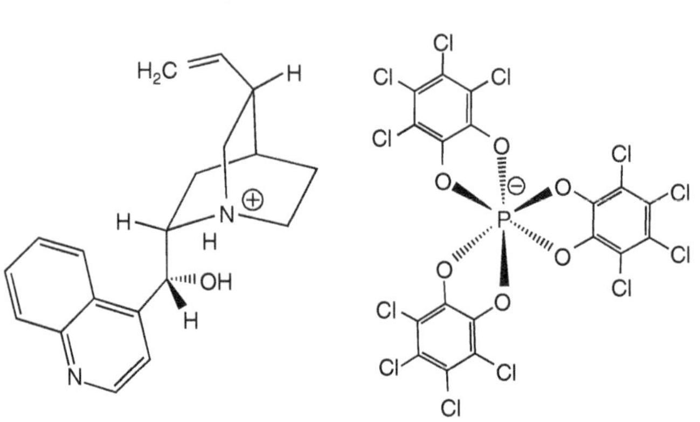

(l)-(**2.37**)   (Δ)-(**2.32**)

Figure 2.60
View of the Cinchonidinium cation (left) and of the Trisphat anion (right).

At this stage an important practical point should be made. To determine the structure of a compound, we have used a single crystal, which could have been selected from among dozens of others. If crystallization leads to two possible diastereomers, there is nothing to say that the crystal that we have chosen is representative of the majority. It is important to check this point carefully, by measuring the CD of the crystal being studied and comparing it with that of the bulk of the crystals, either taken up in solution or powdered. The absence of this check may lead to errors in the assignment of the absolute configuration. We have used the example of a salt, but the rationale is the same for those cases where the chiral elements of a compound are linked by covalent or dative bonds.

In the absence of what may be termed a 'chiral probe' to establish the absolute configuration, it is nevertheless possible in some cases to reach the absolute configuration by an X-ray diffraction study of a single crystal. This is how Bijvoet[80] determined the absolute configuration of sodium rubidium bitartrate (+)-NaRb tartrate **(2.5)-NaRb**.

The anomalous scattering diffraction of X-rays takes into account the phase of the diffused wave. The diffusion factor $f = f' + if''$ contains a real part f' and an imaginary part if''. Atomic absorption is taken into account in the imaginary part, and is dependent on the atoms under consideration. The heaviest atoms give rise to a more significant anomalous scattering. Organometallic and coordination derivatives containing metal atoms are more likely to provide the necessary data to measure the anomalous scattering than organic compounds. The wavelength of the radiation also plays an important role, and should be as close as possible in energy to the atomic absorption being considered.

Today, structural calculation from the diffractogram allows the Flack's parameter[81] to be determined automatically. This is equal, or very close, to zero if the absolute configuration attributed to the molecule is correct, and is equal to unity in the case of the opposite configuration. It is equal or close to 0.5 if the two configurations are present in equal amounts in the crystal.

Just as in the case of the relative configuration, it is important to establish that the monocrystal being studied is representative of the bulk of the separated crystals. In the absence of a monocrystal, powder X-ray diffraction studies can also yield important information. The determination of the space group (chiral or not) and the unit cell parameters give us an indication of the chiral nature, or otherwise, of a molecular material, and in conjunction with other techniques, such as CD, sometimes allow the absolute configuration to be established.

2.4.2.2 Optical Rotation

The optical rotation, the optical dispersion and chiroptical methods in general yield information that is directly correlated to the absolute configuration of an enantiomer. However, this correlation is far from being clear. In practice, and in the absence of other information, it is difficult to predict the absolute configuration of a compound based solely on its optical rotation. Under the same conditions of solvent, temperature and wavelength, the sign of the rotation can be inverted for compounds that are nevertheless very close, such as (Λ)-(+)-$[Cr^{III}(C_2O_4)_3]K_3$ **(2.10)-K$_3$** and (Λ)-(−)-$[Co^{III}(C_2O_4)_3]K_3$ **(2.11)-K$_3$**.

The CD measurement, with a great deal of caution in its interpretation, can allow the absolute configuration to be assigned to a complex belonging to a family, where the configuration of one of the members is already established. In this case, it can be assumed that two molecules of identical configuration must have the same Cotton effect for electronic transitions of the same origin.

There are numerous 'rules' allowing the stereochemistry of a compound under investigation to be related to the sign of one of its Cotton effects. These 'rules' are restricted in their application to a precise type of compound, for example, it is the case for nonconjugated cyclohexanones (octant rule),[82] or for ruthenium (II) complexes of the bis-, tris-bipyridine or bis, tris-phenanthroline type. Bosnich[64,65] demonstrated that in the case of ruthenium (II) complexes with two 2,2'-bipyridine, 1,10-phenanthroline or acetylacetonate (1^-) (acac) ligands, the Cotton effect observed around 249 nm (bis-bpy complexes), 263 nm (bis-phen) or 286 nm (bis-acac) is strongly negative at low energy and positive at high energy if the absolute configuration is (Δ). These Cotton effects are attributed to metal–ligand charge transfer (MLCT).

2.4.3 Determination of the Enantiomeric Excess (*ee*)

We will describe three significant methods based on the measurement of optical rotation, nuclear magnetic resonance and chromatography. However, other methods are possible (infrared, Raman, etc.), while some such as calorimetry are hardly ever used for metallic complexes.

2.4.3.1 Measurement of the Specific Rotation

The specific rotation of a compound is, as we have seen (Section 2.4.1.1), defined by $[\alpha]_{\text{specific}} = [\alpha]_{\text{measured}}/lc$ where $[\alpha]_{\text{measured}}$ is the observed rotation expressed in °, l is the path length in dm and c is the concentration in g cm^{-3} (for a pure material c is the density in g cm^{-3}). If the specific rotation of a pure enantiomer is known, it is possible to calculate the *ee* of a mixture enriched in this enantiomer directly according to the relationship:

$$ee = [\alpha]_{\text{specific enriched}}/[\alpha]_{\text{specific pure enantiomer}}.$$

This method appears simple, yet several precautions are necessary. Apart from the fact of knowing the specific rotation of the pure enantiomer, which is itself not always straightforward, the measurement of the optical rotation must be made under the same conditions of solvent, concentration, temperature and wavelength as were used to obtain the specific rotation of the pure enantiomer. Moreover, the measured rotation is often weak, which increases the error in the measurement. It also happens not infrequently that the complexes being studied are strongly coloured, and as a result the measurement is not possible with a standard polarimeter. The presence of impurities with a large optical rotation may also invalidate the reading.

2.4.3.2 Nuclear Magnetic Resonance (NMR)

This is a technique widely used for the determination of the ee of a mixture of diamagnetic enantiomers. On its own, NMR cannot distinguish between enantiomers, rather it is an indirect method which requires that the enantiomer mixture be converted into a mixture of diastereomers. This can be achieved by:

(i) *a chemical reaction that adds a chiral element to the molecule,*
(ii) *creation of diastereomeric ion pairs,*

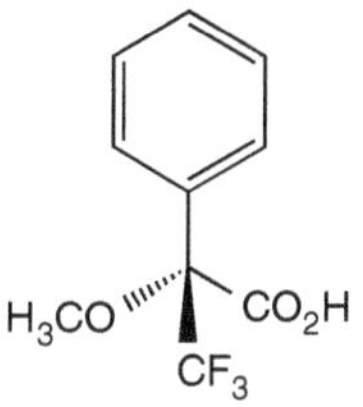

(R)-**(2.39)**

Figure 2.61
View of the (R)-(α-methoxy-α-(trifluoromethyl)phenylacetic acid) (R)-(**2.39**).

(iii) using a chiral chemical shift reagent, or
(iv) using a chiral solvation agent.

In each case it is a matter of obtaining a diastereomeric environment around the nuclei whose resonances can then be distinguished. The *ee* is then calculated by integration of the peaks corresponding to each of the diastereomers. In reality, it is a diastereomeric excess that is calculated, but this can be converted into an *ee* if the chiral auxiliary agent used is itself enantiomerically pure. In order for the precision of the measurement to reach 5%, each peak area must be obtained with an accuracy of 1.25%.

(a) Chemical reaction adding a chiral element to the molecule. The chiral derivatizing agent most widely used for forming a covalent bond is Mosher's acid:[60] (R)- and (S)- (α-methoxy-α-(trifluoromethyl)phenylacetic acid) (**2.39**) (Figure 2.61).

Its use requires that the compound under investigation reacts with an organic acid group. The *ee* can be measured by proton or fluorine NMR. However, certain precautions must be taken to ensure that the measured *ee* is meaningful.

  (i) The reaction between the chiral agent and the enantiomer mixture to be studied must be complete. In simple terms, this means that there must not be any kinetic resolution.

  (ii) There must not be any racemization during the reaction.

  (iii) There must not be any decomposition of the complex being studied.

Having to fulfil all these conditions does not make the use of this type of reagent very easy in the majority of cases of metal complex chemistry.

(b) Creation of diastereomeric ion pairs. Many coordination complexes exist in the form of salts in which one of the ions is chiral and the other not. This is the case for many complexes hexacoordinated by neutral ligands, of ruthenium, osmium, iridium and so on. Replacing the achiral counter-ion with an enantiomerically pure chiral counter-ion leads to two pairs of diastereomeric ions. These exhibit different NMR spectra and so the ee can be measured as described above.

In order to measure the ee of cations of chiral complexes, Lacour[83] proposed a number of enantiomerically pure anions, of which we will mention in particular (Δ)- and (Λ)-Trisphat (**2.32**) and (Λ)-BINPHAT (**2.40**) (Figure 2.62).

These two anions form relatively intimate pairs with organometallic and coordination cations, with certain ammonium and phosphonium cations, for example. The specific interactions which result allow an enantio-differentiation to be made by

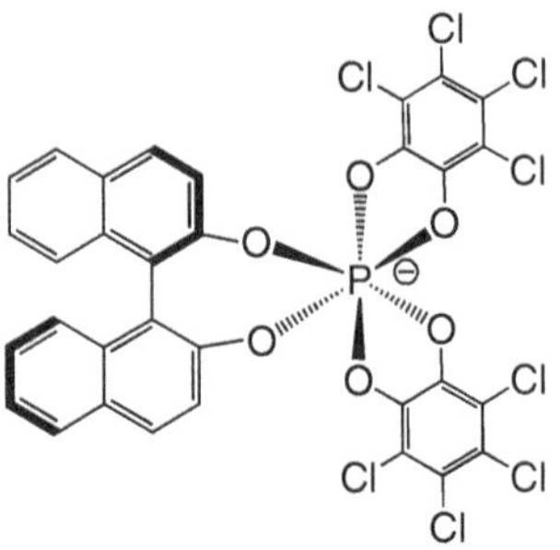

(Λ)-**(2.40)**

Figure 2.62
View of the (Λ)-BINPHAT **(2.40)**.

NMR. These chiral anionic reagents are diamagnetic, thus avoiding broadening of the peaks in the NMR spectra, which is not the case, as we shall see, with certain chiral lanthanide complexes used as chemical shift agents.

(Δ)-(−)-Trisphat-Cinchonidinium **(2.32)**-Cinchonidinium,[84] is particularly interesting. Its synthesis in an enantiomerically pure form ($[\alpha]_D^{20} = -361$ ($c = 0.1$, EtOH)) is easy, its optical purity can be confirmed by phosphorus NMR and it does not have any protons likely to hinder the reading of the NMR spectrum.

In Figure 2.63 several types of cation are shown whose enantiomeric purity has been determined by proton, phosphorus or nitrogen NMR using **(2.32)**.[33,84–86]

Recent work[83] has also shown that (Δ)-Trisphat **(2.32)** can be used to measure the enantiomeric purity of neutral complexes of chromium and palladium (Figure 2.64).

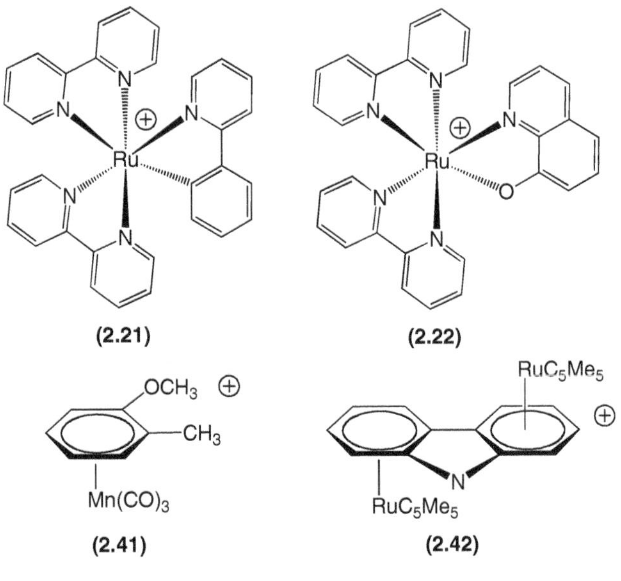

(2.21)   **(2.22)**

(2.41)   **(2.42)**

Figure 2.63
Some examples of chiral complexes whose optical purity was determined by[1]H NMR using (Δ)-Trisphat as chiral shift reagent.

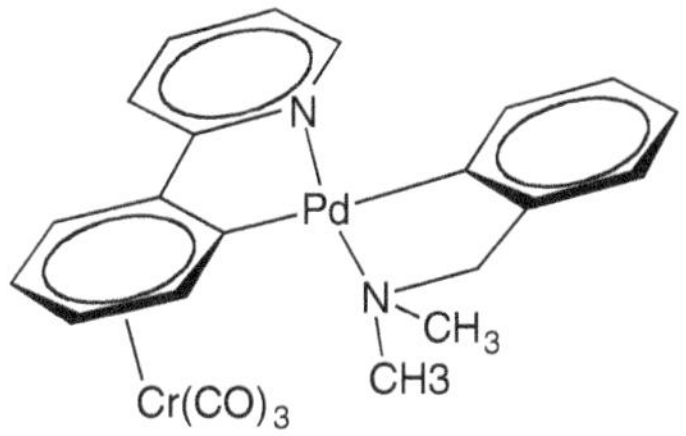

(2.43)

Figure 2.64
View of the palladium complex (**2.43**).

(c) Chiral chemical shift reagent. Chiral chemical shift reagents are principally chiral lanthanide complexes. They act as Lewis acids towards basic compounds (in the Lewis sense), becoming more or less temporary ligands in the metal coordination sphere, and thereby forming two diastereomers in which the Lewis acid and base are more or less associated (Scheme 2.2).

$$(ML_3)^* + \text{Chiral molecule} \Leftrightarrow (ML_2^*, \text{Chiral molecule}) + L^*$$

Scheme 2.2

The most well-known chiral chemical shift reagents are chiral lanthanide complexes (Figure 2.65). The broadening of NMR signals due to the paramagnetic character of a number of these chiral agents reduces the precision of the measurement. It has been shown, incidentally, that (Δ)-Trisphat (**2.32**) is clearly superior to Eu(tfc)$_3$ (**2.44-Eu**) in the determination of the enantiomeric purity of salts containing the [Ru(phen)$_2$py$_2$]$^{2+}$ cation (**2.16**).[87]

(d) Chiral solvation agents, chiral solvents and chiral liquid crystals. The differentiation of two enantiomers by NMR is possible by studying them in the presence of a chiral solvent or in chiral liquid crystals.[88–90] The molecule–solvent or molecule–liquid crystal interaction is sufficiently energetic for the diastereomeric association to have a lifetime longer than the NMR timescale.

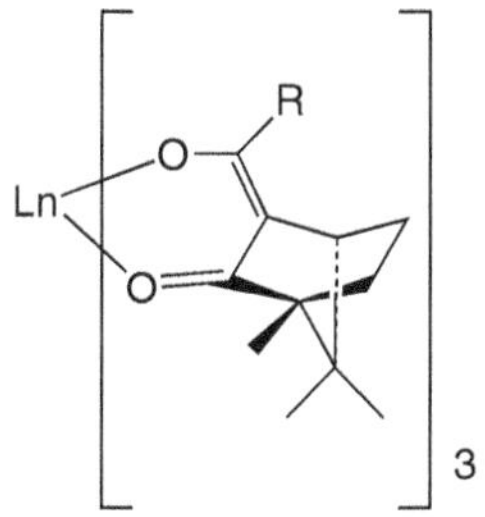

Ln = Eu, Pr, Yb
R = tBu, CF$_3$, C$_3$F$_7$
(2.44)

Figure 2.65
Lanthanide complexes used as chiral shift reagents.

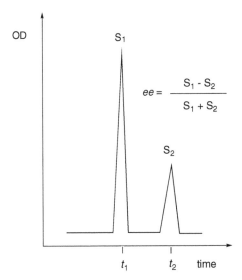

Figure 2.66
Determination of the *ee* of a mixture of enantiomers using chiral chromatography techniques.

2.4.3.3   Chromatography

Chromatographic techniques are certainly among the most precise for the measurement of *ee*. While this assertion may be true for organic compounds, it is less so for organometallic and coordination compounds. In particular, gas phase chromatography, which is simple to put into practice for organic compounds, is practically impossible for metal complexes due to the high vaporization temperatures which lead to decomposition of the product. High performance liquid chromatography (HPLC) is certainly more suitable due to its room temperature operation, the only limitation being the solubility in a solvent or solvent mixture compatible with the stationary phase used.

Each enantiomer carried by the mobile phase interacts in a different manner with the stationary phase which contains an enantiomerically pure chiral element. There again, it is diastereomeric interactions that are translated into different elution rates through the column. In principle, whatever the detector (ultraviolet-visible spectrometer, refractometer, etc.) the response factor is identical for the two enantiomers being analysed. As a result, integration of the peak areas corresponding to each enantiomer leads to a measure of the *ee* by the simple relationship: $ee = (S_1 - S_2)/(S_1 + S_2)$ (Figure 2.66).

To obtain a high degree of precision in the measurement, the chromatographic peaks should be as close as possible to Gaussian and be separated right to their base. It should also be checked that no other product has the same retention time as either of the enantiomers. In effect, even a small quantity of an impurity with a strong ultraviolet-visible absorption (in the case that such a detector is used) can have a significant effect on the measurement.

By using derivatives of cellulose 3,5-dimethylphenyl carbamate (Chiracel OD) as the chiral phase Bergman[91] measured the enantiopurity of complexes of molybdenum **(2.45)** and iridium **(2.46)** (Figure 2.67). With the same type of stationary phase, Gladysz[23] measured the *ee* of rhenium complexes $ReCp(NO)P(C_6H_5)_2Br$ **(2.13)**, and uranyl-salophen[92] complexes of uranium were analysed with a Chiracel ODH column.

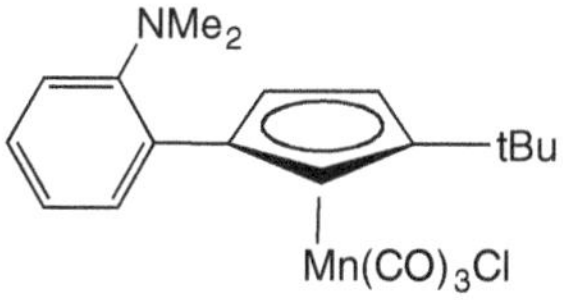

Figure 2.67
View of the manganese and iridium (**2.45**) and (**2.46**) complexes.

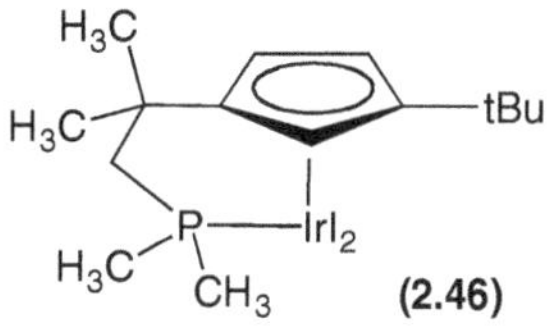

2.5 Principles of Resolution and Preparation of Enantiomers

As stated by Pasteur and quoted by Jacques in his book[93] *La Molécule et son Double*: *'one is completely mistaken in believing that dissymmetry is created when racemic compounds are produced.'*

This signifies that the resolution of a racemic compound into its enantiomers, or their enantioselective synthesis, is the decisive part in the work of the chemist interested in chirality. From this point of view there is a certain ambiguity in the literature as the concept of chirality is used both to describe a molecule possessing this geometric property, and also to describe the enantiomers of a racemic mixture. This confusion is even more pertinent in the description of crystalline materials possessing a chiral space group and resulting from spontaneous resolution during the course of crystallization, since on occasions the authors consider that enantiomeric separation has been achieved whereas in fact it is only a conglomerate that is formed. We consider that the term enantiomers should only be used when their partial or total separation has been carried out and their specific chiro-optic properties described.

2.5.1 Spontaneous Resolution

When a racemic compound in solution crystallizes in an enantiomorphic unit cell, thus forming a conglomerate, several possibilities can arise. In the ideal case, the crystals that are formed are of a reasonable size and the enantiomeric forms have distinct geometries. This was the situation that Pasteur[31] encountered with sodium ammonium tartrate tetrahydrate (**2.5**), and also that which allowed Werner[34] to separate by hand the complex $[Rh^{III}(C_2O_4)_3]K_3$ (**2.24**)-$\mathbf{K_3}$. It must be noted, however, that even in this case the spontaneous resolution must be accompanied by an asymmetric manual sorting operation.

This type of resolution is relatively rare, but in some cases it can allow the first few enantiomerically pure crystals to be obtained which can then be used in the crystallization of one enantiomer by entrainment.

Recently, Kostyanovsky[94] proposed a new methodology to separate enantiomers crystallizing in a chiral space group and forming conglomerates. This methodology is named the 'double conglomerate procedure' and works without an external chiral reagent. The idea is inspired by the works of Pasteur, Gernez[11] and Werner[34] showing that a supersaturated solution of a racemic mixture leads to the crystallization of one enantiomer of the same configuration of the initial external crystal used as seed, in the case where the crystallization occurs with spontaneous resolution.

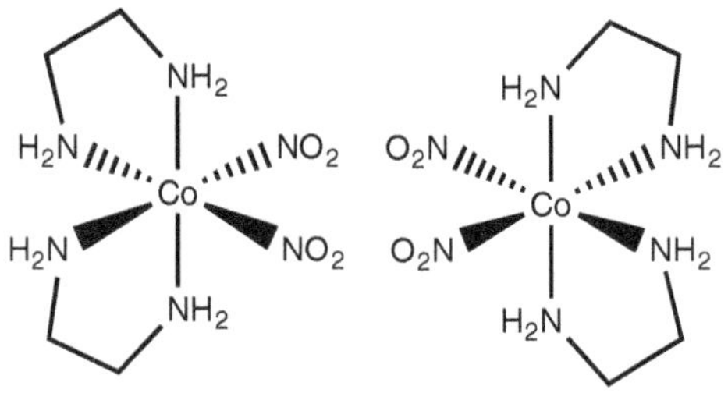

Figure 2.68
View of the cationic part of the two (**2.47**) enantiomers.

This direct separation of enantiomers is named resolution by entrainment. Kostya-novsky[95] uses the term 'conglomerator' for one element (ion) which is able to initiate the crystallization of a conglomerate instead of a racemic crystal. For example Na^+ is a conglomerator for the $NaNH_4{}^{2+}$ tartrate salt (**2.5**) because it is responsible for the formation of a conglomerate while the diammonium tartrate salt crystallizes as a true racemic compound.

In the case of $Co^{III}(eda)_2(NO_2)_2X$ (**2.47**)-**X** (X = Cl, Br) (eda = ethylene diamine) the chloride was found to be more soluble than the bromide and each of them crystallizes as a conglomerate (Figure 2.68).

Adding NaBr to the (**2.47**) chloride solution in water, Kostyanovsky[94] observed the formation of bromide crystals exhibiting optical rotation, whereas the remaining solution has the opposite rotation. The rotation direction of the recovered crystals is not predictable and statistically they are (+) or (−) in a 1:1 ratio. It is possible to imagine that the first crystal formed, resulting from anion exchange, is a single crystal of only one configuration, and able to act as a seed producing internal entrainment. However, this type of resolution is limited by the tendency of racemic solutions to lead to enantiomers which form epitaxy.

2.5.2 Use of a Chiral Auxiliary

We have just seen the theoretical importance of direct resolution by crystallization and the practical limitations of such a method. Moreover, for compounds that crystallize as racemic compounds this route is clearly not allowed. However, the use of a pure chiral auxiliary, leading to diastereomers with different physico-chemical properties, allows us to separate them by a variety of techniques and in particular fractional crystallization. We can schematize the process of formation and separation of the diastereomers in a simple way, as well as the recovery of the chiral auxiliary and the enantiomers (Scheme 2.3).

(+)-, (-)-A + (+)-B → [(+)-A, (+)-B; (-)-A, (+)-B] formation of the diastereomer mixture

[(+)-A, (+)-B ; (-)-A, (+)-B] → [(+)-A, (+)-B] + [(-)-A, (+)-B] separation of the mixture

[(+)-A, (+)-B] → (+)-A, (+)-B separation of the auxiliary (+)-B

[(-)-A, (+)-B] → (-)-A, (+)-B separation of the auxiliary (+)-B

Scheme 2.3

We will give two examples, one involving the formation of a covalent bond, and the other the formation of an ion pair.

(a) Formation of a covalent bond. In the following example, the separation of the (R) and (S) enantiomers of the complex [Mn(C$_5$H$_5$)(NO)COP(C$_6$H$_5$)$_3$]X **(2.12)-X**[21] was achieved by using natural menthol as the chiral auxiliary. The covalent bond so formed could be readily broken without effecting the absolute configuration of the manganese complex.

(b) Formation of an ion pair. This approach is widespread in organometallic and coordination chemistry since many complexes exist as salts. It has allowed the resolution of many hexacoordinated complexes of Cr, Co, Rh, Zn, Ir, Os and Ru. The principle is based on the metathesis of the achiral anion or cation by an enantiomerically pure chiral anion or cation. Among the chiral ions most used are, for example, the following anions: tartrates **(2.5)**, phosphates (Trisphat **(2.32)**, BINPHAT **(2.40)**, and camphosulphonates **(2.3)**, and the cations: [Ni^{II}(phen)$_3$]$^{2+}$ **(2.48)**, or those derived from optically active natural amines such as cinchonine and strychnine.

In all cases, if both of the enantiomers of the chiral auxiliary are available then, according to Marckwald's principle, both of the enantiomers of the racemic mixture to be separated can be obtained. Note that it is always easier to remove pure crystals by filtration, than to extract a pure enantiomer from the filtrate. This method of separation is illustrated by the following example of the resolution of a ruthenium (II) salt[96,97] [Ru(bpy)$_2$py$_2$]Cl$_2$ **(2.16)-Cl$_2$** (Scheme 2.4).

(Δ)-, (Λ)-(2.16)-Cl$_2$ + (d)-dibenzoyl-(R, R)-tartrate Na$_2$ → (Δ)- (2.16), (d)-dibenzoyl-(R, R)-

tartrate (pure crystals)

(Δ)-, (Λ)-(2.16)-Cl$_2$ + (l)-dibenzoyl-(S, S)-tartrate Na$_2$ → (Λ)- (2.16), (l)-dibenzoyl-(S, S)-

tartrate (pure crystals)

Scheme 2.4

2.5.3 Chromatography

Chromatography is the method of choice for the separation of enantiomers or diastereomers on a preparative scale.

(i) Separation of diastereomers on an achiral support using an achiral eluant
Many examples are given in the literature of which we will mention two, one corresponding to helicoidal chirality and the other to the chirality of a tetrahedral centred complex. The separation of the diastereomers [(Δ)Ru(phen)$_2$(CH$_3$CN)$_2$ **(2.49)**, (Δ)-Trisphat] and [(Λ)Ru(phen)$_2$(CH$_3$CN)$_2$**(2.49)**,(Δ)-Trisphat] is possible on a silica column with methylene chloride as the eluant.[98]

(ii) Separation of enantiomers on a chiral support using a chiral eluant
Starting from the fact that hexacoordinated ruthenium salts of the general formula [Ru(bpy)$_2$L$_2$]X$_2$ or [Ru(phen)$_2$L$_2$]X$_2$ are capable of forming specific interactions with double stranded DNA, Strekas[99] suggested that the two (Δ) and (Λ) enantiomers could be separated by using column chromatography with double-stranded DNA immobilized on hydroxyapatite as the stationary phase.

(iii) Separation of enantiomers on an achiral column using a chiral mobile phase
This technique was used to resolve trinuclear iron complexes by using sephadex SP-C25 as the stationary phase and an aqueous solution of (+)-tartratoantimonate (III) as eluant.[100]

2.5.4   Enantioselective Synthesis

In the preceding paragraph we briefly described the methods of separating enantiomers starting from a racemic mixture. We will now consider another approach called '*enantio-selective synthesis*' or '*asymmetric synthesis*'. We can define enantioselective synthesis as a step in the synthetic process during which, with the aid of a man-made or naturally occurring (enantiomerically pure) chiral element (reagent, solvent, catalyst), another chiral element is created or transformed, with the formation of one of the possible enantiomers in either partial or total excess. The concept of asymmetric synthesis was introduced by Marckwald in 1904.[101,102] This concept is ambiguous, since in fact asymmetric syntheses are above all diastereoselective syntheses of enantiomers.

Before describing some examples, it is necessary to define two terms, stereoselectivity and stereospecificity.

(i) Stereoselectivity
A reaction is said to be stereoselective if one stereoisomer is formed in excess with respect to all others. This is termed enantioselectivity in the case of enantiomers, or diastereoselectivity for diastereomers.

(ii) Stereospecificity
A reaction is stereospecific if, starting from compounds of different configurations, stereoisomerically distinct products are formed. The stereochemistry of the products is determined by that of the starting materials. A stereospecific reaction is necessarily stereoselective.

2.5.4.1   Stereoselective Reactions

Only a limited number of examples will be given to illustrate the concept.

(i) Asymmetric cyclopalladation
This reaction, discovered by Sokolov and Troitskaya,[103,104] allows the synthesis of enantiomers of 1,2-disubstituted ferrocenes (**2.50**) (*pR*) or (*pS*) (*ee* = 0.6). The reaction can be directed towards either of the configurations depending on the configuration of the amino acid (N-Acetyl-Leucine) used as catalyst (Figure 2.69).
In a variant of this reaction described by Kagan[105] the chiral inducer is contained within the starting material; in the example shown in Figure 2.70 it orientates the metallation by *t*BuLi leading to the lithium derivative (*pR*)-(**2.51**).

(ii) Diastereoselective radical coupling
An example of a diastereoselective reaction is the radical coupling of dinuclear acetylenic cobalt complexes. This coupling leads to the compound of relative configuration (R, R)* (**2.52**) with a diastereoselectivity of 0.9[106] (Figure 2.71).

(iii) Enzymatic reactions
Enantioselective enzymatic reactions are mostly used in organic chemistry. The first of these to be mentioned was in the reduction of hept-6-yne-2-one to hept-6-yn-2-ol

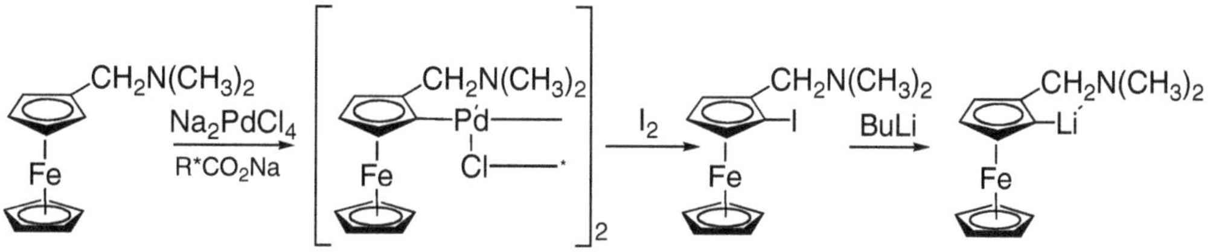

(2.50)

R*CO$_2$Na = N-Acetyl-D-Leucine **(*pS*)-(2.50)** $[\alpha]_D^{20}$ = - 447° (C = 0.53, CH$_2$Cl$_2$)

R*CO$_2$Na = N-Acetyl-L-Leucine **(*pR*)-(2.50)** $[\alpha]_D^{20}$ = + 400° (C = 0.52, CH$_2$Cl$_2$)

Figure 2.69
Asymmetric cyclopalladation of a ferrocenic amine derivative.

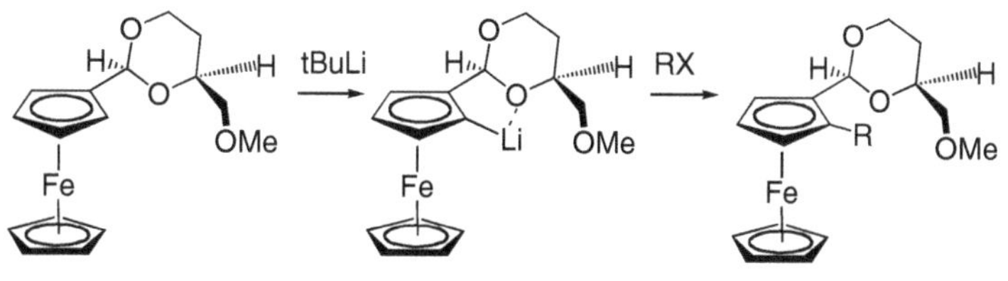

(*pR*) (*pR*)-(**2.51**)

Figure 2.70
Asymmetric functionalization of a ferrocenic compound.

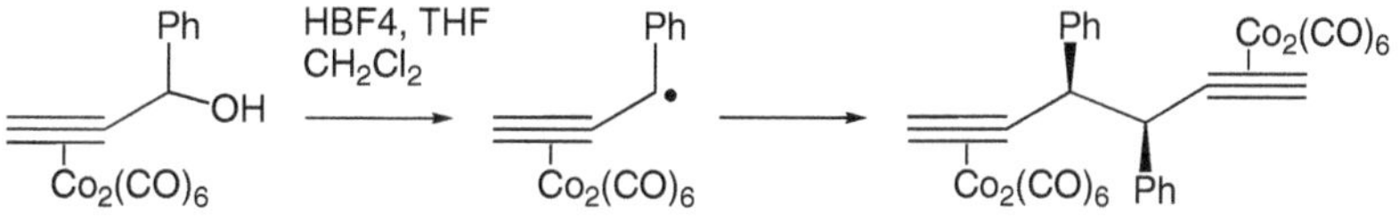

(2.52)

Figure 2.71
Diastereoselective coupling of chiral complexed radicals.

with the configuration (*S*). The reduction reactions of ferrocene and/or benchrotrene ketones to optically active alcohols by enzymes are also known. In the example shown in Figure 2.72, treatment of methyl 1-2 phenyl diacetate chromium carbonyl **(2.53)** (*meso* compound) with pig liver esterase leads in an enantioselective manner to the monosaponified derivative (*pS*)-(**2.54**).[107]

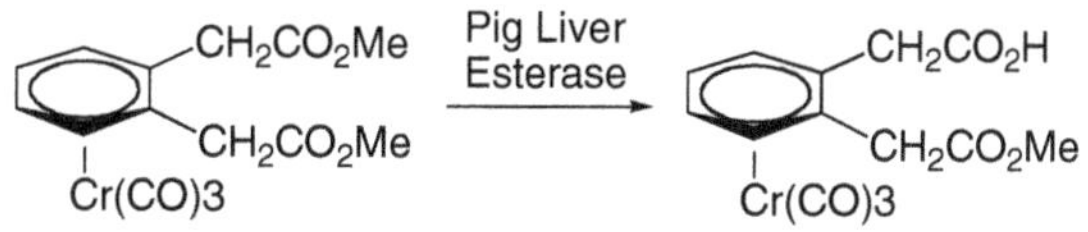

(2.53) (*pS*)-**(2.54)**

Figure 2.72
Enantioselective saponification of an achiral complexed diester using pig liver esterase as chiral reagent.

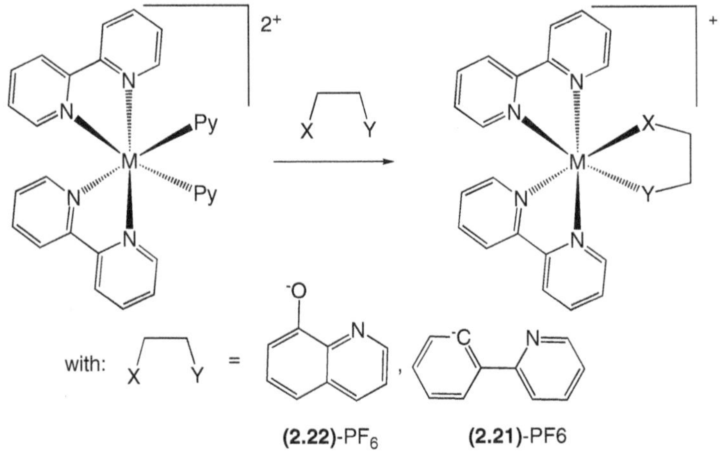

Figure 2.73
Stereospecific ligand substitution.

2.5.4.2 Stereospecific Reactions

(i) *Ligand substitution*
In the synthesis of hexacoordinated ruthenium complexes, the substitution of two monodentate ligands by a third bis(chelating) ligand generally takes place with retention of configuration[33] (Figure 2.73).

(ii) *Enantiospecific auto-assembly*
The formation of a polymetallic helicate of (P) or (M) configuration can be induced by the configuration of the ligand.[108] In Figure 2.74, the ligand of (M) configuration induces the formation of a trimetallic helix (P)-(**2.55**).

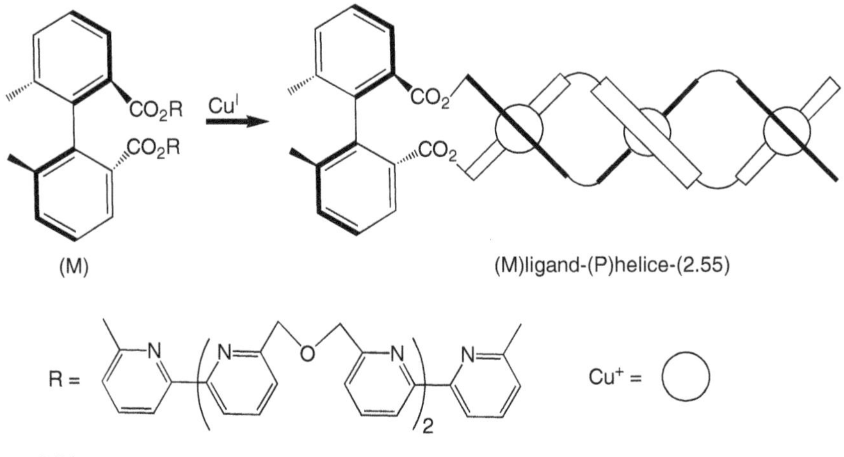

Figure 2.74
The ligand of (M) absolute configuration acts as a chiral inductor in the formation of a pure (P) helix.

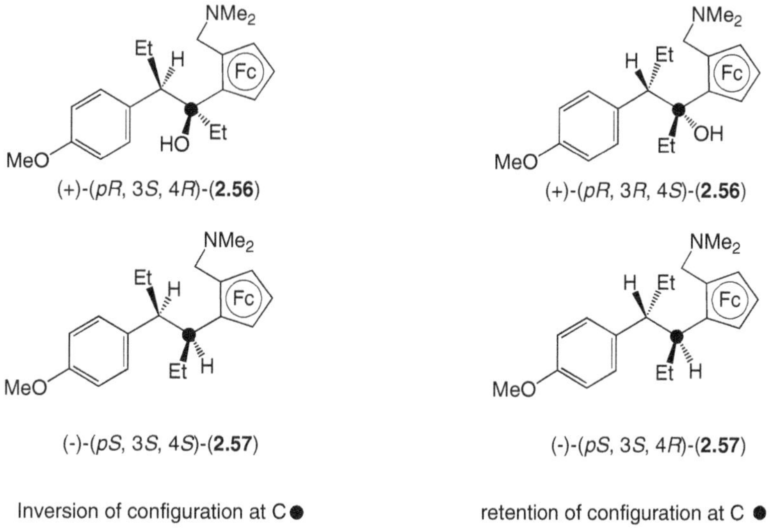

(+)-(pR, 3S, 4R)-(**2.56**)   (+)-(pR, 3R, 4S)-(**2.56**)

(-)-(pS, 3S, 4S)-(**2.57**)   (-)-(pS, 3S, 4R)-(**2.57**)

Inversion of configuration at C●   retention of configuration at C ●

Figure 2.75
Diastereospecific reduction of ferrocenic alcohols by ionic hydrogenation.

(iii) Diastereospecific reduction of ferrocenic alcohols by ionic hydrogenation
This reaction, derived from the Kursanov Parnes reaction,[109] consists of the reduction of an alcohol by a hydride in an acidic medium. In the case of α-ferrocenic alcohols this reaction takes place, in general, with retention of configuration due to the fact that the iron atom stabilizes the carbocation formed. However, diastereospecific reactions are observed when the substituents at the carbon atom bearing the hydroxyl group are bulky. Figure 2.75 shows that there is retention of configuration for one of the diastereomers and inversion for the other.

2.6 Summary

Throughout this chapter we have presented a historical introduction and all the events from Pasteur to Werner who contributed to the birth of the new discipline 'stereo-chemistry'. Starting from the rules of Le Bel and Van't Hoff to Cahn–Ingold and Prelog the universal descriptors were presented to describe chiral objects. To this end several chiral organometallic and coordination complexes were chosen which exhibited different geometrical shapes. Different methods to determine the enantiomeric excesses were presented. Finally the principle of resolution and preparation of enantiomers were discussed. Thus in this chapter historical events, principles and general consideration of chirality were discussed, the following chapters are more precise and will deal with chirality in catalysis, supramolecular and material science.

References

[1] A. Werner and A. Vilmos, *Z. Anorg. Allg. Chem.* **1899**, *21*, 145.
[2] A. Werner, *Ber. Dtsch. Chem. Ges.* **1911**, *44*, 1887.

[3] A. Werner, *Chem. Ber.* **1911**, *44*, 1887.
[4] A. Werner, *Ber. Dtsch. Chem. Ges.* **1914**, *47*, 3087.
[5] F. S. Kipping and W. J. Pope, *J. Chem. Soc., Trans.* **1895**, *67*, 354.
[6] F. S. Kipping and W. J. Pope, *J. Chem. Soc., Trans.* **1893**, *63*, 548.
[7] G. B. Kauffman and I. Bernal, *Pure Appl. Chem.* **1988**, *60*, 1379.
[8] I. Bernal and G. B. Kauffman, *J. Chem. Educ.* **1987**, *64*, 604.
[9] V. L. King, *J. Chem. Educ.* **1942**, 345.
[10] A. Werner, *Ber. Dtsch. Chem. Ges.* **1914**, *47*, 2171.
[11] D. Gernez, *Com. Rend. Acad. Sci.* **1866**, *63*, 843.
[12] J.-A. Le Bel, *Bull. Soc. Chim. Paris* **1874**, *22*, 337.
[13] J. Jacques, *Bull. Soc. Chim.* **1995**, *132*, 5.
[14] L. Pasteur, *Com. Rend. Acad. Sci.* **1848**, *26*, 535.
[15] G. B. Kauffman and R. D. Myers, *J. Chem. Educ.* **1975**, *52*, 777.
[16] L. Kelvin, *Baltimore Lectures on Molecular Dynamics and the Wave Theory of Light*, Cambridge University Press, London, **1904**.
[17] G. P. Moss, *Pure Appl. Chem.* **1996**, *68*, 2193.
[18] J. Simon and P. Bassoul, *Design of Molecular Materials*, John Wiley and Sons, Ltd, Chichester, **2000**.
[19] S. Alvarez, M. Pinsky, M. Llunell and D. Avnir, *Cryst. Eng.* **2001**, *4*, 179.
[20] S. Alvarez, P. Alemany and D. Avnir, *Chem. Soc. Rev.* **2005**, *34*, 313.
[21] H. Brunner, *Angew. Chem., Int. Ed.* **1999**, *38*, 1194.
[22] J. H. Merrifield, C. E. Strouse and J. A. Gladysz, *Organometallics* **1982**, *1*, 1204.
[23] J. A. Ramsden, C. M. Garner and J. A. Gladysz, *Organometallics* **1991**, *10*, 1631.
[24] F. Richter and H. Vahrenkamp, *Chem. Ber.* **1982**, *115*, 3243.
[25] M. Gruselle, R. Thouvenot, B. Malezieux, C. Train, P. Gredin, T. V. Demeschik, L. L. Troitskaya and V. I. Sokolov, *Chem. Eur. J.* **2004**, *10*, 4763.
[26] G. Simonneaux, A. Meyer and G. Jaouen, *J. Chem. Soc., Chem. Commun.* **1975**, 69.
[27] B. Kolp, H. Viebrock, A. von Zelewsky and D. Abeln, *Inorg. Chem.* **2001**, *40*, 1196.
[28] A. von Zelewsky, *Stereochemistry of Coordination Compounds*, John Wiley and Sons, Ltd, Chichester, **1996**.
[29] H. Amouri, R. Thouvenot, M. Gruselle, B. Malézieux and J. Vaissermann, *Organometallics* **2001**, *20*, 1904.
[30] M. C. Biagini, M. Ferrari, M. Lanfranchi, L. Marchio and M. A. Pellinghelli, *J. Chem. Soc., Dalton Trans.* **1999**, 1575.
[31] J. Jacques, A. Collet and S. H. Wilen, *Enantiomers, Racemates and Resolution*, John Wiley and Sons, Inc., New York, **1981**.
[32] J. Breu, H. Domel and S. Alexander, *Eur. J. Inorg. Chem.* **2000**, 2401.
[33] M. Brissard, O. Convert, M. Gruselle, C. Guyard-Duhayon and R. Thouvenot, *Inorg. Chem.* **2003**, *42*, 1378.
[34] A. Werner, *Ber. Dtsch. Chem. Ges.* **1914**, *47*, 1954.
[35] M. Mitscherlisch, *Com. Rend. Acad. Sci.* **1844**, 720.
[36] G. Coquerel, *Top. Curr. Chem.* **2007**, *269*, 1.
[37] G. Coquerel and S. Petit, *J. Cryst. Growth* **1993**, *130*, 173.
[38] S. Decurtins, H. W. Schmalle, P. Schneuwly and H. R. Oswald, *Inorg. Chem.* **1993**, *32*, 1888.
[39] S. Decurtins, H. W. Schmalle, P. Schneuwly, J. Ensling and P. Guetlich, *J. Am. Chem. Soc.* **1994**, *116*, 9521.
[40] R. Andres, M. Brissard, M. Gruselle, C. Train, J. Vaissermann, B. Malezieux, J.-P. Jamet and M. Verdaguer, *Inorg. Chem.* **2001**, *40*, 4633.
[41] P. Hayoz, A. Von Zelewsky and H. Stoeckli-Evans, *J. Am. Chem. Soc.* **1993**, *115*, 5111.
[42] J. Lacour and H.-V. Virginie, *Chem. Soc. Rev.* **2003**, *32*, 373.
[43] F. P. Dwyer and A. M. Sargeson, *J. Phys. Chem.* **1956**, *60*, 1331.
[44] G. B. Kauffman, L. T. Takahashi and N. Sugisaka, *Inorg. Syn.* **1966**, *8*, 207.

[45] E. I. Zhilyaeva, G. V. Shilov, O. A. Bogdanova, R. N. Lyubovskaya, R. B. Lyubovskii, N. S. Ovanesyan, S. M. Aldoshin, C. Train and M. Gruselle, *Mater. Sci.* **2005**, *22*, 565.

[46] E. Fischer, *Berichte* **1891**, *24*, 2683.

[47] E. Fischer, *Berichte* **1891**, *2*, 1836.

[48] R. S. Cahn and C. K. Ingold, *J. Chem. Soc.* **1951**, 612.

[49] R. S. Cahn, C. K. Ingold and V. Prelog, *Experientia* **1956**, *12*, 81.

[50] R. S. Cahn, C. Ingold and V. Prelog, *Angew. Chem., Int. Ed.* **1966**, *5*, 385.

[51] V. Prelog and G. Helmchen, *Angew. Chem., Int. Ed.* **1982**, *94*, 614.

[52] T. Damhus and C. E. Schäffer, *Inorg. Chem.* **1983**, *22*, 2406.

[53] K. F. Purcell and J. C. Kotz, *Inorganic Chemistry* **1977**, W. B. Saunders Co., Philadelphia.

[54] N. G. Connelly, T. Damhus, R. M. Hartshorn and A. T. Hutton (Eds), *Nomenclature of Inorganic Chemistry: IUPAC Recommendations 2005*, **2005**.

[55] J. A. McClevertya nd N. G. Connelly, *Nomenclature of Inorganic Chemistry II: Recommendations 2000*, **2001**.

[56] K. Schloegl, *Top. Stereochem.* **1967**, *1*, 39.

[57] A. Collet, J. Crassous, J.-P. Dutasta and L. Guy, *Molécules Chirales: Stéréochimie et Propriétés*, EDP Sciences/CNRS Editions, Les Ulis, Paris, **2006**.

[58] IUPAC, *Inorg. Chem.* **1970**, *9*, 1.

[59] R. Andres, M. Gruselle, B. Malezieux, M. Verdaguer and J. Vaissermann, *Inorg. Chem.* **1999**, *38*, 4637.

[60] M. Ziegler and A. von Zelewsky, *Coord. Chem. Rev.* **1998**, *177*, 257.

[61] E. Charney, *The Molecular Basis of Optical Activity: Optical Rotatory Dispersion and Circular Dichroism*, **1979**.

[62] G. B. Kauffman, S. Nobuyuki and R. K. Reid (checked by Murmann), *Inorg. Synth.* **1989**, *25*, 139.

[63] R. Caspar, H. Amouri, M. Gruselle, C. Cordier, B. Malezieux, R. Duval and H. Leveque, *Eur. J. Inorg. Chem.* **2003**, 499.

[64] B. Bosnich, *Inorg. Chem.* **1968**, *7*, 178.

[65] B. Bosnich, *Inorg. Chem.* **1968**, *7*, 2379.

[66] A. J. McCaffery, S. F. Mason and B. J. Norman, *J. Chem. Soc. A* **1969**, 1428.

[67] A. J. McCaffery and S. F. Mason, *Proc. Chem. Soc.* **1963**, 211.

[68] N. Das, A. Ghosh, O. M. Singh and P. J. Stang, *Org. Lett.* **2006**, *8*, 1701.

[69] L. A. Nafie, *Annu. Rev. Phys. Chem.* **1997**, *48*, 357.

[70] L. A. Nafie, *Appl. Spectrosc.* **1996**, *50*, 14A.

[71] L. Hecht, A. Phillips and L. D. Barron, *J. Raman Spectrosc.* **1995**, *26*, 727.

[72] L. D. Barron, L. Hecht, A. F. Bell and G. Wilson, *Appl. Spectrosc.* **1996**, *50*, 619.

[73] L. D. Barron, A. R. Gargaro, L. Hecht, P. L. Polavarapu and H. Sugeta, *Spect. Acta, Part A: Mol. Biomol. Spect.* **1992**, *48A*, 1051.

[74] G. L. J. A. Rikken and B. A. van Tiggelen, *Nature* **1996**, *381*, 54.

[75] G. L. J. A. Rikken and E. Raupach, *Nature* **2000**, *405*, 932.

[76] G. L. J. A. Rikken and E. Raupach, *Nature* **1997**, *390*, 493.

[77] L. D. Barron, *Nature* **2000**, *405*, 895.

[78] E. Raupach, G. L. J. A. Rikken, C. Train and B. Malezieux, *Chem. Phys.* **2000**, *261*, 373.

[79] J. Lacour, C. Ginglinger, C. Grivet and G. Bernardinelli, *Angew. Chem., Int. Ed.* **1997**, *36*, 608.

[80] J. M. Bijvoet, A. F. Peerdeman and A. J. van Bommel, *Nature* **1951**, *168*, 271.

[81] H. D. Flack and G. Bernardinelli, *Acta Cryst. A* **1999**, *A55*, 908.

[82] W. Moffitt, R. B. Woodward, A. Moscowitz, W. Klyne and C. Djerassi, *J. Am. Chem. Soc.* **1961**, *83*, 4013.

[83] J. Lacour and R. Frantz, *Org. Biomol. Chem.* **2005**, *3*, 15.

[84] J. Lacour, C. Ginglinger, F. Favarger and S. Torche-Haldimann, *Chem. Commun.* **1997**, 2285.

[85] M. Brissard, H. Amouri, M. Gruselle and R. Thouvenot, *CR Chimie* **2002**, *5*, 53.

[86] H. Amouri, R. Caspar, M. Gruselle, C. Guyard-Duhayon, K. Boubekeur, D. A. Lev, L. S. B. Collins and D. B. Grotjahn, *Organometallics* **2004**, *23*, 4338.

[87] G. Bruylants, C. Bresson, A. Boisdenghien, F. Pierard, K.-D. Mesmaeker, J. Lacour and K. Bartik, *New J. Chem.* **2003**, *27*, 748.

[88] J. Courtieu, P. Lesot, A. Meddour, D. Merlet and C. Aroulanda, *Encyclopedia of Nuclear Magnetic Resonance* **2002**, *9*, 497.

[89] A. Meddour, J. Uziel, J. Courtieu and S. Juge, *Tetrahedron: Asymmetry* **2006**, *17*, 1424.

[90] L. Beguin, J. Courtieu, L. Ziani and D. Merlet, *Magn. Reson. Chem.* **2006**, *44*, 1096.

[91] S. N. Paisner and R. G. Bergman, *J. Organomet. Chem.* **2001**, *621*, 242.

[92] A. Dalla Cort, L. Mandolini, C. Pasquini and L. Schiaffino, *J. Org. Chem.* **2005**, *70*, 9814.

[93] J. Jacques, *La Molécule et son Double*, Hachette, Paris, **1992**.

[94] R. G. Kostyanovsky, V. Y. Torbeev and K. A. Lyssenko, *Tetrahedron: Asymmetry* **2001**, *12*, 2721.

[95] V. Krstic, S. Roth, M. Burghard, K. Kern and G. L. J. A. Rikken, *J. Chem. Phys.* **2002**, *117*, 11315.

[96] X. Hua and A. von Zelewsky, *Inorg. Chem.* **1995**, *34*, 5791.

[97] X. Hua and A. Lappin Graham, *Inorg. Chem.* **1995**, *34*, 992.

[98] M. Gruselle, R. Thouvenot, R. Caspar, K. Boubekeur, H. Amouri, M. Ivanov and K. Tonsuaadu, *Mendeleev Comm.* **2004**, 282.

[99] A. D. Baker, R. J. Morgan and T. C. Strekas, *J. Am. Chem. Soc.* **1991**, *113*, 1411.

[100] B. Hasenknopf and J. M. Lehn, *Helv. Chim. Acta* **1996**, *79*, 1643.

[101] W. Marckwald, *Ber. Dtsch. Chem. Ges.* **1904**, *37*, 1368.

[102] W. Marckwald, *Ber. Dtsch. Chem. Ges.* **1904**, *37*, 349.

[103] V. I. Sokolov, L. L. Troitskaya and O. A. Reutov, *J. Organomet. Chem.* **1979**, *182*, 537.

[104] V. I. Sokolov, *Pure Appl. Chem.* **1983**, *55*, 1837.

[105] O. Riant, O. Samuel and H. B. Kagan, *J. Am. Chem. Soc.* **1993**, *115*, 5835.

[106] G. G. Melikyan, F. Villena, A. Florut, S. Sepanian, H. Sarkissian, A. Rowe, P. Toure, D. Mehta, N. Christian, S. Myer, D. Miller, S. Scanlon, M. Porazik and M. Gruselle, *Organometallics* **2006**, *25*, 4680.

[107] J. A. S. Howell, M. G. Palin, G. Jaouen, B. Malezieux, S. Top, J. M. Cense, J. Salauen, P. McArdle, D. Cummingham and M. O'Gara, *Tetrahedron: Asymmetry* **1996**, *7*, 95.

[108] C. R. Woods, M. Benaglia, F. Cozzi and J. S. Siegel, *Angew. Chem., Int. Ed.* **1996**, *35*, 1830.

[109] B. Malezieux, M. Gruselle, L. L. Troitskaya, V. I. Sokolov and J. Vaissermann, *Organometallics* **1994**, *13*, 2979.

3 Some Examples of Chiral Organometallic Complexes and Asymmetric Catalysis

In this chapter, we would like to cover the key role of chirality in the organometallic chemistry of mononuclear complexes since the resolution of the first organometallic chiral complex $[CpMn(CO)(NO)(PPh_3)][PF_6]$[1] up to the present time and to show the fast evolution of this field. In this context we will discuss the synthesis and resolution of some chiral-at-metal complexes, followed by their ability to promote chiral induction upon coordination to organic substrates. Moreover the use of some chiral-at-metal complexes as auxiliaries to promote stoichiometric enantioselective reactions in organic synthesis will be discussed. Then a summary of the most important asymmetric catalyses will be presented, because of the importance of this field to academic and industrial applications.[2] However, we note that there are specialized books devoted to this important subject of 'asymmetric catalysis', and a reader may consult the cited references.[3–5] As stated in Chapter 2, the asymmetric carbon atom has dominated stereochemistry since its discovery by van 't Hoff and Le Bel in 1874.[6,7] Another important event in chirality apart from asymmetric carbon is the octahedral tris-chelating complexes in coordination chemistry such as $[Co(en)_3]X_3$ which have a D_3 symmetry and are therefore chiral. The first octahedral complex was prepared in optically active form by A. Werner in 1911 and proved that an asymmetric carbon is not a prerequiste for chirality.[8] Organometallic chemistry is intermediate between organic and inorganic chemistry because it looks at the interaction between inorganic metal ions and organic molecules. Organometallic chemistry plays a key role in most examples in asymmetric organic synthesis and in metal-based catalytic reactions.

3.1 Chirality at Metal Half-sandwich Compounds

3.1.1 Chiral Three-legged Piano Stool: the $CpMnL^1L^2L^3$ Model

3.1.1.1 Synthesis and Resolution of the First Organometallic Compound

Half-sandwich compounds with three-legged piano stool geometry with different substituents are archetypal examples of optically active chiral-at-metal complexes. In 1969 Brunner *et al.* prepared and resolved the first chiral organometallic complex with four

Chirality in Transition Metal Chemistry: Molecules, Supramolecular Assemblies and Materials   H. Amouri and M. Gruselle
© 2008 John Wiley & Sons, Ltd

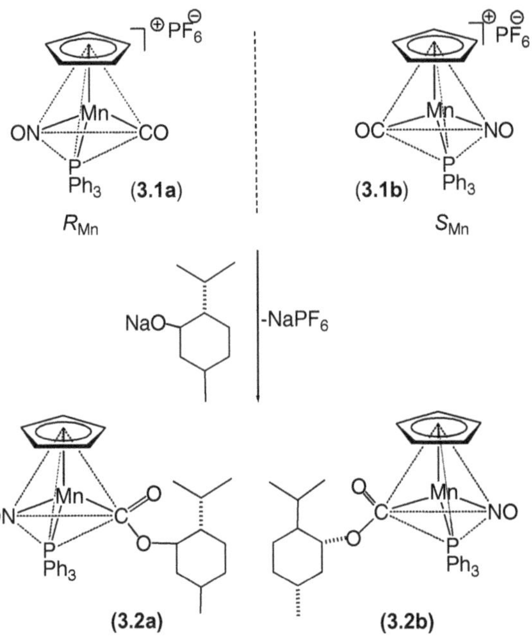

Scheme 3.1
Resolution of the first organometallic compound **3.1**.

different substituents (*R/S*)-[CpMn(CO)(NO)(PPh₃)][PF₆] (**3.1a/3.1b**) by converting the pair of enantiomers into a pair of diastereomers **3.2a/3.2b** using the optically active alcohol (1*R*,3*R*,4*S*)-*l*-menthol (Scheme 3.1).[1,9,10] The neutral *l*-menthyl esters **3.2a/3.2b** differed only in the configuration of the manganese atom and had different solubility properties; hence they could be separated.

To complete the resolution, the optically active auxiliary was removed from the separated diastereomers by bubbling HCl through solutions of **3.2a** and **3.2b** in benzene. During this reaction the C—O$_{menthyl}$ bond was cleaved. Menthol was formed and the carbonyl group at the metal centre was restored. The counter anion Cl⁻ was then replaced by PF₆⁻. Thus starting from diastereomer **3.2a** ($[\alpha]_{579}^{25} = +460$ ($c = 0.1$, benzene)),[1] the PF₆⁻ salt **3.1a** ($[\alpha]_{579}^{20} = +375$)[9] was formed; while the diastereomer **3.2b** ($[\alpha]_{579}^{25} = -450$ ($c = 0.1$, benzene))[1] was transformed into the PF₆⁻ salt **3.1b** ($[\alpha]_{579}^{20} = -386$).[9] The CD spectra of the diastereomeric manganese complexes of **3.2a** and **3.2b** are shown in Figure 3.1. Unlike the diastereomers **3.2a** and **3.2b** which epimerize in solution, the salts **3.1a** and **3.1b** were found to be configurationally stable;[11] no change in optical purity was observed for solutions in CH₂Cl₂ over a period of several days. Complexes **3.1a** and **3.1b** represented the first resolved, optically active organometallic compound with four different substituents. The stereogenic manganese atom was the only source of chirality.

3.1.1.2   Configurational Stability

An important issue in these chiral complexes is the configurational stability at the metal centre. A complex may be *configurationally stable* over weeks while others epimerize or racemize rapidly even at low temperatures.[12] Thus in one case complexes may be

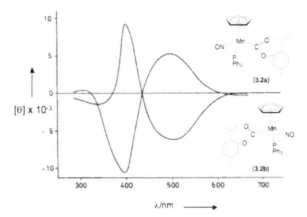

Figure 3.1
CD spectra of **3.2a** and **3.2b** (reproduced with permission from reference[10]).

classified as *configurationally stable* if it is possible to isolate and manipulate them faster than they epimerize, while other compounds are *configurationally labile* and cannot be obtained in a stereochemically defined ratio in solution due to the rapid process of epimerization.

Unlike cationic carbonyl complexes **3.1a/3.1b** the related neutral benzoyl species **3.3a/3.3b** racemized quickly with half-lives of the order of minutes and hours. at 20°C.[12] The neutral complexes **3.3a/3.3b** were obtained by treating **3.1a/3.1b** with PhLi. Brunner *et al.* investigated the steric and electronic factors on the rate of racemization of these manganese complexes and showed that racemization proceeds via dissociation of the PAr_3 ligand. Further studies carried out by the same author varied the *para* substituents X and Y on the aryl groups in **3.3**, with X, Y = NMe_2, OMe, F, CF_3 and others (Figure 3.2). It appeared that racemization slowed down in the case of electron releasing groups Y on the tertiary phosphine and/or electron withdrawings groups X in the para position of the benzoyl substituent. On the other hand, steric factors were also shown to have an effect. For example, in the case of small substituents L = CO, $P(OEt_3)_3$, $P(n$-$Bu)_3$, irrespective whether they were of good π-acceptors or donors, the chiral complexes CpMn(COPh)(NO)(L) were configurationally stable, whereas the presence of the bulkier ligand PPh_3 led to steric crowding in **3.3** and hence faster dissociation and more rapid epimerization.[13]

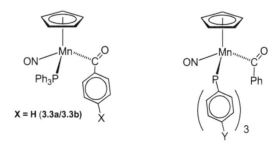

Figure 3.2

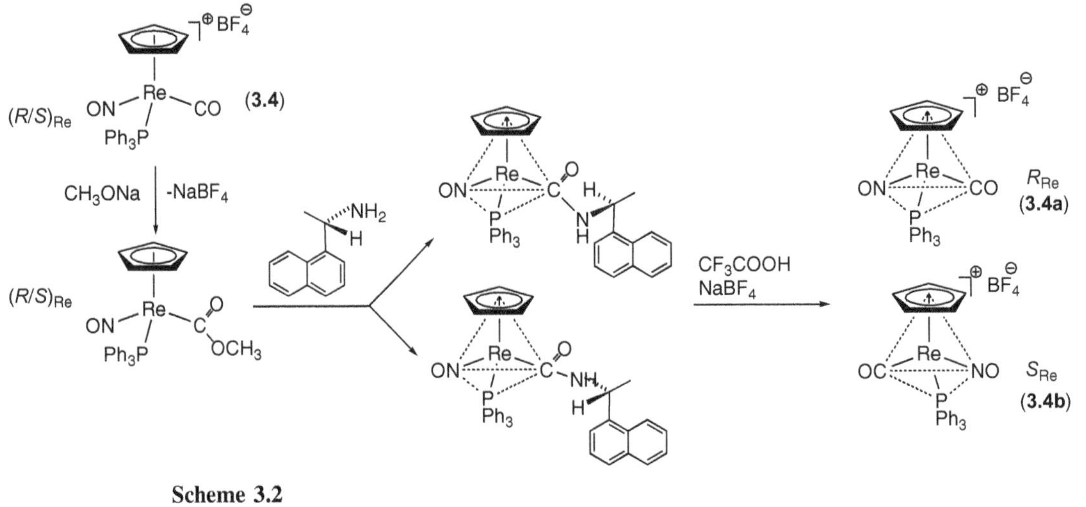

Scheme 3.2
Resolution of the rhenium ccomplex [CpRe(CO)(NO)PPh$_3$)][BF$_4$] (**3.4**).

3.1.2 Chiral Three-legged Piano Stool: the CpReL^{1}L^{2}L^3 Model

3.1.2.1 Synthesis, Resolution and Configurational Stability

Gladysz and coworkers made a great contribution to modern chirality in organometallic chemistry by designing the Re chiral model CpRe(PPh$_3$)(NO)(CH$_3$) (**3.6**).[14] The latter was obtained optically pure from the resolved metal carbonyl [CpRe(CO)(NO)(PPh$_3$)] [BF$_4$] complex (**3.4**) by reduction with NaBH$_4$.[15] It is worth mentioning that [CpRe(CO) (NO)(PPh$_3$)][BF$_4$] (**3.4**), which is isostructural to Brunner's chiral complex [CpMn(CO) (NO)(PPh$_3$)][PF$_6$] (**3.1**), was resolved using a different method which involved the formation of 'amides' [CpRe(NO)(PPh$_3$){CONHCH(CH$_3$)C$_{10}$H$_7$}] (**3.5**) using optically pure α-(1-naphthyl)-ethylamine (Scheme 3.2). Upon protonation by CF$_3$COOH followed by anion metathesis using NaBF$_4$, the optically pure rhenium complexes (R)-[CpRe(CO) (NO)PPh$_3$)][BF$_4$] (**3.4a**) and/or (S)-[CpRe(CO)(NO)PPh$_3$)][BF$_4$] (**3.4b**) were obtained.

Upon protonation of **3.6** at −80°C in CH$_2$Cl$_2$ or PhCl, new chiral Lewis acids **3.7** and **3.8** were formed with solvent weakly coordinated via the chlorine centre (Scheme 3.3).[16]

Gladysz has investigated the solution behaviour and configurational stability of these chiral rhenium complexes; further, he extensively studied the chiral recognition process that results from the enantioface binding of alkenes, aldehydes and ketones on these rhenium chiral Lewis acids **3.7** and/or **3.8**. There is a huge body of papers and reviews on this chemistry, and the reader may consult the referenced review.[17]

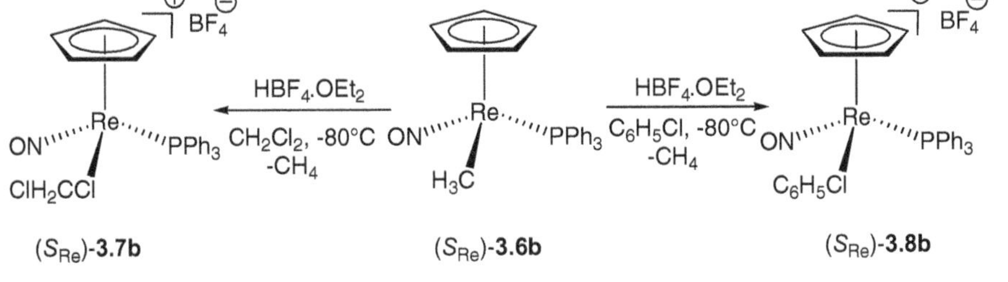

Scheme 3.3

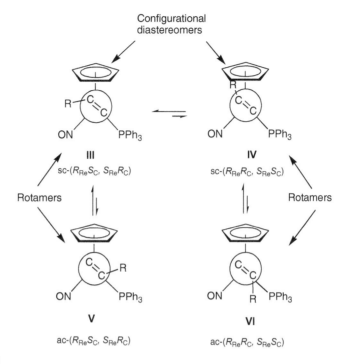

Scheme 3.4
III, more stable rotamer of more stable diastereomer; IV, more stable rotamer of less stable diastereomer, V, less stable rotamer of more stable diastereomer; VI less stable rotamer of less stable diastereomer (reprinted with permission from reference[18], copyright 1990, American Chemical Society).

Thus the interaction of the chiral-at-metal Lewis acid **3.7** and/or **3.8** with unsaturated π-ligands was thoroughly examined.[18] Scheme 3.4 shows the possible diastereomers and rotamers of the product $[CpRe(NO)(PPh_3)(R-CH=CH_2)][BF_4]$ (**3.9**) with unsymmetrical olefins. In general, a mixture of $(R_{Re}S_C, S_{Re}R_C)/ (S_{Re}S_C, R_{Re}R_C)$ diastereomers were formed that differed in the alkene enantioface bound to rhenium. The $(R_{Re}S_C, S_{Re}R_C)$ diastereomers were more stable than the $(S_{Re}S_C, R_{Re}R_C)$ diastereomers. In general, the alkene substituent projected toward the small nitrosyl ligand and opposite the bulky PPh$_3$ group. In these complexes the absolute configuration of rhenium is given first and the stereogenic carbon of the olefin second.

For each diastereomer two Re—(C=C) rotamers are possible ($sc = $ synclinal; $ac = $ anticlinal). A synclinal conformer is one in which the highest priority substituent on the Re (η^5-Cp) and the (C=C) centroid (=CHR) define a $60 \pm 30°$ torsion angle.[18] An anticlinal conformer is one in which the highest priority substituents define a $120 \pm 30°$ torsion angle. The torsion angles in idealized structures III/IV and V/VI (Scheme 3.4) are 45° and 135° respectively.

3.1.2.2. Enantioface Binding Selectivity

Gladysz *et al.* have investigated the kinetic and thermodynamic binding selectivities of η^2 π-ligands (π-ligands = alkenes, aldehydes and ketones) to the rhenium Lewis acids **3.7** and **3.8**.[19,20] The studies with monosubstituted alkenes and the dichloromethane

Table 3.1 Binding selectivity for diastereomeric π aromatic and aliphatic aldehyde complexes [CpRe(NO)(PPh$_3$)(η^2-O=CHR)][BF$_4$] (**3.10**).

T	R = aromatic groups	$(R_{Re}S_C, S_{Re}R_C)/$ $(S_{Re}S_C, R_{Re}R_C)$	R = aliphatic groups	$(R_{Re}S_C, S_{Re}R_C)/$ $(S_{Re}S_C, R_{Re}R_C)$
173 K	4-C$_6$H$_4$OCH$_3$	74:26	CH$_3$	99.0/1.0
173 K	4-C$_6$H$_4$CH$_2$CH$_3$	79:21	CH$_2$CH$_3$	99.8/0.2
173 K	4-C6H4CH3	76:24	CH$_2$CH$_2$CH$_3$	99.5/0.5
173 K	C$_6$H$_5$	78:22	CH(CH$_3$)$_2$	>99.9/< 0.1
173 K	4-C$_6$H$_5$Cl	83:17	C(CH$_3$)$_3$	>99.9/< 0.1
173 K	4-C$_6$H$_5$CF$_3$	89:11		
173 K	C$_6$F$_5$	97:3		

complex **3.7** show that the *kinetic diastereoselectivity* for the formation of the diastereomers $(R_{Re}S_C, S_{Re}R_C)/$ $(S_{Re}S_C, R_{Re}R_C)$ are a modest 2:1, and this ratio is 84:16 in the case of the bulkier *t*-butyl-ethylene. However, the thermodynamic selectivities are considerably higher; for example, when a mixture of $(R_{Re}S_C, S_{Re}R_C)/$ $(S_{Re}S_C, R_{Re}R_C)$ is heated in C$_6$H$_5$Cl at 95–100°C they equilibrate to between $\sim$ 95:5 and 99:1 mixtures of diastereomers.

Aldehydes and monosubstituted alkenes differ only at one X = C terminus (X = O vs X = CH$_2$) and are approximately isosteric. Thus rhenium Lewis acid **3.7** with aldehydes [(η^5-Cp)Re(NO)(PPh$_3$)(η^2-O=CHR)][BF$_4$] (**3.10**) give configurational diastereomers and Re-(X=C) rotamers analogous to those of the complexes with monosubstituted alkenes. In brief, equilibrium constants are similar; however, diastereomers of [(η^5-Cp)Re(NO) (PPh$_3$)(η^2-O=CHR)][BF$_4$] (**3.10**) rapidly interconvert below room temperature. The isomerization mechanism involves intermediate σ-complexes – a type of energy minimum not available to alkene complexes. At 173 or 183K, the aldehyde complexes [(η^5-Cp)Re(NO)(PPh$_3$)(η^2-O=CHR)][BF$_4$] give binding selectivities $(R_{Re}S_C, S_{Re}R_C)/$ $(S_{Re}S_C, R_{Re}R_C)$ higher than those observed for alkenes (see Table 3.1). A profound electronic effect is obvious. Electron withdrawing aryl substituents, which enhance π-acidity give higher selectivities (up to 97:3). In contrast electron-donating aryl substituents that diminish π-acidity and enhance σ-basicity give diminished selectivities (as low as 74:26).[21] On the other hand, aliphatic aldehyde complexes exhibit a high ratio of chiral recognition relative to the aromatic aldehydes. In this family of complexes steric effects were proposed to influence the enantioface binding selectivities, for example $(R_{Re}S_C, S_{Re}R_C)/$ $(S_{Re}S_C, R_{Re}R_C)$ ratios increase as the sizes of the O=CHR substituents from methyl (99.0:1.0) to *n*-alkyl (99.5-99.8: 0.5-0.2) to *sec*- and *tert*-alkyl (> 99.9:< 0.1).

3.1.2.3 Key Structural Data and Chiral Recognition

It has been thought that the binding selectivities obtained within the chiral rhenium complexes might be amenable to a structural rationalization. Hence olefin as well as aldehyde complexes of **3.7** were subjected to an extensive series of crystallizations. In general each complex crystallized as the more stable $(R_{Re}S_C, S_{Re}R_C)$ diastereomer with =CHR *trans* to the bulky PPh$_3$ group and pointing towards the small nitrosyl ligand. Importantly the distances between the rhenium and the carbon stereocentres Re—C* increased as the $(R_{Re}S_C, S_{Re}R_C)/$ $(S_{Re}S_C, R_{Re}R_C)$ ratios decreased (see Table 3.2). Stronger π accepting aldehydes gave shorter bonds and higher chiral recognition.[21]

Table 3.2 Key structural parameters for aromatic and aliphatic aldehyde complexes (*RS*, *SR*)-[CpRe(NO)(PPh$_3$)(η^2-O=CH*R*)][BF$_4$] (**3.10**).

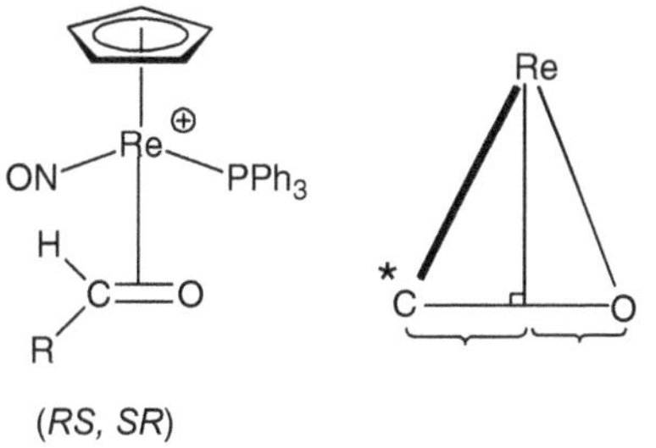

(*RS*, *SR*)

R	Re—C*	Slippage	OC-R bend-back-angle	Re-O=C plane with Re-P bond	O=C bond
C$_6$F$_5$	**2.157(5)** Å	20.1%	20.5°	2.5°	1.317 (6)Å
C$_6$H$_4$Cl	**2.172(4)** Å	29.9%	19.3°	20°	1.336 (6)Å
C$_6$H$_5$	**2.182(6)** Å	30.7%	17.5°	12.8°	1.324 (8)Å
Et-C$_6$H$_5$	**2.184(5)** Å	29.5%	17.5°	6.6°	1.307 (6)Å
Et	**2.15(1)** Å	24%	19.2°	17°	1.35 (1) Å
n-Pr	**2.150(4)** Å	20%	19.3°	20.5°	1.338 (5)Å

3.1.3 Other Related Complexes with Chiral-at-Metal Centre

3.1.3.1 Half-sandwich Complexes with CpML^{1}L^{2}L^3 M = Fe, Mo

Several chiral complexes were prepared with a variety of transition metals.[22] For example the iron complex CpFe(CO)(COR)(PPh$_3$) (**3.11**) was prepared and resolved.[23] Such a complex is configurationally stable, and its application to stoichiometric

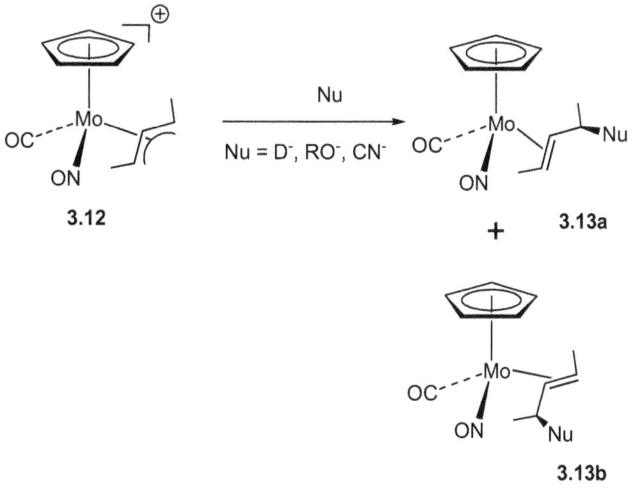

Scheme 3.5

asymmetric synthesis of organic compounds was studied by Davies, Liebeskind and coworkers (see Section 3.2). Furthermore Faller and coworkers[24] studied nucleophilic attack at the diastereotopic terminal carbons (Scheme 3.5) of the allyl ligand of the chiral complex [CpMo(CO)(NO)(CH₃CH=CH-CH-CH₃)]$^{+1}$ (**3.12**). The additions showed high selectivity, and this was explained by electronic asymmetry of the complex. Although the CO and NO ligands were almost indistinguishable, they exerted distinct electronic *trans* effects leading to different degrees of reactivity at allylic termini.

3.1.3.2   Half-sandwich Complexes with Chiral Cp Equivalents

Another method that allows enantioselective synthesis of half-sandwich complexes with a chiral metal centre is to replace the 'η^5-C₅H₅' ligand with a chiral *Cp*- equivalent. Such an approach introduces a second element of chirality in a precursor complex besides the stereogenic metal centre and may influence the control of configuration of the metal atom during the synthesis. Thus chiral *Cp* equivalents of the type 'η^5-(R^*-C₅H₄)' are reported with $R^* = $ menthyl, neomenthyl and fused-pinene. In particular a diastereomeric ratio of 82:18 has been achieved for the chiral complexes η^5-(R^*-C₅H₄)Ru(CO)(Cl)(Pi-Pr₃) (**3.14a/3.14b**) where $R^* = $ fused-pinene (Figure 3.3). It is believed that the selectivity increases with increasing steric bulk of the phosphine.[25]

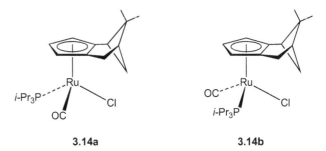

3.14a   3.14b

Figure 3.3

3.1.3.3   Half-sandwich Complexes with Chiral Chelating Ligands

The use of chiral chelating ligands to control the metal configuration in diastereomeric half-sandwich complexes have been widely employed with success. It is believed that chiral chelating ligands may control the configuration of a metal centre relative to a chiral monodentate ligand better. Most of the examples reported in the literature involve '(arene)Ru' precursors either [(p-cymene)Ru(μ-Cl)₂]₂ or the related [(benzene)Ru (μ-Cl)₂]₂ and chiral chelating ligands.

Indeed high diastereomeric ratios of 86:14 were obtained for the cationic (benzene)Ru complexes **3.15a/3.15b** prepared from the reaction of [(C₆H₆)Ru(μ-Cl)₂]₂ with the sodium salt of anion **3.16** (Figure 3.4).[26] The X-ray crystal structure of one of the diastereomers was determined. The ^{1}H NMR of **3.15** recorded at −80°C showed the same 86:14 ratio of the two diastereomers, comparable to that of the crude material at room temperature. These data suggested that the configurational stability of the metal centre is very low. Thus the thermodynamic ratio was established rapidly even at low temperature.

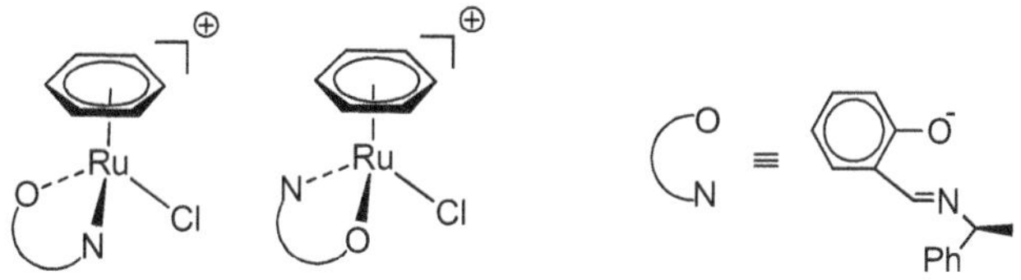

Figure 3.4

On the other hand, the reaction of $[(p\text{-cymene})Ru(\mu\text{-Cl})_2]_2$ with the chiral chelating ligands **3.17** and **3.18** remarkably gave the respective cationic complexes **3.19** and **3.20** as single disatereomers (Figure 3.5).[27,28] As stated in the beginning of this section chiral chelation has powerful control on the metal configuration in diastereomeric half-sandwich complexes.

However, the most employed half-sandwich complexes with chiral chelating ligands were reported by Noyori *et al.*[29] who designed a new type of active and robust Ru(II) catalyst precursors based on complexes of monotosylated diamines. The structure of one of these catalyst precursors **3.21** was determined by X-ray analysis and showed that the metal has (*R*) configuration (see Figure 3.6).

More recently a new ruthenium 'zipper catalyst' **3.22** which is chiral-at-metal centre and bears a PN chelating ligand was reported by Grotjahn and coworkers (Figure 3.7).[30] This catalyst can isomerize a variety of alkene derivatives, including diallyl ether, geraniol and *t*-butylpent-4-enyloxydimethylsilane, to form high purity (E)-isomer products. Depending on the alkene selected, the catalyst can shift the double bond as many as 30 positions along the alkyl chain.

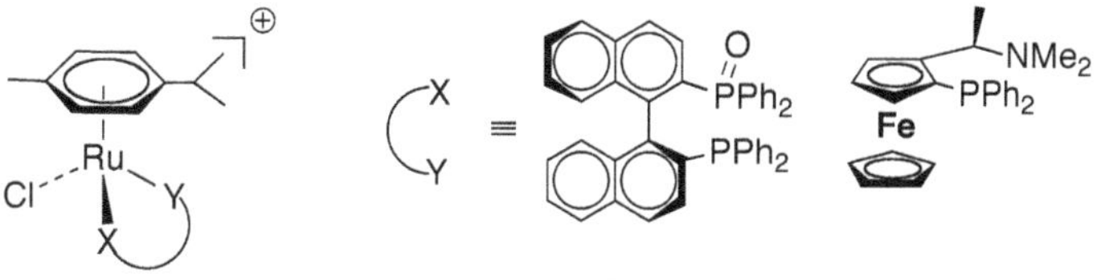

3.19, X-Y = 3.17
3.20, X-Y = 3.18

Figure 3.5

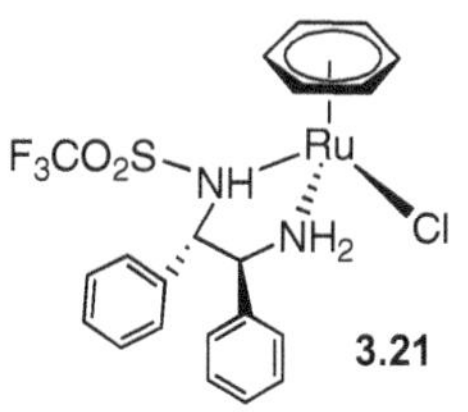

3.21

Figure 3.6

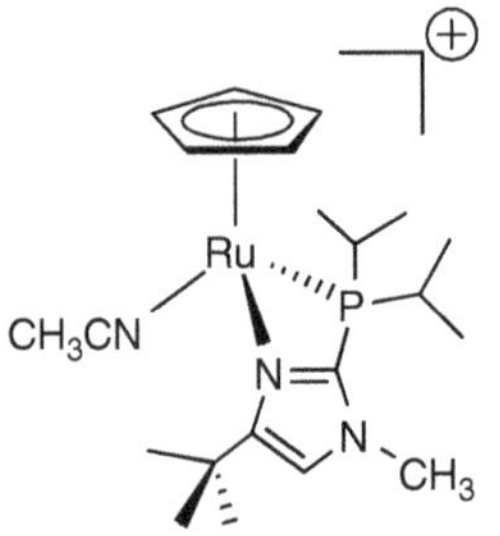

Figure 3.7

3.1.3.4 Half-sandwich Complexes with Tethered Chiral Ligands

Another approach to control the metal configuration of complexes with chiral-at-metal centres is to anchor the π-bonded *Cp* ligand to a chiral donor group that will also coordinate to the metal centre. In such a system, the tethered chiral donor ligand is expected to place the metal centre in a rigid defined chiral environment even if the analogous *Cp*-complex without a tether would be configurationally labile. Thus *Cp*-anions with different chiral phosphane donor functions have been synthesized and treated with [Ru(PPh$_3$)$_3$Cl$_2$] to yield the related half-sandwich complexes **3.23–3.25** (Figure 3.8). The observed diastereomer ratios range from 59:41 up to $> 99 :< 1$.[31–33] High isomer ratios are expected, if the chiral information transmitted to the metal centre is located nearby.

More recently a different synthetic method was used to prepare tethered CpRu complexes. In these examples **3.26–3.27** (Figure 3.9) the Cp ligand is trisubstituted, including the tethered phosphine ligand. The best diastereo-selectivities were found for the sterically demanding derivative with R $= t$-Bu and bulky phosphines like PPh$_3$ and PBu$_3$, where only one diastereomeric product could be observed by NMR spectroscopy.[34,35]

In 2005, M. Wills and coworkers reported the synthesis of a Ru(II) tethered version of Noyori's catalysts **3.28**[36,37] (Figure 3.10). These complexes have a rigid structure and are stereochemically well defined; moreover they are very active catalysts for asymmetric hydrogenation of ketones.

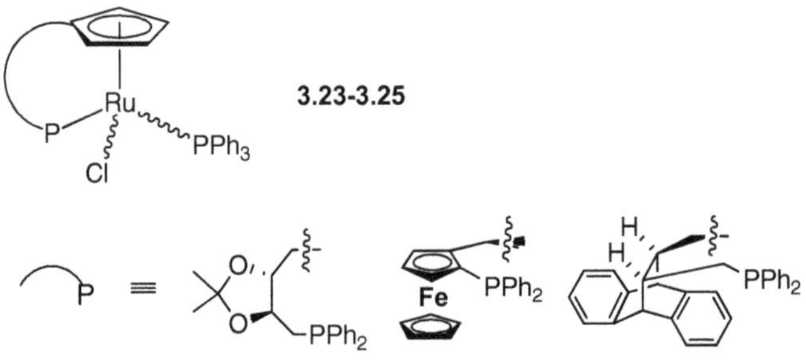

Figure 3.8

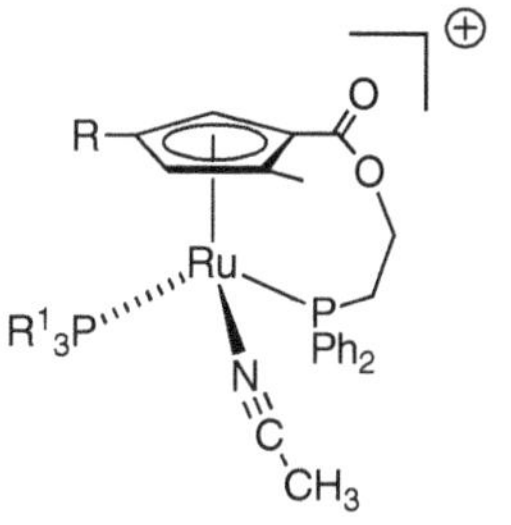

R^1 = -Ph, **3.26**
R^1 = -Bu, **3.27**

Figure 3.9

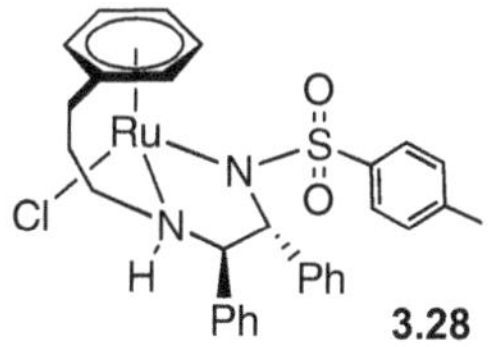

Figure 3.10

Chiral-at-metal Complexes in Organic Synthesis

3.2.1 The Chiral Acyl–Iron Complex

The acyl–iron complex CpFe(CO)(COCH$_3$)(PPh$_3$) (**3.29**) is the most used chiral-at-metal centre species for organic synthesis.[38] This complex adopts a conformation that places the acetyl oxygen *anti* to carbon monoxide with one of the triphenylphosphine phenyl groups approximately parallel to the plane of the acetyl ligand, shielding one face of the acetyl (Figure 3.11).

  This complex has several interesting characteristics: (i) it is easy to prepare and handle, (ii) it is chiral-at-iron and can be resolved, and (iii) the protons α to the acyl group are acidic and the corresponding metal acyl enolate undergoes a variety of transformations including alkylations, aldol reactions, conjugate addition reactions and Diels–Alder reactions (Scheme 3.6).

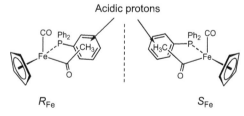

Acidic protons

R_{Fe} S_{Fe}

Figure 3.11
Chiral acetyl-iron [(Cp)Fe(CO)(COMe)(PPh$_3$)] (**3.29**).

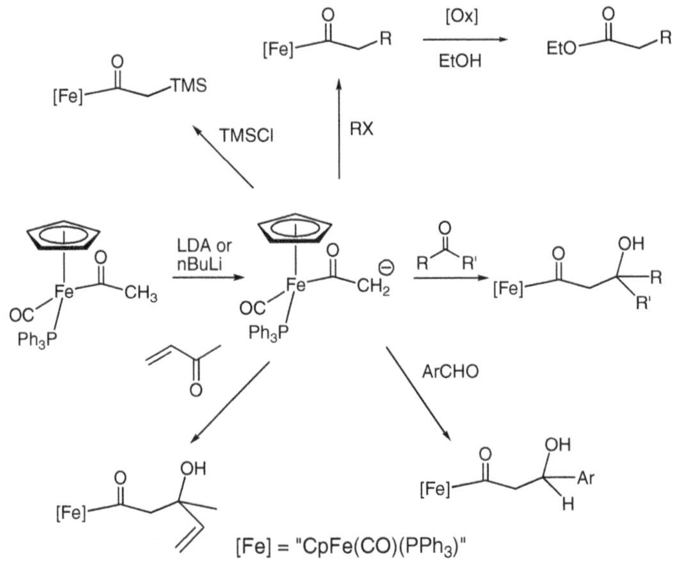

Scheme 3.6

Liebeskind[39] and Davies[40] have independently developed the use of this chiral iron complex for enantioselective organic syntheses, particularly of a variety of optically active molecules. The reasons for this behaviour are that the complex is chiral-at-iron and one face is hindered by the PPh$_3$ ligand; reaction of these acyl–iron enolates occur with very high stereoselectivity (Scheme 3.7).

Thus Liebeskind[41] studied the stereoselective aldol condensation with iron acyl enolates and showed that the choice of counter ion had a dramatic effect on the stereoselectivity of the two diastereomers **3.30** and **3.31**. With i-Bu$_2$Al$^+$ adducts of the enolate, diastereomer **3.30** predominated with **3.30** and **3.31** in a ratio as high as 8.2:1. When SnCl$^+$ is used instead, the stereoselectivity switched completely to favour the disatereomer **3.30** in a ratio of 1:11 (Scheme 3.8).[42]

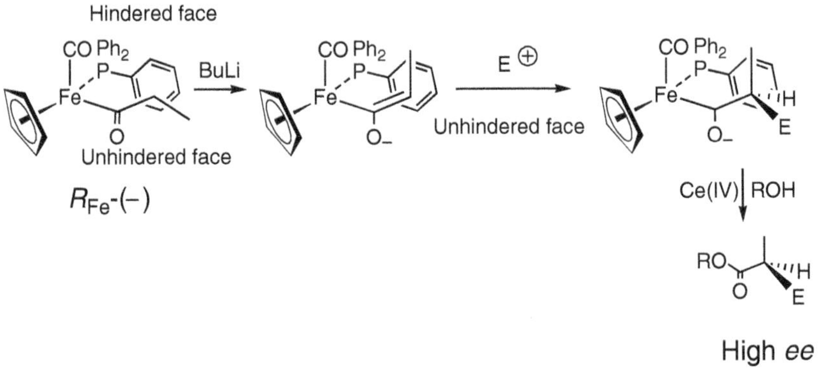

Scheme 3.7

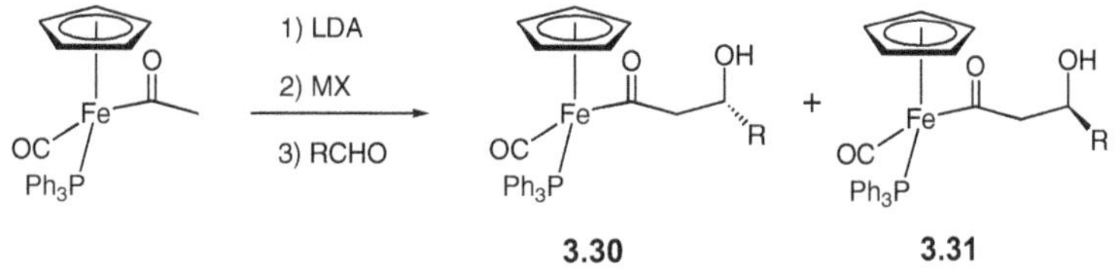

Scheme 3.8

Davies examined the same reaction.[43,44] He found that when using an excess of Et_2Al^+ adducts of the enolate, the diastereoselectivity ratio became > 100:1. Further, if Cu(I) was used as counter ion the opposite stereochemistry was obtained.[45] Moreover both Davies and Liebsekind used this chiral iron auxiliary in a stereoselective synthesis of β-lactams.[46,47] Liebeskind reported that chiral iron enolate complex condensed with imines in the presence of Et_2Al^+ counter ion to give two isomers with a ratio up to 20:1. Oxidation with I_2/R_3N produced the racemic β-lacatms (Scheme 3.9)

Davies reported the use of resolved (S) iron crotonyl complex **3.32**.[48,49] Treatment with lithium benzylamide afforded the (S_{Fe},3S)-β amino complex-**3.33** in 90% yield as a single diastereoisomer, which upon oxidative decomplexation with Br_2 afforded the optically active β-lactam (S)-**3.34**. Furthermore, conjugate addition of lithiumbenzyla-mide to **3.32** and tandem methylation gave (S_{Fe}, 2R, 3S)-β-amino-α-methyl complex **3.35** in 91% yield as a single diastereomer; subsequent oxidative decomplexation yielded the cis-β-lactam (3R, 4S)-**3.36** (Scheme 3.10)

Davies employed the iron chiral auxiliary to establish unambiguously the absolute configuration of Winterstein's acid (3-N,N-dimethylamino-3-phenylpropanoic acid) as (R).[50] Winterstein's acid (3-N,N-dimethylamino-3-phenyl-propanoic acid) had been identified as the carboxylate side chain of the diterpene taxine B.[51,52] This acid had also been prepared by resolution; however, its absolute configuration was not established unambiguously. The previous methodology was applied to prepare the optically pure (R) and (S) ester derivatives of Winterstein's acid (Scheme 3.11). Conjugate addition of lithium dimethylamide to the chiral iron cinnamoyl complexes (S_{Fe},E) and (S_{Fe},Z)-[Cp]Fe (CO)(PPh₃)(COCH=CHPh) **3.37** and **3.38** proceeded with high diasterteoselectivity to

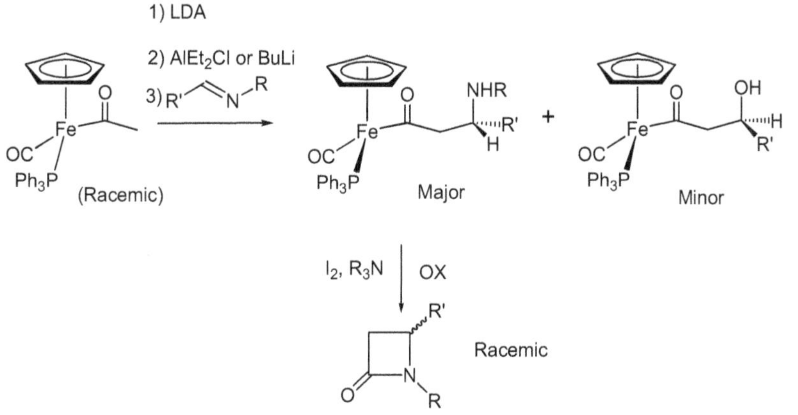

Scheme 3.9

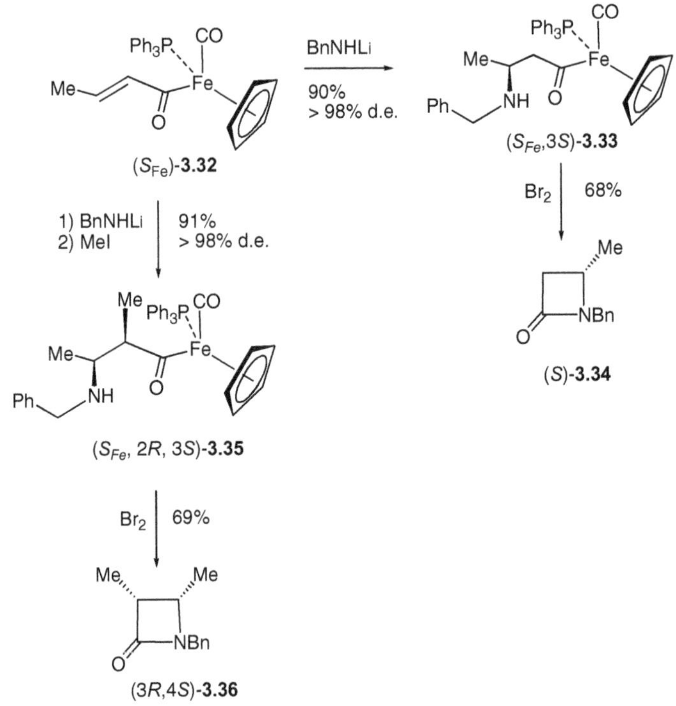

Scheme 3.10
Formation of *cis*-β-lactam (3*R*, 4*S*)-**3.36**.

give (S_{Fe}, 3*R*)- and (S_{Fe}, 3*S*)-β-amino complexes **3.39** and **3.40**. Oxidative decomplexation of the latter complexes by NBS followed by reaction work up provided the (*R*)-**3.41** and (*S*)-**3.41** ethylester derivatives of Winsterstein's acid. The specific rotation of each enantiomer was measured {(*S*)-**3.41**; $[\alpha]_D^{20} + 13.9(c = 0.5, \text{ CHCl}_3), [\alpha]_{436}^{20} + 31.5$ ($c = 0.5, \text{CHCl}_3$); for (*R*)-**3.41** ; $[\alpha]_{436}^{20} - 31.1$ ($c = 0.5, \text{CHCl}_3$)}. Comparison with the ethyl ester derived from Winterstein's acid { $[\alpha]_D^{18} - 15.5$ (neat), $[\alpha]_{436}^{20} - 9.55(c = 9.0,$

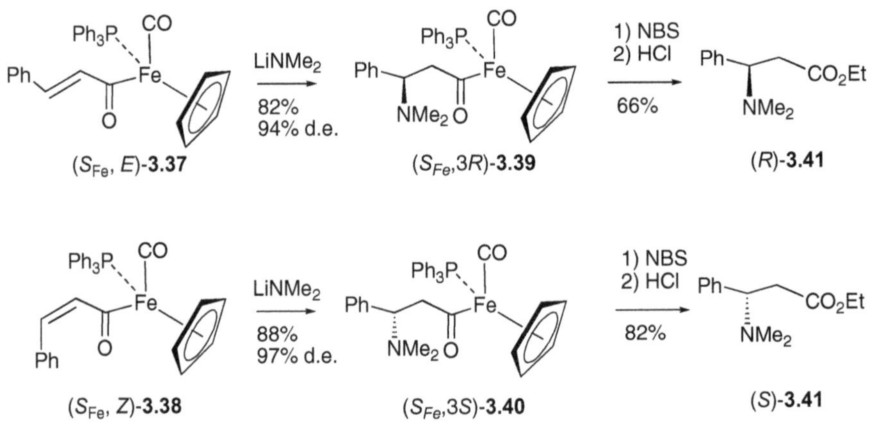

Scheme 3.11
Asymmetric synthesis of ethyl ester derivatives of Winterstein's acid **3.41**.

$CHCl_3$)} unambiguously established the absolute configuration of Winterstein's acid as (R).

3.2.2 The Chiral Cyclic Acyl–Cobalt Complex

Bergman has prepared a chiral cyclic acyl–cobalt complex **3.42** (Figure 3.12) and showed that the corresponding enolate complex underwent asymmetric, high-yielding aldol and alkylation reactions.[53] Moreover, this chiral acyl–cobalt complex was resolved using an optically pure phosphine. Starting with the related enolate cobalt complex, it was possible to prepare the enantiopure compound. Thus alkylation of the cyclic enolate complex with pivaldehyde, followed by oxidative decomplexation, produced the chiral cyclobutatone **3.43** optically pure in 70% yield. The purity of **3.43** was confirmed by ^{1}H NMR using chiral shift reagents (Scheme 3.12).

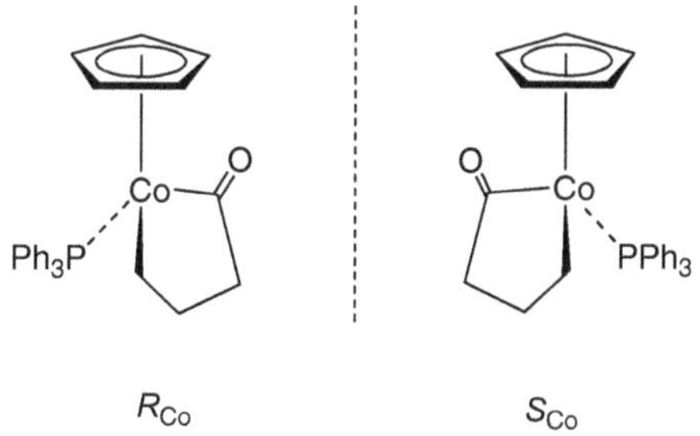

Figure 3.12
Cyclic acyl–cobalt complex **3.42**.

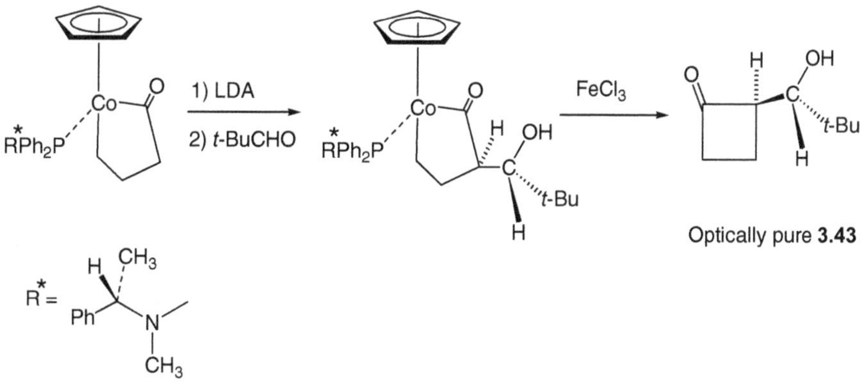

Scheme 3.12

3.2.3 The Lewis Acid–Rhenium Complex

The Lewis acid rhenium complex described in Section 3.1.2, shows high enantioface binding towards alkenes, aldehydes and ketones. For example aldehyde substrates bind to the (S)-[CpRe(NO)(PPh$_3$)]$^+$ Lewis acid via an η^2 coordination mode to give the (R_C S_{Re}) aldehyde complex with the bulky RCH= terminus *anti* to the PPh$_3$ ligand, and the alkyl group R *syn* to the small NO ligand.[54] The aldehyde prefers to have the small proton rather than the bulky R group pointing toward the medium-sized η^5-C$_5$H$_5$ ring. The

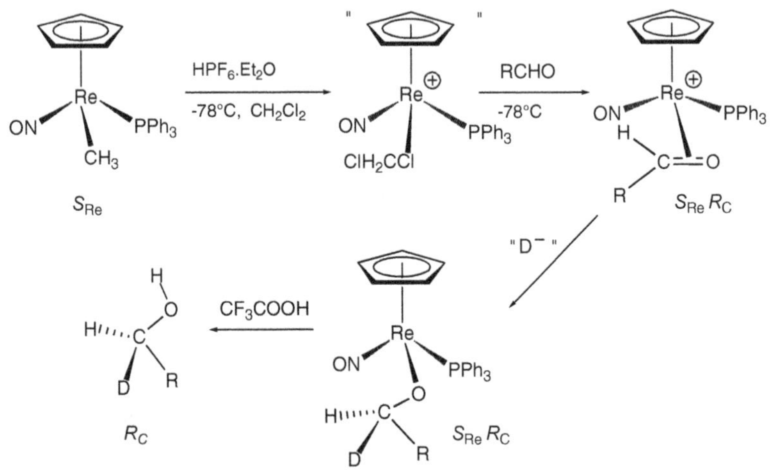

Scheme 3.13

synthesis of optically active alcohols is successfully performed using optically active aldehyde complexes. Thus nucleophilic D^- addition occurs at only one of the two faces of the aldehyde to give the deuterated alkoxide complex. Subsequent acidification with CF_3COOH provides a single enantiomer of the alcohol product (Scheme 3.13).[55] It is worth noting that nuleophiles attack the aldehyde complexes by σ-aldehyde complex intermediates from the side opposite to the bulky PPh_3,[56,57] so that the selectivity is not a direct function of which face of the aldehyde is bound; however the alkene complexes undergo attack from direction opposite to the π-bound face.[58]

3.3 Asymmetric Catalysis by Chiral Complexes

Enantiomeric compounds react with other chiral molecules at intrinsically different rates. In fact this is extremely important for human metabolism since human proteins are composed of *l*-amino acids. As a consequence, the need to prepare compounds efficiently in enantiopure form is increasing dramatically. For example, in the drug industry nine of the top 10 pharmaceutical drugs have chiral active ingredients, seven of which are enantiopure.[59] For these reasons, a great deal of interest has been devoted to asymmetric synthesis. Moreover, stereoselective syntheses that are based on asymmetric catalysis can be advantageous from the standpoints of both efficiency and economy, since a small amount of a chiral catalyst can produce a large quantity of enantiopure active molecules. For their contributions to developing 'asymmetric catalysis', William S. Knowles, Ryoji Noyori and K. Barry Sharpless were awarded the 2001 Nobel Prize for Chemistry. [60]

3.3.1 Asymmetric Hydrogenation

3.3.1.1 Chiral Phosphine Ligands

An alkene with two different substituents at the same carbon is prochiral. This implies that the two faces of the alkene are different. The addition of an H_2 molecule to the

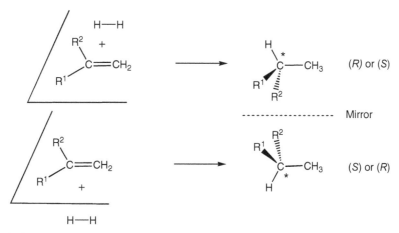

Figure 3.13

double bond from one face produces one enantiomer while from the opposite face the other enantiomer is formed (see Figure 3.13). Thus if we could control the H_2 docking to one face rather than the other, we would have an asymmetric synthesis.

Transition metal compounds bind to prochiral olefins to give two enantiomers, despite the fact that neither the metal complex nor the olefin is chiral. In fact this is understandable since the initially prochiral carbon (with asterisk) becomes coordinated to four different substituents, one of which is the metal (Figure 3.14).

As shown previously in this chapter if one of the ligands L^* of the metal M is optically pure with S configuration, then two diastereomers will be formed (see Figure 3.15). In general diastereomers have different chemical properties; thus the rate of hydrogenation of the two diastereomers will be different, and eventually one of the enantiomers of the product will be obtained.

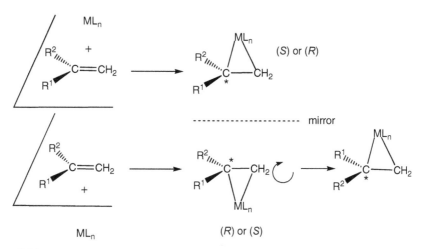

Figure 3.14

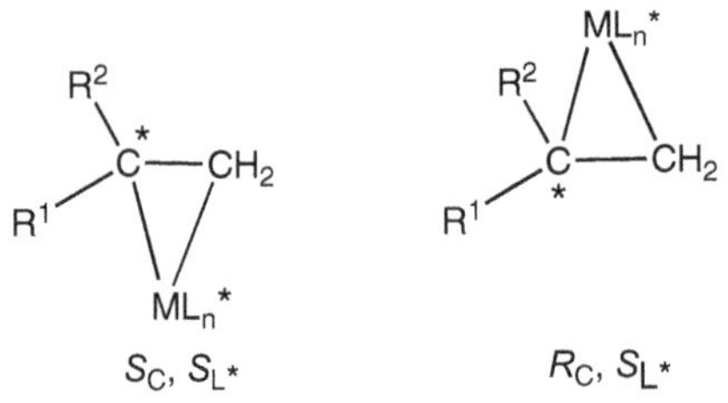

S_C, S_{L^*} R_C, S_{L^*}

Figure 3.15

Thus asymmetric catalysis is important for both academic and industrial applications, since we need to use only small amounts of a resolved asymmetric catalyst to obtain large quantities of optically pure active molecules. We can also compare asymmetric catalysts to enzymes, since they are also efficient catalysts in biological processes.

The first efficient asymmetric hydrogenation catalyst was developed by Kagan[61] based on the Schrock–Osborn catalyst $[L_2RhS_2]^+$ (L = phosphine, S = Solvent).[62–65] Instead of the monodentate phosphines, the chiral ligand DIOP (**3.44**), prepared from tartaric acid, chelates the rhodium centre. The resolved ligand contains two chiral carbon centres and has C_2 symmetry (Figure 3.16). Upon coordination to the rhodium centre a five-membered ring is formed with helical chirality which of course can be either right-handed or left-handed. Such a configuration forces the phenyl groups on the phosphorus to adopt a propeller-like arrangement, and hence the chiral information is transferred to the binding site of the metal.

Eventually the docking of the prochiral ligand to the metal centre will favour one face relative to the other one. During the catalytic cycle the hydrogen addition will produce one of the two enantiomers preferentially.

Halpern and coworkers[66] investigated the mechanism of these asymmetric hydrogenation reactions with the resolved (*S,S*)-chiraphos ligand and showed that the minor diastereomer was more active than the major diastereomer as observed by NMR techniques during the catalytic cycle and thus produced the major hydrogenated *R*-product (**3.45**) (Scheme 3.14).

An important industrial application of asymmetric hydogenation was developed by Willam S. Knowles. The drug L-DOPA (**3.46**), known for the treatment of Parkinson's disease, was prepared on a large scale using a rhodium catalyst but with the DIPAMP (**3.47**) chelating ligand (Scheme 3.15). In this ligand the chirality was centred at the phosphorus atom.[67] The inversion barrier of the tricoordinated phosphorus was above 30 kcal/mole, which meant that the configurational stability of the phosphorus atom was

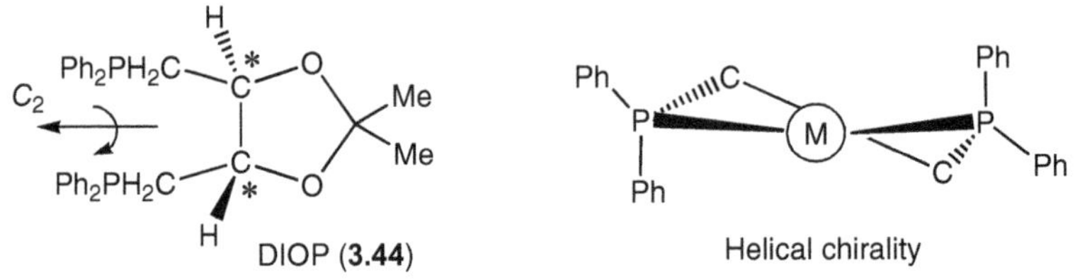

DIOP (**3.44**)   Helical chirality

Figure 3.16

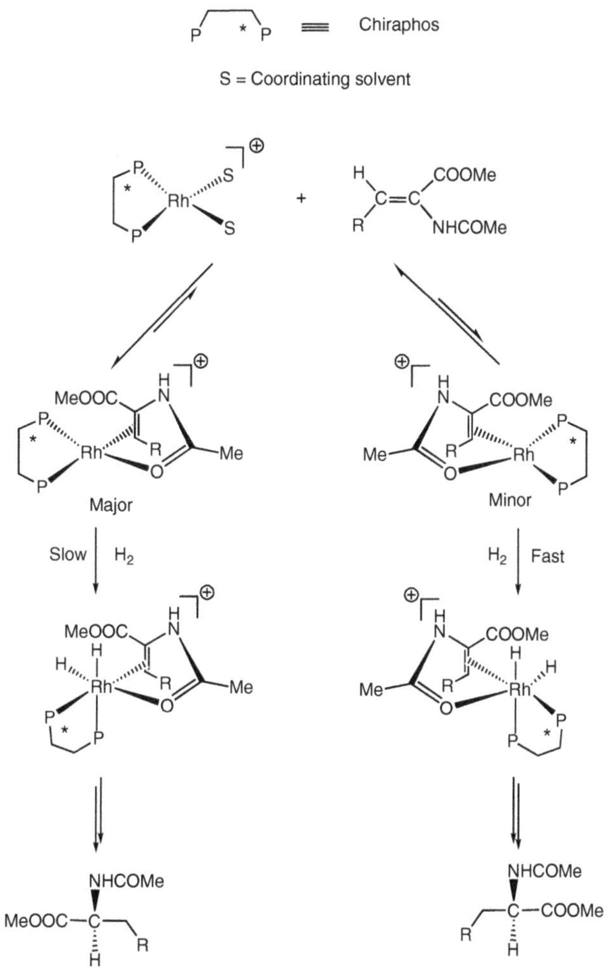

Scheme 3.14

high even at 100°C. Another important feature for the success of this industrial application as noted by Knowles in his Nobel lecture was the ease in separation of the pure enantiomer from the partly racemic product formed during the catalytic process. Because in the L-DOPA process the intermediate formed was a conglomerate rather than a racemate (see definition in Chapter 2), separation of the L-form from the LD form was efficient. For instance a 90% ee value provided easily 90% of a pure enantiomer. In his Nobel lecture Knowles[68] indicated that this lucky break was most welcome and stated that *'If one thinks back, ours was the same luck Pasteur encountered in his classical tartaric acid separation, 150 years ago.'*

Another important development in this area comes from Noyori who designed the new chiral chelating ligand BINAP (**3.48**) (Figure 3.17).[69] Unlike preceding examples, which displayed central chirality, the BINAP ligand possesses 'axial chirality' (atropisomerism). Due to steric hindrance, the two binaphthyl groups are distorted along the

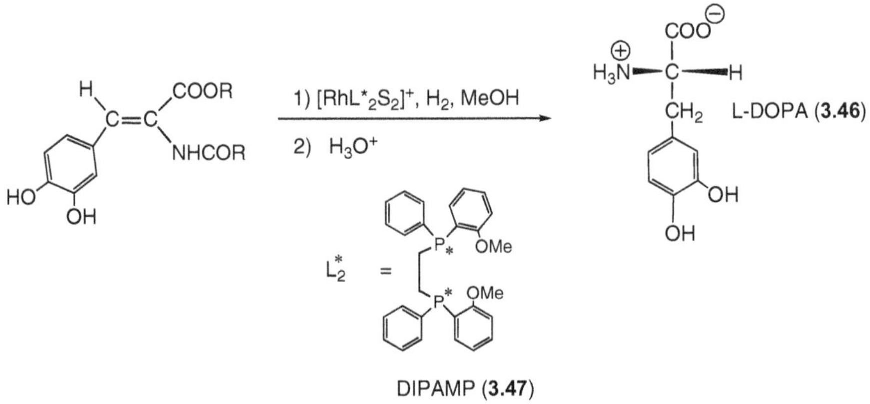

Scheme 3.15

C—C bridge. This phosphine is a remarkable ligand for asymmetric hydrogenation of olefins with cationic rhodium catalysts and ketones with ruthenium (II) catalysts.

Most notable was use of the cationic BINAP-Rh complex in asymmetric isomerization of allylic amines, which permitted an industrial synthesis of (−)-menthol from myrcene (see Scheme 3.16).[70]

More recently Noyori developed asymmetric hydrogenation of simple ketones with BINAP/diamine-ruthenium complexes.[71,72] In this system the catalytic process contrasted with the conventional mechanism of asymmetric hydrogenation of unsaturated bonds which requires metal-substrate π-complexation. In fact with BINAP/diamine-ruthenium neither the ketone substrate nor the alcohol product interacted with the metallic centre during the catalytic cycle. The enantiofaces of the prochiral ketones were differentiated on the molecular surface of the coordinatively saturated RuH intermediate.

Other chiral diamine-(η^6-arene)ruthenium catalysts were developed by Noyori where the chirality was centred at the metal (see Figure 3.18). These complexes were effective catalysts for asymmetric transfer hydrogenation of carbonyl compounds and a mechanism involving a metal–ligand bifunctional process was proposed.[73–76]

Later, in the 1990s a remarkable asymmetric catalyst with the chiral chelating phosphine DuPHOS (**3.49**) (Figure 3.19) was developed by M. J. Burk at DuPont.[77] This cationic DuPHOS/rhodium complex showed very high ee values of 99% in asymmetric hydrogenation of enamides. As can be seen the choice of chiral phosphine chelating ligand had a dramatic effect on the ee values.

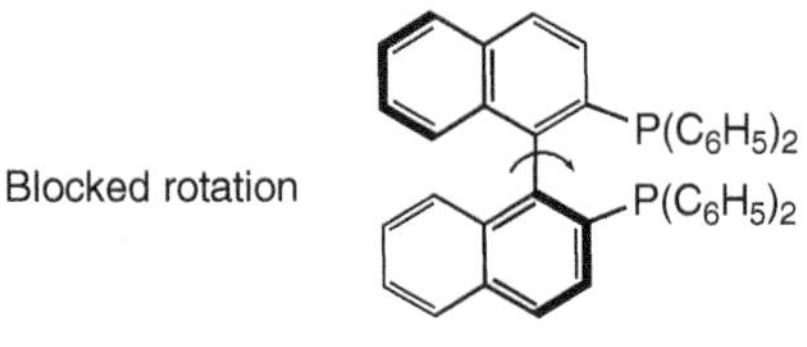

(*S*)- BINAP (**3.48**)

Figure 3.17

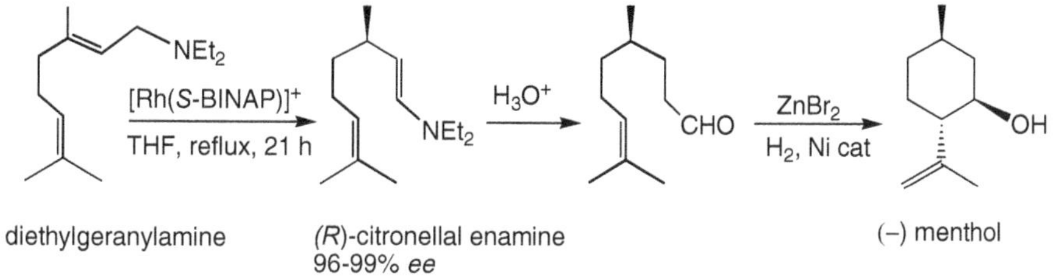

diethylgeranylamine   (*R*)-citronellal enamine   (−) menthol
96-99% *ee*

Scheme 3.16

In the search for new ligands for asymmetric catalysis Ben L. Feringa and coworkers developed the synthesis of chiral monodentate phosphoramidites (PipPhos)[78] (Figure 3.20) which were excellent ligands for asymmetric hydrogenation of aromatic enol acetates, and enol carbamates with high *ee* values up to 98%.

More recently and in the course of new efficient catalysts for asymmetric hydrogenations, efforts were directed to the use of optically active chelating ligands but with ferrocene backbones.

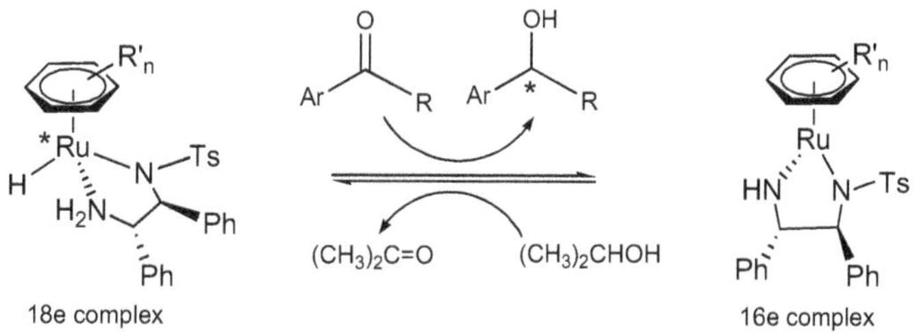

Figure 3.18

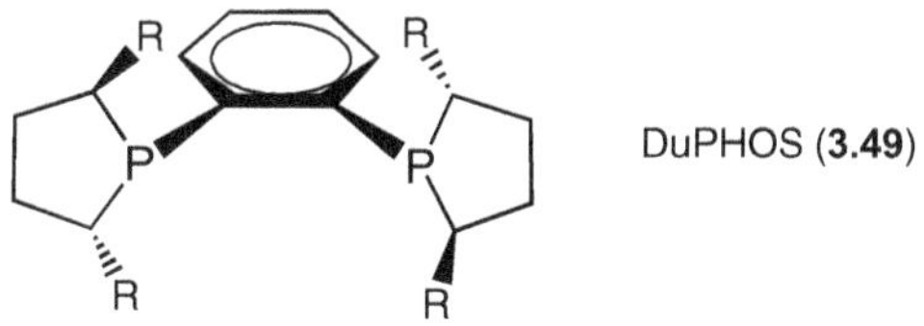

DuPHOS (**3.49**)

Figure 3.19

PipPhos

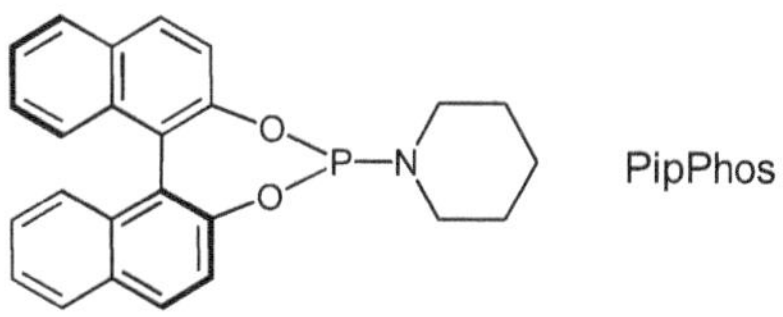

Figure 3.20

3.3.1.2   Chiral Ferrocene Ligands in Asymmetric Hydrogenation

Ferrocene has a rigid sandwich structure, which is highly stable and easily available. There are several methods for its functionalization at different positions of the sandwich

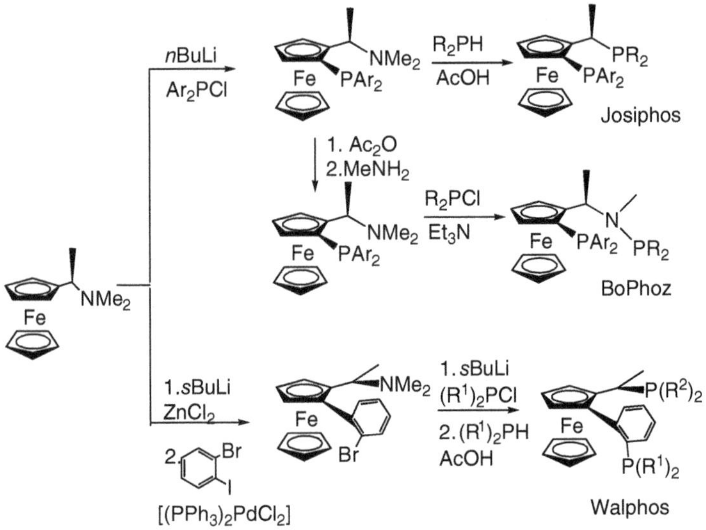

Scheme 3.17
Ugi's method.

structure, which allow the easy introduction of a vast variety of substituents – usually coordinating phosphorus, nitrogen and sulfur atoms. All of these characteristics make ferrocene an excellent scaffold for the design of chiral ligands for asymmetric catalysis. There are several procedures for the functionalization of ferrocene; the most widely used strategies for the preparation of enantiopure 1,2 disubstituted ferrocene ligands are the Ugi amine method[79] (Scheme 3.17), the oxazoline approach and the sulfoxide approach described by Kagan (Scheme 3.18).[80,81]

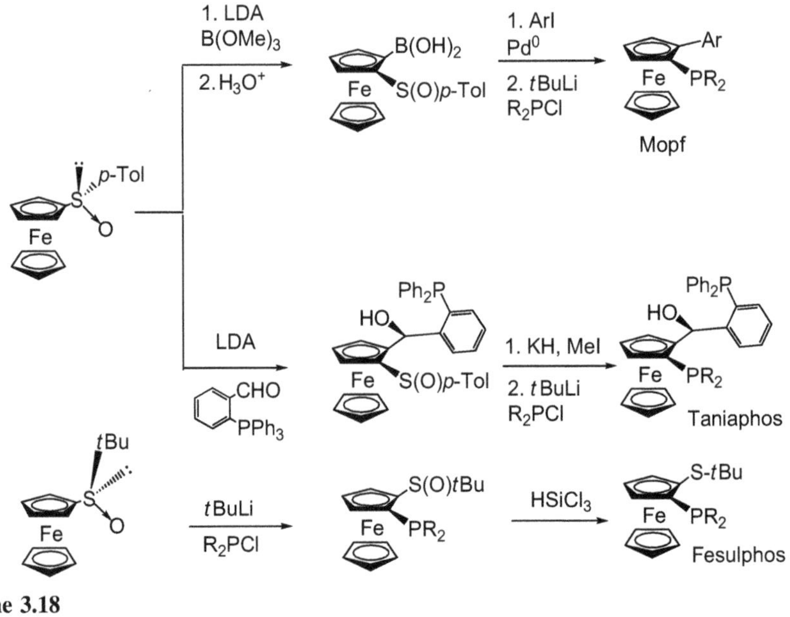

Scheme 3.18
Kagan's method.

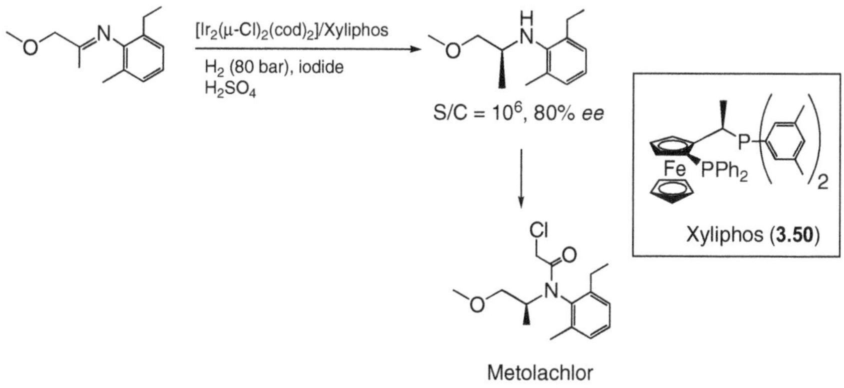

Scheme 3.19
Xyliphos/Ir-catalysed enantioselective hydrogenation of the imine derivative and synthesis of Metolachlor.

In recent years an amazing number and variety of chiral ferrocene ligands have been used in asymmetric catalysis. A quite remarkable example of the great utility of chiral ferrocene ligands is the synthesis of a precursor of the herbicide (1*S*)-metolachlor by an Ir-Xyliphos-catalysed asymmetric imine hydrogenation reaction (Scheme 3.19).

The catalyst is generated *in situ* by mixing [Ir(μ-Cl)(Cod)]$_2$, Xyliphos (**3.50**), iodide (NaI or NBu$_4$I) and a Bronsted acid (CH$_3$COOH or H$_2$SO$_4$). This catalyst is extremely reactive, being one of the fastest homogeneous systems known (S/C = 10^6) and it has become an extremely effective industrial process and constitutes the largest enantiose-lective catalytic process in industry (turnover numbers (TONs) of 2×10^6 and turnover frequencies (TOFs) of around $6 \times 10^5 \, h^{-1}$, at more than 10^4 tons per annum).[82,83]

On the other hand, hydrogenation of the more challenging simple ketones has recently been reported by using the [RuCl$_2$(PPh$_3$)$_3$]/Fc-Fox system where Fc-Fox = ferrocenyl phosphine-oxazolene ligands (**3.51**) (Scheme 3.20). The Ru/Ferrocenyl based ligands are extremely effective catalysts with remarkable enantioselectivities (up to 99% ee) and excellent S/C ratios (up to 10 000–50 000).[84]

During the last few years a large number of articles and reviews have described the use of chiral ferrocene-based ligands for an amazing variety of asymmetric catalytic reactions involving hydrogenation of alkenes, ketones and imines, hydrosilylation of ketones, 1,2 addition to carbonyl compounds, conjugate additions and a diversity of metal mediated C—C coupling reactions. Among the ferrocene-based ligands used, were the Josiphos family of bisphosphine ferrocene ligands reported by Togni *et al.*,[85] the

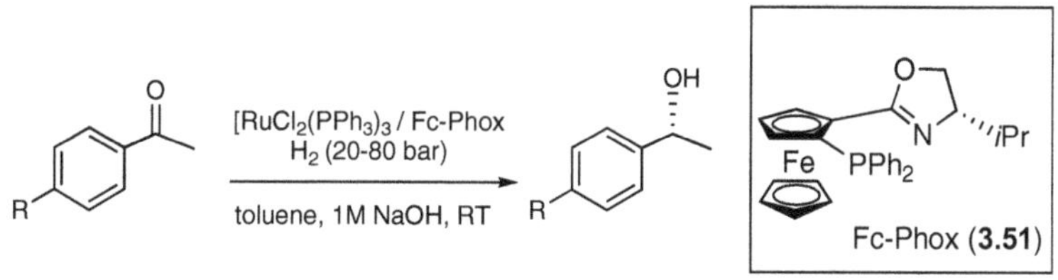

Scheme 3.20

ferrocenylphosphino-oxazoline (Fc-Phox) ligands reported by the groups of Richards,[86] Sammakia[87] and Uemara,[88] the development of the 1,5-bisphosphine Taniaphos by Knochel and coworkers,[89] and more recently by Fu for the preparation of chiral heterocycles, ferrocene-based ligands in asymmetric catalysis.[90] For a specialized recent review concerning applications of *'chiral ferrocene ligands in asymmetric catalysis'*, the reader may consult a review by Carretero *et al.*[91]

3.3.2 Asymmetric Epoxidation and Dihydroxylation

Asymmetric oxidation of alkenes is a useful reaction in organic synthesis. Sharpless and coworkers made an outstanding contribution in this field. The first practical method of asymmetric epoxidation of allylic alcohols was reported by Sharpless in 1980.[92] The epoxidation agent was *tert*-butylhydroperoxide, and the asymmetric catalyst was composed of diethyl tartrate (DET) D-(−) or L-(+) and titanium tetraisopropoxide.

Use of D-(+) DET leads to addition of the epoxide oxygen from the bottom, while when L-(−) DET is employed the epoxide oxygen is added from the top (Scheme 3.21). Enantiomeric excesses for this process are very high ($\geq 90\%$).

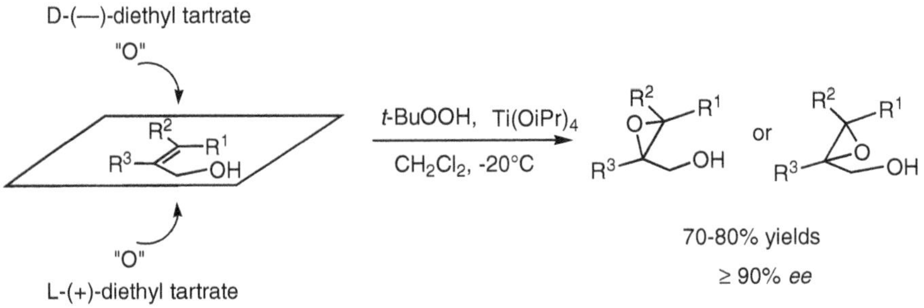

Scheme 3.21

The great value of the asymmetric 'Sharpless epoxidation' lies in both the variety of applications for the products of the reaction and the variety of allylic alcohols that will give a product in good yield and high enantiomeric excess, as well as the simplicity of the reagents used to carry out the above reaction.

Sharpless and his group also developed several other asymmetric additions to the carbon–carbon double bond. Among these reactions, perhaps the best developed was reported in 1988 and concerned asymmetric dihydroxylation (AD-mix) which led to 1,2 diols in good yields and high enantiomeric excess. In this system the oxidizing agent was N-methymorpholine-N-oxide in aqueous acetone, and the catalyst was osmium tetra-oxide.[93] Once again the choice of the optically active chiral ligand played an important role to induce the asymmetric addition.[94] Thus when the dihydroquinidine ligand (DHQD)$_2$PHAL (**3.53**) (Figure 3.21) was used for asymmetric hydroxylation of *trans*-stilbene, the *S,S*-enantiomer of the chiral diol was produced in almost quantitative yield. The use of the other optically pure ligand dihydroquinine (DHQ)$_2$PHAL (**3.52**) (Figure 3.21) provided a 99.8% of the *R,R* diol.[95]

Another interesting asymmetric catalyst for epoxidation of simple olefins such as *cis*-PhCH=CHMe was reported by Jacobsen and coworkers. In this example the catalyst was a chiral manganese salen-based complex (**3.54**) (Figure 3.22), and the oxidant used was NaOCl; high yields and ee of > 90% were achieved.[96]

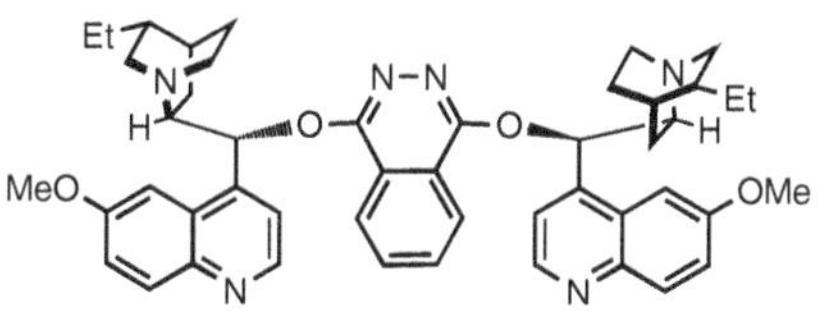

(DHQ)₂PHAL (**3.52**)

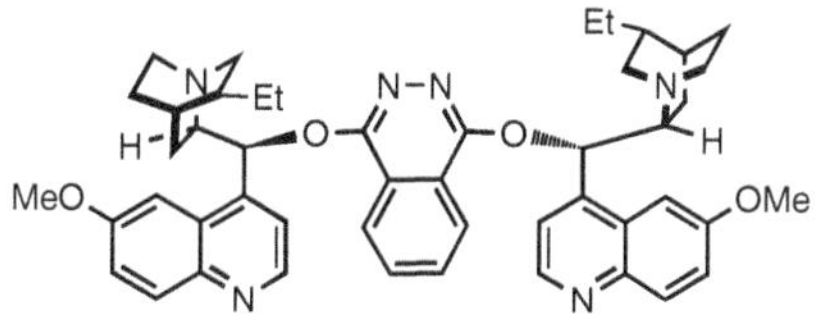

(DHQD)₂PHAL (**3.53**)

Figure 3.21

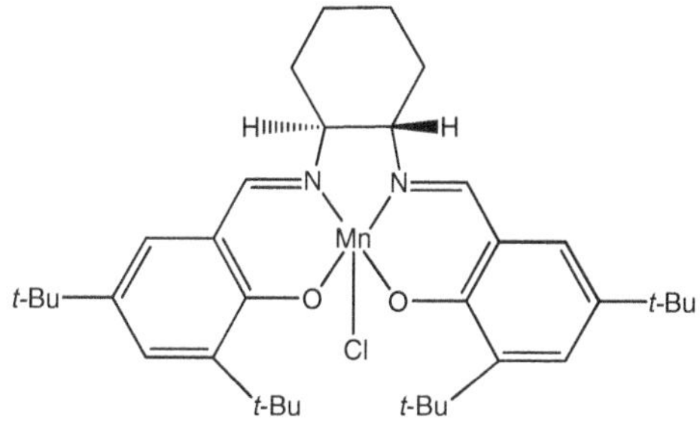

t-Bu

t-Bu (**3.54**)

Cl

t-Bu   *t*-Bu

Figure 3.22

3.3.3 Gold Complexes in Asymmetric Catalysis

Transition metal compounds containing gold were once considered as being catalytically inert; just in the past few years gold has emerged as a powerful homogeneous catalyst for the electrophilic activation of alkynes towards a variety of nucleophilic additions.[97,98] The strong Lewis acidity of cationic Au(I) and/or Au(III) as well as their ability to stabilize cationic reaction intermediates imparts unique reactivity to such catalysts and makes them important tools in the development of new synthetic methods in organic chemistry.[99] Thus there are an impressive number of reports on the use of gold catalysts to carry out a variety of catalytic transformations of alkynes towards: (a) nucleophiles, (b) alkenes or (c) cyclization reaction of enynes,[100–105] as well as (d) the preparation of functionalized arenes [106] and phenols.[107] These have been reported and the reader may consult the reviews.[108,109] However, we note that only a few examples of gold complexes in asymmetric catalysis have been reported.

In 1986 Ito and Hayashi pioneered the use of Au(I) homogeneous catalysts in asymmetric organic synthesis.[110] Thus, the chiral ferrocenylphosphine/Au(I) catalyst precursor (**3.55/3.56**) formed *in situ*, catalysed asymmetric aldol reactions of an isocyanoacetate with aldehydes to produce optically active substituted oxazolines with high enantio- and diastereoselectivity (Scheme 3.22). The author suggested that the use of gold is essential for the high selectivity, a silver or copper catalyst being much less selective.

More recently, Echavarren and coworkers performed a detailed investigation of the enantioselective alkoxycyclization of functionalized enynes with gold (I) complexes in the presence of a variety of chiral ligands such as BINAP, chiral 1,2-ferrocenyl phosphines and chiral 1,2-ferrocenyl-P^N bidentate ligands.[111] Interestingly the authors

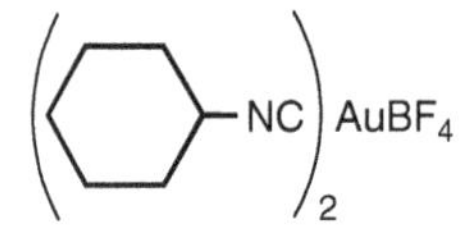

(3.55)

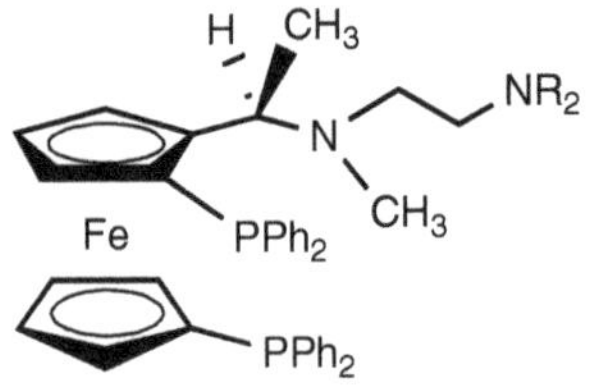

R = Et, **3.56a**
R = Me, **3.56b**

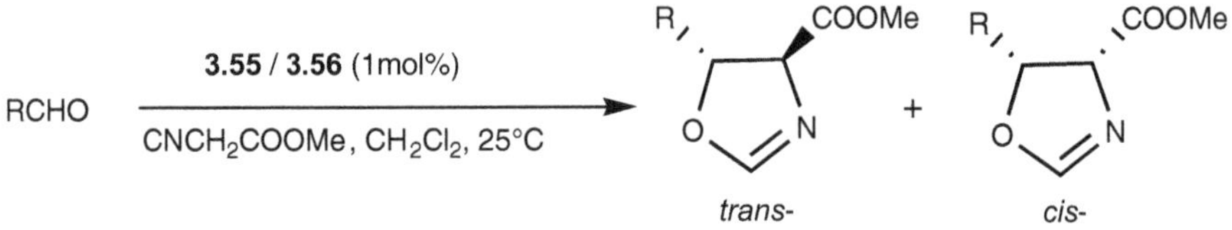

Scheme 3.22
Hayashi-Ito asymmetric Aldol reaction with chiral ferrocenylphosphine–gold complex.

noticed that L*Au(I) based catalysts (L* = chiral ligand) were more active than the related L*Pt(II) catalysts. For instance alkoxycyclization reactions occurred at room temperature with Au(I) catalysts while for Pt(II) based catalysts, a higher temperature was required. Remarkably the (*R*)-(TolBINAP)$(AuCl)_2$ (**3.57**) catalysed at 24°C the asymmetric methoxycyclization reaction of enyne (**3.58**) in MeOH to give the methoxycyclized compound (−)-**3.59** in moderate yield but with high enantioselectivity of 94% ee (Scheme 3.23).

The use of gold complexes with chiral phosphines to promote asymmetric cyclization reactions remains the most studied approach of several groups.[112,113] However, Toste and coworkers pushed this chemistry further, when they combined the additive effects of a chiral phosphine ligand and a chiral counter ion which forms an ion pair with the active species ≪ $[L^*Au]^+X^{-*}$ ≫ (L^* = chiral phosphine ligand ; X^* = chiral counter ion). Thus a dramatic increase in Au(I)-catalysed enantioselective intramolecular hydroalkylation of

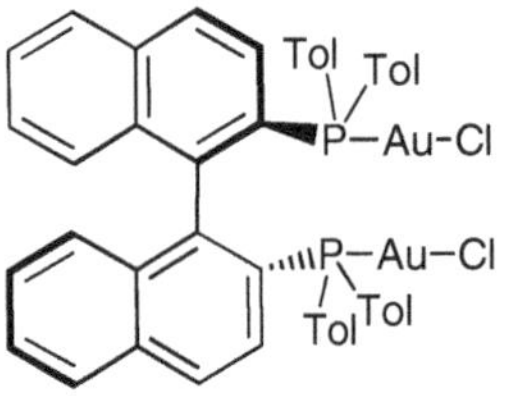

(*R*)-TolBINAP$(AuCl)_2$ (**3.57**)

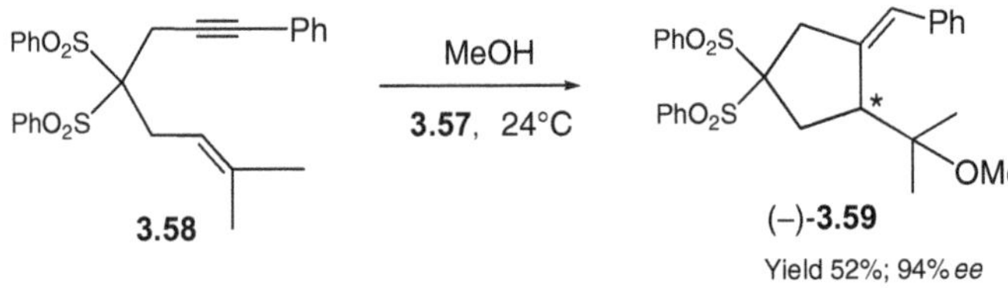

Scheme 3.23
Enantioselective selective alkoxycyclization of enynes by (*R*)-TolBINAP$(AuCl)_2$ complex.

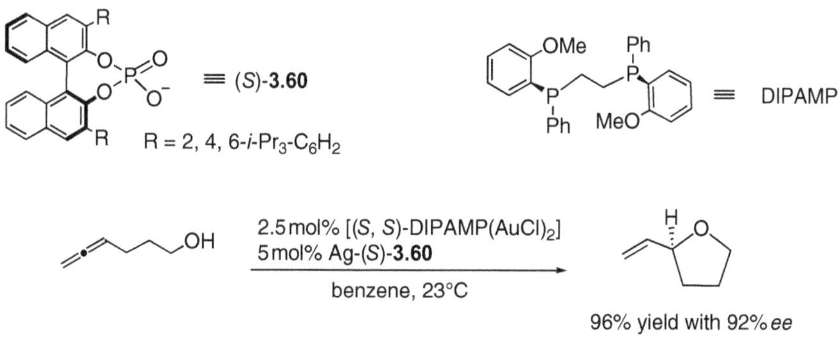

R = 2, 4, 6-*i*-Pr₃-C₆H₂ ≡ (*S*)-**3.60**

DIPAMP

2.5 mol% [(*S*, *S*)-DIPAMP(AuCl)₂]
5 mol% Ag-(*S*)-**3.60**
benzene, 23°C

96% yield with 92% *ee*

Scheme 3.24
Counter ion mediated enantioselective intramolecular hydroalkylation.

allenols was obtained with [(*S*,*S*)-DIPAMP-(AuCl)₂] in the presence of chiral binaphtol-derived phosphate counter ion (*S*)-**3.60** in benzene to give monosubstituted tetrahydrofuran products in good yields with greater than 90% ee (Scheme 3.24).[114]

Interestingly the combination of (*S*,*S*)-DIPAMP and (*S*)-**3.60** showed the right match for high enantioselectivity, in contrast to that observed with (*S*,*S*)-DIPAMP and (*R*)-**3.60**. The same reaction carried out with a chiral phosphine ligand, rather than a chiral *counter* ion gave poor enantioselectivities. The chiral counter anion strategy coupled with the chiral phosphine ligand also worked well for Au(I)-catalysed enantioselective intramolecular hydroamination of allenes.

In the case of Au(I)-catalysed asymmetric hydrocarboxylation of allenes a high ee of 82% was observed when (*S*)-BINAP(AuCl)₂ was combined with the counter ion of opposite chirality, that is *R*-**3.60**. In the presence of the *S*-**3.60** counter ion the reaction produced nearly racemic product. These examples clearly illustrate the powerful effect of the chiral counter ion on the enantioselectivity.

Krause and coworkers used Au(I) and/or Au(III) catalysts for stereoselective cycloisomerization of various *β*-hydroxyallenes to the corresponding chiral dihydropyrans. In these reactions, the authors showed that chirality transfer to the cyclized products has occurred.[115] As mentioned in the beginning of this section, although the use of gold complexes in catalytic transformations has proved to be an efficient tool in organic synthesis where a variety of chemical transformation occurs to give the desired products, yet only few examples in which gold catalysts have been used in asymmetric synthesis were reported. It is expected that this area will grow rapidly in the near future.

3.3.4 Asymmetric Nucleophilic Catalysis

In the preceding examples, the asymmetric catalyst is a Lewis acid and hence the catalytic processes reported so far involve electrophilic activation by a metal-centred chiral Lewis acid. There is another strategy, although less explored, which consists of designing chiral Lewis bases for *nucleophilic catalysis*. It is well known that Lewis bases such as nitrogen heterocycles and tertiary phosphines and amines catalyse a variety of important chemical processes. For instance 4-(dimethylamino)pyridine (DMAP) catalyses the acylation of alcohols by anhydrides; the mechanism by which DMAP accelerates this process provides an instructive illustration of how nucleophiles can

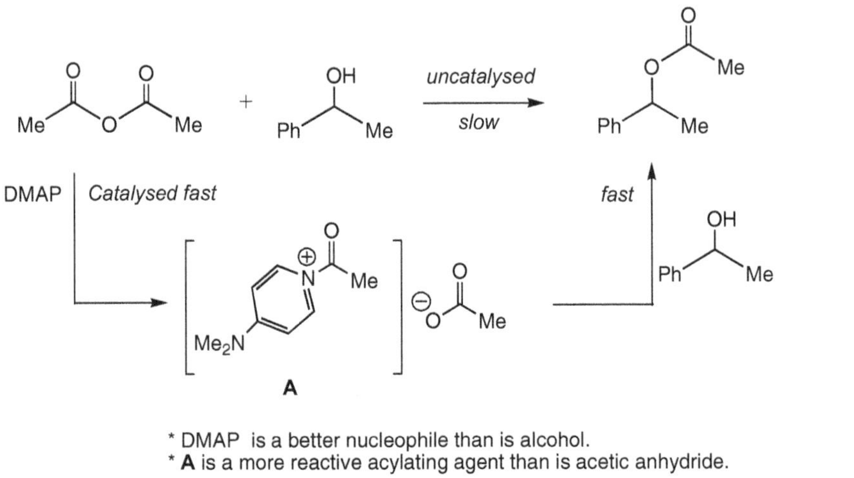

* DMAP is a better nucleophile than is alcohol.
* **A** is a more reactive acylating agent than is acetic anhydride.

Scheme 3.25

catalyse chemical transformations (Scheme 3.25).[116] In this field a limited number of asymmetric catalysts have been reported with planar chirality and having a ferrocene scaffold.

On this topic, several outstanding contributions were reported by Fu and cow-orkers,[90,117] in which new asymmetric nucleophilic catalysts based on chiral ferro-cene-type heterocycles were designed. To this end the planar-chiral PPY ferrocene complex (PPY = 4-(pyrrolidino)-pyridine (**3.61**) was prepared and resolved. Complex **3.61** catalysed the enantioselective rearrangement of A-acylated Azlactones to give C-acylated isomers with high yields and ee of 82–90%.[118] The powerful effect of the chiral ferrocene scaffold was clearly evident if compared to the same reaction with the organic catalyst DMAP reported in 1970 by Steglich and Höfle where only racemic compounds were formed (Scheme 3.26).

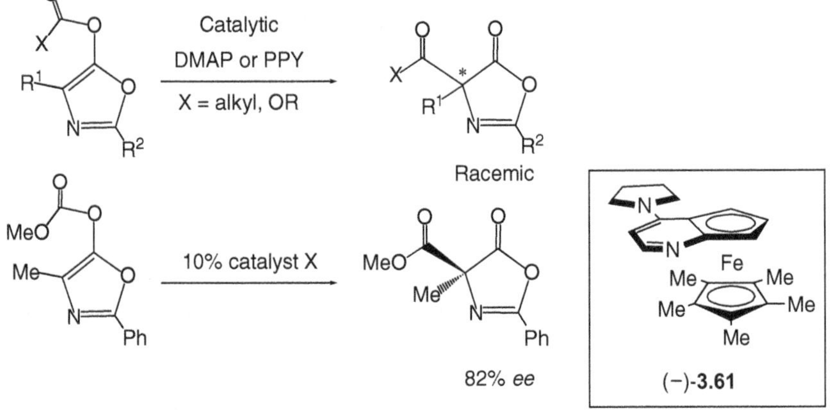

Scheme 3.26

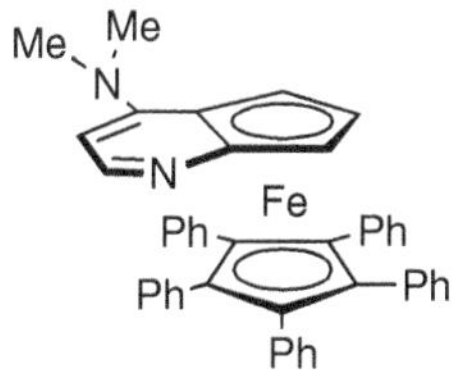

Figure 3.23

Further modification of the chiral environment of the catalyst by replacing the Cp* group of **3.61** by the bulkier C_5Ph_5 group allowed the development of new catalysts (−)-**3.62** for the kinetic resolution of secondary alcohols during acylation of alcohols with anhydrides (Figure 3.23).

In kinetic resolution, the key parameter is the selectivity factor *s*, which measures the relative rate of reaction of the two enantiomers. A selectivity factor greater than 10 is required for a kinetic resolution to be synthetically useful.

Fu and coworkers[119] showed that acylations catalysed by **3.62** in the presence of *tert*-amyl alcohol allowed the resolution of a wide array of arylalkyl carbinols with excellent stereoselection (Table 3.3).

Other nucleophile catalysed reactions using chiral ferrocene scaffolds have been investigated by different groups but with lower enantioselectivity compared with those

Table 3.3

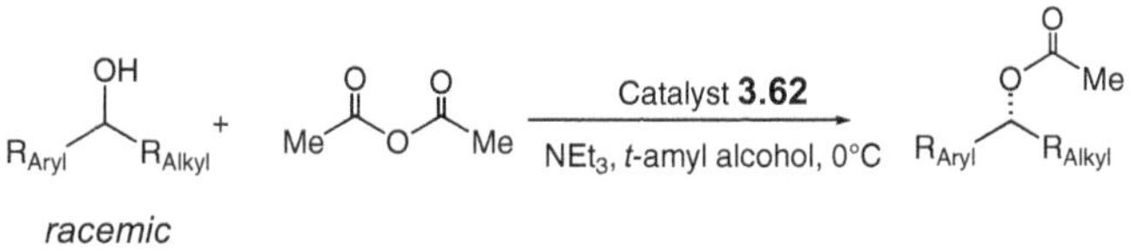

Entry	Unreacted alcohol major enantiomer		s	% ee (% conversion)
1 2 3 4		R = Me Et i-Pr t-Bu	43 59 87 95	99(55) 99(54) 97(52) 96(51)
5	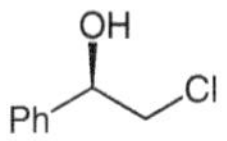		32	98(56)
6	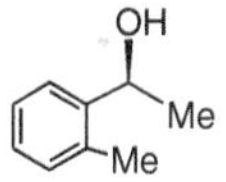		71	99(53)
7	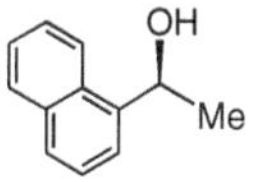		65	95(52)
8	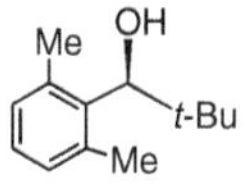		>200	99(51)

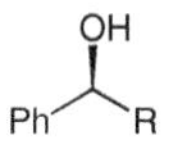

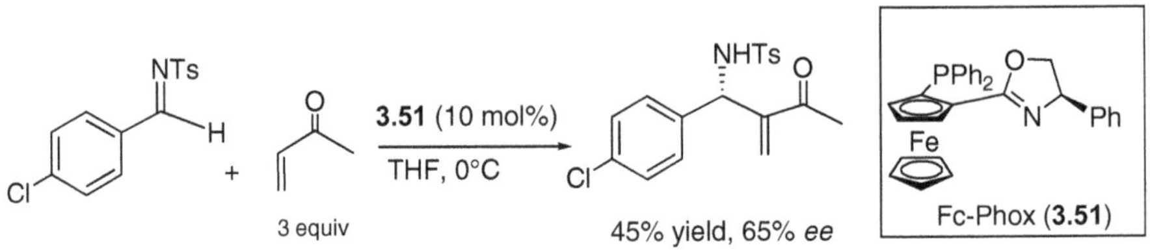

Scheme 3.27

obtained by Fu and coworkers. For instance planar-chiral ferrocenyl dialkylphosphines have been tested in the Baylis–Hillman reaction between aromatic aldehydes and acrylates with Mandyphos providing up to 65% *ee*.[120] In a similar study the Fc-Fox (**3.51**) catalyst gives 65% *ee*[121] for the aza-Baylis–Hillman reaction of the *N*-tosyl aldimine of *p*-chlorobenzaldehyde with methylvinylketone (Scheme 3.27).

Furthermore Kobayashi and coworkers reported the enantioselective reaction of N-acylhydrazones with allyltrichlorosilanes furnishing the corresponding homoallylic amines with high diastereo- and enantioselectivities.[122] Other groups investigated the same reaction using chiral ferrocenyl sulfoxides.[123] In spite of the great progress achieved in recent years on this topic, much effort is still required for the development of new asymmetric Lewis base catalysts with high ee values.

3.4   Summary

As stated before, we have tried in this chapter to present an overview of chirality in modern organometallic chemistry. The resolution of the first organometallic compound [CpMn(CO)(NO)PPh₃)][PF₆] gave an impetus to the development of chirality in this area and hence the configurational stability of such compounds was examined, as well as the phenomenon of chiral induction or chiral recognition upon binding of a prochiral organic substrate to a metal centre. The use of these chiral organometallic complexes in organic synthesis was then developed, and finally, and most importantly, asymmetric catalyses remain the most active area due to their importance in academic and industrial applications. We feel a complete coverage of all aspects of chirality in organometallic chemistry remains a difficult task, but we hope that we have been able to provide a guide to the most recent developments in this area.

References

[1] H. Brunner, *Angew. Chem. Int. Ed.* **1969**, *8*, 382.
[2] A. M. Thayer, *Chem. Eng. News* **2005**, *5*, 40.
[3] E. N. Jacobsen, A. Pfaltz and H. Yamamoto (Eds), *Comprehensive Asymmetric Catalysis, Vol. 1–3*, Springer, New York, **1999**.
[4] I. Ojima, (Ed.), *Catalytic Asymmetric Synthesis*, VCH, New York, **2000**.
[5] H. U. Blaser and E. Schmidt (Eds) *Asymmetric Catalysis on Industrial Scale*, Wiley-VCH, New York, **2004**.
[6] J. H. van't Hoff, *Arch. Neerl. Sci. Exactes Nat.* **1874**, *9*, 445.
[7] J. A. Le Bel, *Bull. Soc. Chim. Fr.* **1874**, *22*, 337.

[8] A. Werner, *Ber. Dtsh. Chem. Ges.* **1911**, *44*, 1887.

[9] H. Brunner and H.-D. Schindler, *J. Organomet. Chem.* **1970**, *24*, C7.

[10] H. Brunner, *Angew. Chem. Int. Ed.* **1999**, *38*, 1194.

[11] H. Brunner, *Angew. Chem. Int. Ed.* **1971**, *10*, 249.

[12] H. Brunner, *Eur. J. Inorg. Chem.* **2001**, 905.

[13] H. Brunner, *J. Organomet. Chem.* **1975**, *94*, 189.

[14] J. H. Merrifield, C. E. Strouse and J. A. Gladysz, *Organometallics* **1982**, *1*, 1204.

[15] F. Agbossou, E. J. O'Connor, C. M. Garner, N. Q. Mendez, J. M. Fernandez, A. T. Patton, J. A. Ramsden and J. A. Gladysz, *Inorg. Synth.* **1992**, *29*, 211.

[16] J. H. Merrifield, J. M. Fernandez, W. E. Buhro and J. A. Gladysz, *Inorg. Chem.* **1984**, *23*, 4022.

[17] J. A. Gladysz and B. J. Boone, *Angew. Chem. Int. Ed.* **1997**, *36*, 551.

[18] G. S. Bodner, T. S. Peng, A. M. Arif and J. A. Gladysz, *Organometallics* **1990**, *9*, 1191.

[19] T.-S. Peng, A. M. Arif and J. A. Gladysz, *Helv. Chim. Acta* **1992**, *75*, 442.

[20] Y. Wang and J. A. Gladysz, *Chem. Ber.* **1995**, *128*, 213.

[21] B. J. Boone, D. P. Klein, J. W. Seyler, N. Q. Mendez, A. M. Arif and J. A. Gladysz, *J. Am. Chem. Soc.* **1996**, *118*, 2411.

[22] C. Ganter, *Chem. Soc. Rev.* **2003**, *32*, 130.

[23] H. Brunner and E. Schmidt, *J. Organomet. Chem.* **1972**, *36*, C18.

[24] J. W. Faller, M. R. Mazzieri, J. T. Nguyen, J. Parr and M. Tokunaga, *Pure Appl. Chem.* **1994**, *66*, 1463.

[25] B. Pfister, R. Stauber and A. Salzer, *J. Organomet. Chem.* **1997**, *533*, 131.

[26] H. Brunner, R. Oeschey and B. Nuber, *J. Chem. Soc., Dalton Trans.* **1996**, 1499.

[27] J. W. Faller, B. J. Grimmond and D. G. D'Alliessi, *J. Am. Chem. Soc.* **2001**, *123*, 2525.

[28] C. Standfest-Hauser, C. Slugovc, K. Mereiter, R. Schmid, K. Kirchner, L. Xiao and W. Weissensteiner, *J. Chem. Soc., Dalton Trans.* **2001**, 2989.

[29] S. Hashiguchi, A. Fujii, J. Takehara, T. Ikariya and R. Noyori, *J. Am. Chem. Soc.* **1995**, *117*, 7562.

[30] D. B. Grotjahn, C. R. Larsen, J. L. Gustafson, R. Nair and A. Sharma, *J. Am. Chem. Soc.* **2007**, *129*, 9592.

[31] Y. Kataoka, Y. Saito, K. Nagata, K. Kitamura, A. Shibahara and K. Tani, *Chem. Lett.* **1995**, 833.

[32] Y. Nishibayashi, I. Takel and M. Hidai, *Organometallics* **1997**, *16*, 3091.

[33] B. M. Trost, B. Vidal and M. Thommem, *Chem. Eur. J.* **1999**, *5*, 1055.

[34] N. Dodo, Y. Matsushima, M. Uno, K. Onitsuka and S. Takahashi, *J. Chem. Soc., Dalton Trans.* **2000**, 35.

[35] K. Onitsuka, N. Dodo, Y. Matsushima and S. Takahashi, *Chem. Commun.* **2001**, 521.

[36] F. K. Cheung, A. M. Hayes, J. Hannedouche, A. S. Y. Yim and M. Wills, *J. Org. Chem.* **2005**, *70*, 3188.

[37] A. M. Hayes, D. J. Morris, G. J. Clarkson and M. Wills, *J. Am. Chem. Soc.* **2005**, *127*, 7318.

[38] S. G. Davies, *Aldrichimica Acta* **1990**, *23*, 31.

[39] L. S. Liebeskind and M. E. Welker, *Organometallics* **1983**, *2*, 194.

[40] G. J. Baird and S. G. Davies, *J. Organomet. Chem.* **1983**, *248*, C1.

[41] L. S. Liebeskind and M. E. Welker, *Tetrahedron Letts.* **1984**, *25*, 4341.

[42] L. S. Liebeskind, M. E. Welker and R. W. Fengl, *J. Am. Chem. Soc.* **1986**, *108*, 6328.

[43] S. G. Davies, I. M. Dordor, J. C. Walker and P. Warner, *Tetrahedron Lett.* **1984**, *25*, 2709.

[44] S. G. Davies, I. M. Dordor and P. Warner, *J. Chem. Soc., Chem. Commun.* **1984**, 956.

[45] P. W. Ambler and S. G. Davies, *Tetrahedron Lett.* **1985**, *26*, 2129.

[46] K. Broadly and S. G. Davies, *Tetrahedron Lett.* **1984**, *25*, 1743.

[47] L. S. Liebeskind, M. E. Welker and V. Goedken, *J. Am. Chem. Soc.* **1984**, *106*, 441.

[48] S. G. Davies, I. M. Dordor-Hedgecock, K. H. Sutton and J. C. Walker, *Tetrahedron Lett.* **1986**, *32*, 3787.

[49] S. G. Davies, I. M. Dordor-Hedgecock, K. H. Sutton, J. C. Walker, R. H. Jones and K. Prout, *Tetrahedron Lett.* **1986**, *42*, 5123.

[50] S. G. Davies, J. Dupont, R. J. C. Easton, O. Ichihara, J. M. McKenna, A. D. Smith and J. A. A. de Sousa, *J. Organomet. Chem.* **2004**, *689*, 4184.

[51] R. W. Miller, *J. Nat. Prod.* **1980**, *43*, 425.

[52] E. Graf, S. Weinandy, B. Koch and E. Breitmaier, *Liebigs Ann. Chem.* **1986**, *7*, 1147.

[53] K. H. Theopold, P. N. Becker and R. G. Bergman, *J. Am. Chem. Soc.* **1982**, *104*, 5250.

[54] C. M. Garner, N. Quiros Mendez, J. J. Kowalczyk, J. M. Fernandez, K. Emerson, R. D. Larsen and J. A. Gladysz, *J. Am. Chem. Soc.* **1990**, *112*, 5146.

[55] J. M. Fernandez, K. Emerson, R. H. Larsen and J. A. Gladysz, *J. Am. Chem. Soc.* **1986**, *108*, 8268.

[56] D. M. Dalton, C. M. Garner, J. M. Fernandez and J. A. Gladysz, *J. Org. Chem.* **1991**, *56*, 6823.

[57] D. P. Klein and J. A. Gladysz, *J. Am. Chem. Soc.* **1992**, *114*, 8710.

[58] T.-S. Peng, A. M. Arif and J. A. Gladysz, *J. Chem. Soc., Dalton Trans.* **1995**, 1857.

[59] A. M. Rouhi, *Chem. Eng. News* **2004**, *June 14*, 51.

[60] A. Ault, *J. Chem. Edu.* **2002**, *79*, 572.

[61] H. B. Kagan and P. Dang Tuan, *J. Am. Chem. Soc.* **1972**, *94*, 6429.

[62] R. R. Schrock and J. A. Osborn, *J. Chem. Soc. D: Chem. Comm.* **1970**, 567.

[63] J. A. Osborn and R. R. Schrock, *J. Amer. Chem. Soc.* **1971**, *93*, 2397.

[64] R. R. Schrock and J. A. Osborn, *J. Am. Chem. Soc.* **1976**, *98*, 2134.

[65] R. R. Schrock and J. A. Osborn, *J. Am. Chem. Soc.* **1976**, *98*, 4450.

[66] A. S. C. Chan, J. J. Pluth and J. Halpern, *J. Am. Chem. Soc.* **1980**, *102*, 5952.

[67] W. S. Knowles, *Acc. Chem. Res.* **1983**, *16*, 106.

[68] W. S. Knowles, *Angew. Chem. Int. Ed.* **2002**, *41*, 1998.

[69] R. Noyori and H. Takaya, *Acc. Chem. Res.* **1990**, *23*, 345.

[70] R. Noyori, *Angew. Chem. Int. Ed.* **2002**, *41*, 2008.

[71] T. Ohkuma, H. Ooka, S. Hashiguchi, T. Ikariya and R. Noyori, *J. Amer. Chem. Soc.* **1995**, *117*, 2675.

[72] R. Noyori and T. Ohkuma, *Angew. Chem. Int. Ed.* **2001**, *40*, 40.

[73] A. Fujii, S. Hashiguchi, N. Uematsu, T. Ikariya and R. Noyori, *J. Am. Chem. Soc.* **1996**, *118*, 2521.

[74] K. Murata, K. Okano, M. Miyagi, H. Iwane, R. Noyori and T. Ikariya, *Org. Lett.* **1999**, *1*, 1119.

[75] R. Noyori, M. Yamakawa and S. Hashiguchi, *J. Org. Chem.* **2001**, *66*, 7931.

[76] M. Yamakawa, I. Yamada and R. Noyori, *Angew. Chem. Int. Ed.* **2001**, *40*, 2818.

[77] M. J. Burk, *Acc. Chem. Res.* **2000**, *33*, 363.

[78] (a) B. L. Feringa, *Acc. Chem. Res.* **2000**, *33*, 346. (b) L. Panella, B.L. Feringa, J.G. de Vries and A. J. Minnaard, *Org. Lett.* **2005**, *7*, 4177.

[79] D. Markarding, H. Klusacek, G. Gokel, P. Hoffmann and I. Ugi, *J. Am. Chem. Soc.* **1970**, *92*, 5389.

[80] F. Rebière, O. Riant, L. Ricard and H. B. Kagan, *Angew. Chem. Int. Ed.* **1993**, *32*, 568.

[81] O. Riant, O. Samuel and H. B. Kagan, *J. Am. Chem. Soc.* **1993**, *115*, 5835.

[82] H.-U. Blaser, *Adv. Synth. Catal.* **2002**, *344*, 17.

[83] H.-U. Blaser, W. Brieden, B. Pugin, F. Spindler, M. Studer and A. Togni, *Adv. Synth. Catal.* **2002**, *19*, 3.

[84] F. Naud, C. Malan, F. Spindler, C. Ruggeberg, A. T. Schmidt and H.-U. Blaser, *Adv. Synth. Catal.* **2006**, *348*, 47.

[85] A. Togni, C. Breutel, A. Schnyder, F. Spindler, H. Landert and A. Tijani, *J. Am. Chem. Soc.* **1994**, *116*, 4062.

[86] C. J. Richards, T. Damalidis, D. E. Hibbs and M. B. Hursthouse, *Synlett.* **1995**, 74.

[87] T. Sammakia, H. A. Latham and D. R. Schaad, *J. Org. Chem.* **1995**, *60*, 10.

[88] Y. Nishibayashi and S. Uemura, *Synlett.* **1995**, 79.

[89] T. Ireland, G. Grossheimann, C. Wieser-Jeunesse and P. Knochel, *Angew. Chem. Int. Ed.* **1999**, *38*, 3212.

[90] G. C. Fu, *Acc. Chem. Res.* **2000**, *33*, 412.

[91] R. G. Arrayas, J. Adrio and J. C. Carretero, *Angew. Chem. Int. Ed.* **2006**, *45*, 7674.

[92] T. Katsuki and K. B. Sharpless, *J. Am. Chem. Soc.* **1980**, *102*, 5974.

[93] K. B. Sharpless, W. Amberg, Y. L. Bennani, G. A. Crispino, J. Hartung, K. S. Jeong, H. L. Kwong, K. Morikawa, Z. M. Wang, *et al.*, *J. Org. Chem.* **1992**, *57*, 2768.

[94] G. A. Crispino, K. S. Jeong, H. C. Kolb, Z. M. Wang, D. Xu and K. B. Sharpless, *J. Org. Chem.* **1993**, *58*, 3785.

[95] H. C. Kolb, M. S. V. Nieuwenhze and K. B. Sharpless, *Chem. Rev.* **1994**, *94*, 2483.

[96] E. N. Jacobsen, W. Zhang, A. R. Muci, J. R. Ecker and L. Deng, *J. Am. Chem. Soc.* **1991**, *113*, 7063.

[97] J. H. Teles, S. Brode and M. Chabanas, *Angew. Chem. Int. Ed.* **1998**, *37*, 1415.

[98] A. S. K. Hashmi, L. Schwarz, J.-H. Choi and T. M. Frost, *Angew. Chem. Int. Ed.* **2000**, *39*, 2285.

[99] D. J. Gorin and F. D. Toste, *Nature* **2007**, *446*, 395.

[100] C. Nieto-Oberhuber, S. Lopez and A. M. Echavarren, *J. Am. Chem. Soc.* **2005**, *127*, 6178.

[101] N. Marion, P. de Fremont, G. Lemiere, E. D. Stevens, L. Fensterbank, M. Malacria and S. P. Nolan, *Chem. Comm.* **2006**, 2048.

[102] G. Lemiere, V. Gandon, N. Agenet, J.-P. Goddard, A. de Kozak, C. Aubert, L. Fensterbank and M. Malacria, *Angew. Chem. Int. Ed.* **2006**, *45*, 7596.

[103] G. Lemiere, V. Gandon, K. Cariou, T. Fukuyama, A.-L. Dhimane, L. Fensterbank and M. Malacria, *Org. Lett.* **2007**, *9*, 2207.

[104] E. Genin, L. Leseurre, P. Y. Toullec, J.-P. Genet and V. Michelet, *Synlett.* **2007**, 1780.

[105] L. Leseurre, P. Y. Toullec, J.-P. Genet and V. Michelet, *Org. Lett.* **2007**, *9*, 4049.

[106] A. S. K. Hashmi, S. Schaefer, M. Woelfe, C. Diez Gil, P. Fischer, A. Laguna, M. C. Blanco and M. C. Gimeno, *Angew. Chem. Int. Ed.* **2007**, *46*, 6184.

[107] A. S. K. Hashmi, T. M. Frost and J. W. Bats, *J. Am. Chem. Soc.* **2000**, *122*, 11553.

[108] E. Jimenez-Nunez and A. M. Echavarren, *Chem. Commun.* **2007**, 333.

[109] A. Furstner and W. Davies Paul, *Angew. Chem. Int. Ed.* **2007**, *46*, 3410.

[110] Y. Ito, M. Sawamura and T. Hayashi, *J. Am. Chem. Soc.* **1986**, *108*, 6405.

[111] M. P. Munoz, J. Adrio, J. C. Carretero and A. M. Echavarren, *Organometallics* **2005**, *24*, 1293.

[112] A. D. Melhado, M. Luparia and F. D. Toste, *J. Am. Chem. Soc.* **2007**, *129*, 12638.

[113] R. L. LaLonde, B. D. Sherry, E. J. Kang and F. D. Toste, *J. Am. Chem. Soc.* **2007**, *129*, 2452.

[114] G. L. Hamilton, E. J. Kang, M. Mba and F. D. Toste, *Science* **2007**, *317*, 496.

[115] B. Gockel and N. Krause, *Org. Lett.* **2006**, *8*, 4485.

[116] E. F. V. Scriven, *Chem. Soc. Rev.* **1983**, *12*, 129.

[117] G. C. Fu, *Acc. Chem. Res.* **2006**, *39*, 853.

[118] J. C. Ruble and G. C. Fu, *J. Am. Chem. Soc.* **1998**, *120*, 11532.

[119] J. C. Ruble, J. Tweddell and G. C. Fu, *J. Org. Chem.* **1998**, *63*, 2794.

[120] S. I. Pereira, J. Adrio, A. M. S. Silva and J. C. Carretero, *J. Org. Chem.* **2005**, *70*, 10175.

[121] M. Shi, L.-H. Chen and C.-Q. Li, *J. Am. Chem. Soc.* **2005**, *127*, 3790.

[122] S. Kobayashi, C. Ogawa, H. Konishi and M. Sangiura, *J. Am. Chem. Soc.* **2003**, *125*, 6610.

[123] I. Fernandez, V. Valdivia, B. Gori, F. Alcudia, E. Alvarez and N. Khiar, *Org. Lett.* **2005**, *7*, 1307.

4 Chiral Recognition in Organometallic and Coordination Compounds

Chiral recognition is an important biological process in the field of medical and pharmaceutical sciences. This is because many drugs consist of chiral molecules and usually only one of the enantiomers is pharmaceutically active while the other one often exerts severe side effects. A historic example was the withdrawal of thalidomide from the market because of its unexpected teratogenic side effects.[1,2] Thalidomide was clinically used as a sedative in the late 1950s. It was later suggested that one of the enantiomers of thalidomide, exhibited anti-emetic activity while the other one can cause fetal damage. Chiral molecular recognition is well manifested in our daily life; for example, the difference between orange and lemon odours that we perceive is related to R-(+)- and S-(−)-limonene enantiomers (**4.1**) that exist in orange and lemon respectively (Figure 4.1).[3] Hence enantiospecific receptor-substrate binding is of the utmost importance in biochemical and chemical systems and can be described by the term '*chiral recognition*'.

As a result a great deal of interest has been devoted to the preparation of chiral compounds as enzyme mimics and models, as well as chiral receptors, in the hope of separating biologically active molecules through a chiral recognition process. For example, Peacock *et al.* have prepared and resolved chiral corands featuring two binaphthyl (Binap) groups **4.2** (Figure 4.2).[4]

The chiral recognition ability of **4.2** was examined by two-phase liquid–liquid extraction experiments of racemic mixtures of amino acids and ester guests. Cram and coworkers developed an enantiomer-resolving machine based on such chiral corand receptors and found that the R,R-host was selectively able to extract the D-enantiomer of amino acids.[5]

More recently Rebek and coworkers invented the concept of chiral soft ball receptors capable of recognizing guest molecules in solution and exhibiting chiral memory.[6–12] Unlike the chiral corands, chiral soft balls (**4.3**)$_2$ are formed by dimerization of achiral monomers **4.3** through weak noncovalent interactions in the presence of guest species (Figure 4.3).[13] Although these examples presented in brief are beyond the scope of this chapter they illustrate the power of the *chiral molecular recognition process* occurring between a receptor and a substrate.

The aim of this chapter is to discuss the recent examples of chiral recognition in organometallic and coordination compounds. Most of the examples presented are related to octahedral tris(bidentate) compounds with helical chirality. The homo-chiral

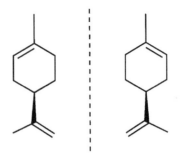

Figure 4.1
R-(+)- and *S*-(−)-limonene (**4.1**) as mirror image molecules responsible for the different orange and lemon odours that we perceive.

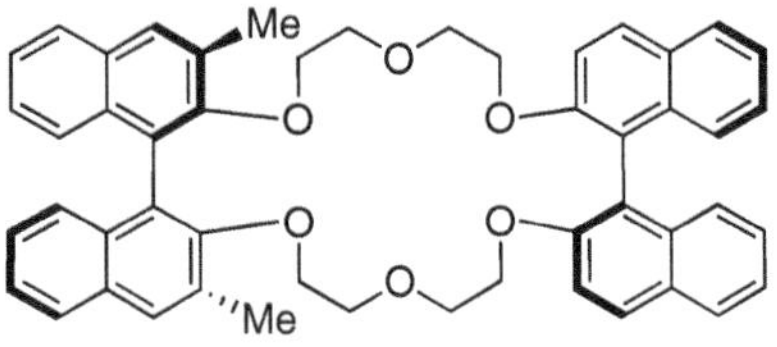

(*R,R*)-**4.2**

Figure 4.2
Example of chiral corands (*R,R*)-**4.2**.

or hetero-chiral recognition exhibited by these compound will be illustrated. Moreover, the employment of chiral cyclodextrins to resolve helical metal complexes will also be discussed. The chiral anion strategy, which involves the use of optically pure anions such as Δ-TRISPAT {TRISPHAT: tris(tetrachlorobenzenediolato)phosphate (V)}, BNP

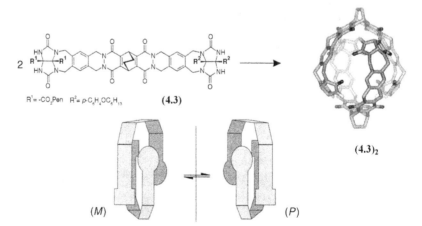

Figure 4.3
Chiral homodimeric soft ball capsule (**4.3**)$_2$ from achiral component (**4.3**) bearing two different glycoluril residues (reprinted with kind permission of Springer Science and Business Media from reference[13] page 32).

{BNP: 1,1' Binaphtyl-2,2'-diyl phosphate}, BNB {BNB: Bis(Binaphthol)-Borate} in the resolution or differentiation of chiral organometallic and coordination complexes through the formation of ion pairs with the appropriate enantiomer molecule, will be highlighted.

Finally a brief introduction to chiral recognition of DNA (a chiral biomolecule) by optically pure octahedral metal complexes will be presented.

4.1 Octahedral Metal Complexes with Helical Chirality

4.1.1 Heterochiral Recognition

In 1984, Yamatera and coworkers showed by ^{1}H-NMR studies that the complex $[Ru(phen)_3]^{2+}$ (**4.4**) displays self-association in solution.[14] In aqueous media this species gives rise to NMR spectra which are significantly dependent on concentration and display features consistent with π-stacking interactions between cations to form dimers (Figure 4.4).

More recently Kol *et al.* demonstrated by NMR and X-ray crystallography that the octahedral eilatin complexes $[M(L-L)_2(eilatin)]^{2+}$ (M = Ru, Os; L-L = bpy, phen, 2,9-dimethyl-phen and 2, 2' biquinole), where eilatin (**4.5**) is a heptacyclic aromatic ligand with strong π-character, dimerize in solution *via* π–π stacking between the eilatin units of each complex.[15] In particular the X-ray crystallography showed that a '*heterochiral association*' is formed between Δ-$[M(L-L)_2(eilatin)]^{2+}$ and Λ-$[M(L-L)_2(eilatin)]^{2+}$ cations (Figure 4.5). The ^{1}H-NMR studies showed that the heterochiral dimer structure (**4.6**) is maintained in solution through a chiral recognition process manifested solely by π–π interactions. Furthermore, the extent of this association is mainly controlled by the steric hindrance introduced by the L–L ligands.

Moreover, the same group found that in solution enriched samples of $[Ru(bpy)_2(\pi$-extended ligand)][PF$_6$]$_2$ (π-extended ligand = eilatin, isoeilatin and tpphz {tetrapyrido [3,2-a:2',3'-c:3'',2''-h:2''',3'''-j]phenazine}) showed the presence of a major and minor set of peaks attributed to two species.[16] The origin of this phenomenon is attributed to a fast equilibrium between monomers and discrete dimers held together by π–π interactions. These dimers result from heterochiral recognition between Δ-$[M(L-L)_2(\pi$-extended ligand)]$^{2+}$ and Λ-$[M(L-L)_2(\pi$-extended ligand)]$^{2+}$ cations. The authors suggest that a

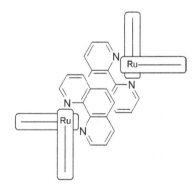

Figure 4.4

Schematic representation of $[Ru(phen)_3]^{2+}$ (**4.4**) showing π-stacking interactions between bidentate phenanthroline ligands, giving rise to dimers in solution.

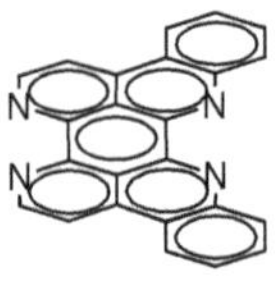

Eilatin (**4.5**)

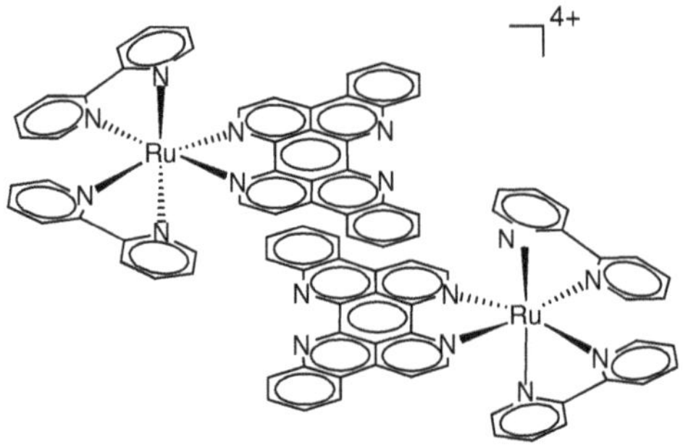

Λ-Δ heterochiral dimer (**4.6**)

Figure 4.5
Eilatin (**4.5**) and the heterochiral association **4.6** with L-L = bpy.

possible explanation for the preferred formation of a heterochiral dimer is a reduced steric repulsion in comparison to the homochiral dimer. This behaviour contrasts with that observed when a chiral anion is employed (see following paragraph).

4.1.2 Homochiral Recognition

Octahedral hexa-coordinated ruthenium(II) complexes bearing 2,2′-bipyridine ligands (bpy) have received considerable attention because they have numerous applications; for instance, as chiral building blocks for supramolecular assemblies, or as chiral probes for biological molecules (polynucleotides, DNA).[17,18] Hence efforts have been devoted to the study of the stereochemistry of these complexes as a result of the different interactions that may arise when enantiomeric forms {Δ− or Λ−} interact with a variety of chiral substrates in solution. Such chiral recognition is of pivotal importance in many biological and chemical systems. Thus the determination of their enantiomeric purity is of prime importance. Lacour and coworkers were the first to report the use of Δ–TRISPHAT (**4.7**) {TRISPHAT: tris(tetrachlorobenzenediolato)phosphate (V)} as a chiral anion shift reagent to determine the enantiopurity of ruthenium polypyridyl complexes.[19,20] Further investigations carried out by Lacour on $Ru(bpy)_3^{2+}$ (**4.8**) have also shown that homochiral ion-pairing mode of association (Δ−, Δ−) is favoured with Δ-TRISPHAT over heterochiral association (Δ−, Λ−) in a solvent of low polarity and proceeds with high selectivity.[21] It should be mentioned that the formation of ion-pairs of transition metal complexes in solution has been the focus of recent reviews by Pregosin *et al.*[22] and Macchioni and coworkers.[23]

Other studies complemented the previous results and have also shown that a homochiral ion-pairing is favoured for $Ru(DEAS-bpy)_3^{2+}$ complex (**4.9**) {DEAS-bpy = 4,4′-bis(diethylamino-styryl)-[2,2′]-bipyridine}[24] and for the compounds [Ru(bpy)₂(L-L)] [PF₆]₂ {L-L = cmbpy = 4-carboxy-4′-methyl-2,2′-bipyridine (**4.10**) ; L-L = dcbpy = 4, 4′-dicarboxy-2,2′ bipyridine (**4.11**)}.[25,26] Interestingly, in the previous two examples the octahedral metal complexes have D_3 symmetry as Δ–TRISPHAT (**4.7**) while in the last example complex **4.10** possesses a C_1 symmetry while **4.11** displays a C_2-symmetry (Figure 4.6). These results suggest that a homochiral association between Δ-TRISPHAT and the octahedral ruthenium complexes is not dependent on a prerequisite D_3-symmetry of the metal complex.

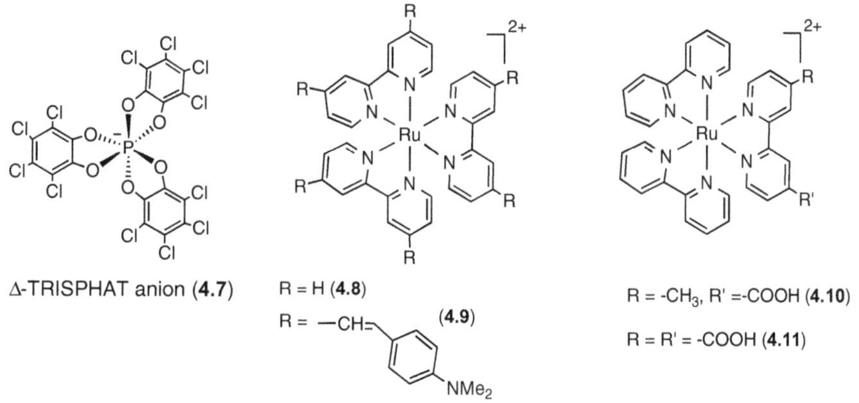

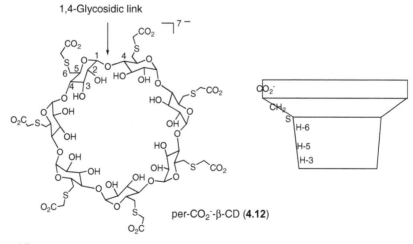

Figure 4.6
Schematic drawing of Δ-TRISPHAT (**4.7**) and some octahedral ruthenium complexes that show (Δ, Δ) homochiral recognition.

The strategy of using Δ–TRISPHAT to resolve or differentiate other organometallic or coordination compounds has been investigated by several groups with success. These studies are presented in Section 4.2.

4.1.3   Chiral Recognition Using Modified Cyclodextrins

The absolute configuration of helical metal complexes $M(phen)_3^{n+}$ (M = Ru(II), Rh(III), Fe(II), Co(II) and Zn(II) and phen = 1,10-phenanthroline was also recognized by modified heptaanion (per-CO_2^--β-cyclodextrins) (**4.12**) in D_2O (Figure 4.7).[27,28]
Cyclodextrins (CDs) are cyclic oligosaccharides composed usually of six to eight D-glucopyranoside units, linked by 1,4-glycosidic bond. The three most important

Figure 4.7
Schematic drawing of per-CO_2^--β-CD (**4.12**) and the bucket-type model of this heptaanion cyclodextrin.

members of the cyclodextrin family are α-cyclodextrin (α-CD), β-cyclodextrin (β-CD) and γ-cyclodextrin (γ-CD) which possess respectively six, seven and eight glucopyranoside units. The shape of a cyclodextrin is often represented as a truncated funnel and, like the upper and lower rims of calixarenes, there are two different faces to the cyclodextrins, referred to as the primary and secondary faces. The primary face is the narrow end of the funnel and comprises the primary hydroxyl groups $-CH_2OH$. The wider secondary face contains the secondary hydroxyl groups $-CR_2OH$ groups. Previous studies using cyclodextrins (CDs) have shown that these hosts recognize chirality of organic substrates.[29–31] Since this chapter is entitled '*Chiral recognition in organometallic and coordination compound*' only modified anionic CDs showing chiral recognition towards metal helical complexes will be reviewed here. On the other hand, for a review on cyclodextrins the reader may consult the reference.[32]

Upon using the capillary zone electrophoresis technique, the native β-CD host did not show any chiral discrimination towards enantiomers of $Ru(phen)_3^{2+}$. In contrast the per-CO_2^--β-CD exhibited stronger association with the Δ-$Ru(phen)_3^{2+}$ enantiomer relative to the Λ-enantiomer and this is reflected by the longer retention time ($t_1\Lambda = 6.64$ min; $t_2\Delta = 7.44$ min). ^{1}H-NMR analysis carried out on the $Ru(phen)_3^{2+}$/per-CO_2^--β-CD system in D_2O allowed the determination of the binding constants with $K_\Delta = 1250\,M^{-1} \pm 50$ and $K_\Lambda = 590\,M^{-1} \pm 40 (K_\Delta/K_\Lambda = 2.12)$. These data suggest that the modified CD host recognizes enantioselectively the Δ-enantiomer twice as much as that of the Λ-enantiomer. To know the binding sites involved in this recognition process, the authors examined the saturated complexation-induced shifts ($\Delta\delta_{sat}$) of per-CO_2^--β-CD (**4.12**) complexed with Δ- and Λ-$Ru(phen)_3^{2+}$. The results show that the binding site of both enantiomers is the -$SCH_2CO_2^-$ precisely the H-5, H-6 and the methylene linkages (Figure 4.7). In addition the signals of the Δ-enantiomer complex shift more than the Λ-enantiomer complex. Moreover the ROESY NMR analysis of the system allowed the detection of correlation peaks between the H^3 and H^5 of the guest and H-5 of the host for the Δ-enantiomer. On the other hand, in the Λ-enantiomer/per-CO_2^--β-CD system, all correlation peaks were much weaker than those for the Δ-enantiomer/per-CO_2^--β-CD system and no correlation peak between H^3 and H-5 was observed. All these results suggest that the Δ-enantiomer is firmly associated with per-CO_2^--β-CD host (**4.12**) and forms a *tight inclusion* complex while the Λ-enantiomer forms a *shallow inclusion* complex (Figure 4.8).

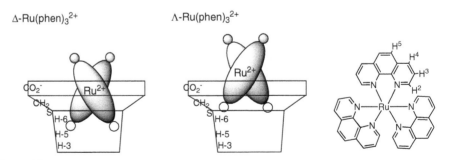

Figure 4.8
Simple models showing *tight inclusion* for Δ-$Ru(phen)_3^{2+}$/per-CO_2^--β-CD (host–guest) and *shallow inclusion* for Λ-$Ru(phen)_3^{2+}$/per-CO_2^--β-CD (host–guest); the atom numbering system of $Ru(phen)_3^{2+}$ show hydrogens involved in the recognition process (adapted with permission from reference[28], copyright 2001, American Chemical Society).

The Δ-enantiomer, having the right-handed helix configuration fits well the primary OH group 'SCH$_2$CO$_2{}^-$ side' of per-CO$_2{}^-$-β-CD. It is worth noting that the hexakis(2,3,6-tri-*O*-methyl)-α-cyclodextrin (TM-α-CD) exhibited enantioselective recognition towards the Λ-enantiomer of Ru(phen)$_3{}^{2+}$ $K_\Delta = 54\,\mathrm{M}^{-1}$ and $K_\Lambda = 108\,\mathrm{M}^{-1}$ $(K_\Delta/K_\Lambda = 0.5)$. The authors suggest that the Ru(phen)$_3{}^{2+}$ is shallowly included into the cavity of the neutral host through van der Waals interactions where the Λ-enantiomer, having a left-handed helix configuration, is preferably bound to the secondary OH group side of the TM-α-CD.

Interestingly when a racemic mixture of the helical complexes M(phen)$_3{}^{2+}$ [M = Fe(II), Co(II) and Zn(II)] was added to the per-CO$_2{}^-$-β-CD host, the circular dichroism spectra showed induced Cotton effects which in the case of the iron complex indicates that the per-CO$_2{}^-$-β-CD enriches the Δ-enantiomer of Fe(phen)$_3{}^{2+}$. This behaviour, known as the Pfeiffer effect,[33–35] occurs when a chiral complex with labile configuration interacts with optically active species to form a pair of disatereomers with one being more stable and hence interconversion between the Δ- and the Λ-enantiomers easily occurs.

More recently MacDonnell, Armstrong and coworkers reported a highly efficient method for the resolution of [Ru(diimine)$_3$]$^{2+}$ complexes using HPLC with cyclodextrin chiral stationary phase. This technique allowed the separation of several monomeric and also dinuclear ruthenium polypyridyl compounds.[36]

4.2 Chiral Recognition Using the Chiral Anion Strategy

4.2.1 Tris(tetrachlorobenzenediolato) Phosphate Anion (TRISPHAT)

The chiral anion TRISPHAT (**4.7**) was first reported in its racemic form as diethylammonium salt [TRISPHAT][Et$_2$NH$_2$] by Schmutzler and coworkers (Scheme 4.1).[37]

Lacour and coworkers made a great contribution to the field of modern chirality by preparing this salt via a new synthetic procedure but in its *enantiopure* form (Scheme 4.2).[19] They also investigated its use as a chiral shift reagent for various organic, organometallic and coordination compounds.[38] Moreover, a family of optically pure hexacoordinated phosphate anions was developed by the same group.[39–41]

In the past decade many groups[42–47] from all over the world were interested in the use of the Δ-TRISPHAT anion to differentiate chiral racemic compounds by ^{1}H-and

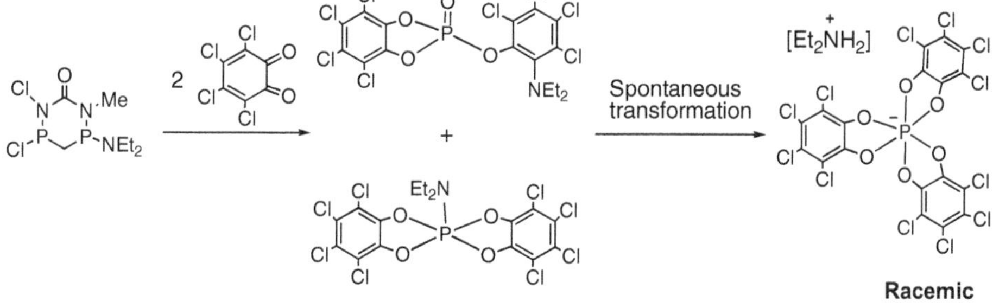

Scheme 4.1
Schmutzler method to racemic TRISPHAT.

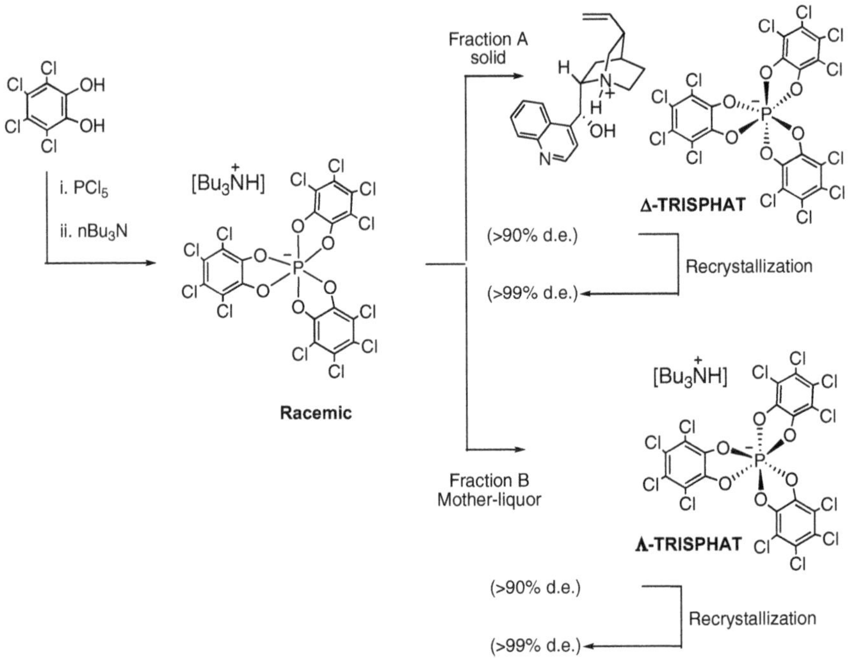

Scheme 4.2
Lacour's method to enantiopure Δ- or Λ-TRISPHAT anion (**4.7**).

^{31}P- NMR techniques, or by resolving them through anion metathesis. We have outlined most of these examples in Table 4.1.

To avoid an exhaustive description and repetition for every single species displayed in Table 4.1, we have selected some compounds of different type and geometry where an appropriate explanation of the chiral recognition process between the optically pure anion and the chiral compound will be presented.

Although it is not surprising that Ru-tris(diimine)$^{2+}$ complexes can easily be differentiated and/or resolved using Δ-TRISPHAT as chiral auxiliary, (see Section 4.1.2), yet the resolution of the dinuclear ruthenium complex *trans*-[bis(Cp*Ru)carbazolyl][PF$_6$] (**4.18**) which displayed a C_2-symmetry is more challenging.

Conversion of this racemic material to a mixture of diastereomeric salts began with anion metathesis. Complex **4.18** in the presence of an excess of [Cinchonidinium][Δ-TRISPHAT] was applied to a neutral aluminium column. Subsequent elution by CH$_2$Cl$_2$ provided the first yellow band which was collected and identified as *trans*-[(Sp,Sp))-bis (Cp*Ru)carbazolyl][Δ-TRISPHAT] (**4.18a**) and *trans*-[(Rp,Rp)-bis(Cp*Ru)carbazolyl] [Δ-TRISPHAT] (**4.18b**) (Figure 4.9). These complexes were separated through fractional crystallization from CHCl$_3$/ether to give samples 1 and 2.

The structure of the major diastereomer **4.18a** in sample 1 was determined by X-ray diffraction analysis. Complex **4.18a** crystallizes in the chiral space group P2$_1$, a view of the complex is shown in Figure 4.10. The absolute configuration of the molecules in the structure was confirmed by refining the Flack's x parameter and was equal to $-0.02(1)$, attesting to the enantiopure character of the crystal.[51] This complex possesses planar chirality and the absolute configuration of the metal centres in the cationic species

Table 4.1 Various metal complexes differentiated and/or resolved using the chiral anion Δ-TRISPHAT.

Compound	Differentiation using NMR techniques	Resolution using Δ-TRISPHAT	Type of symmetry or chirality	References
related. (**4.13**)	Yes	No	Planar chirality	Ref[42]
and derivatives. (**4.14**)	Yes	No	Planar chirality	Ref[44]
related. (**4.15**)	Yes	Yes	Planar chirality	Ref[48]
	Yes	No	Planar chirality	Ref[49]
	Yes	No	Planar chirality	Ref[50]

Table 4.1 *(Continued)*

Compound	Differentiation using NMR techniques	Resolution using Δ-TRISPHAT	Type of symmetry or chirality	References
(4.18)	Yes	Yes	Planar chirality C_2-symmetry	Ref[51]
(4.19)	Yes	Yes	Helical C_2-symmetry	Ref[43]
(4.20)	Yes	Yes	Helical	Ref[52]
(4.21)	Yes	Yes	Helical C_2-symmetry	Ref[46,53,54]

Table 4.1 (*Continued*)

Compound	Differentiation using NMR techniques	Resolution using Δ-TRISPHAT	Type of symmetry or chirality	References
(4.22)	Yes	No	Helical C_2-symmetry	Ref[47]
(4.23)	Yes	No	Helical D_3-symmetry	Ref[55]
(4.24)	Yes	No	Helical C_3-symmetry	Ref[56]
(4.25)	Yes	Yes	Metal centred chirality C_3-symmetry	Ref[57,58]
(4.26)	Yes	Yes	Helical symmetry	Ref[59]

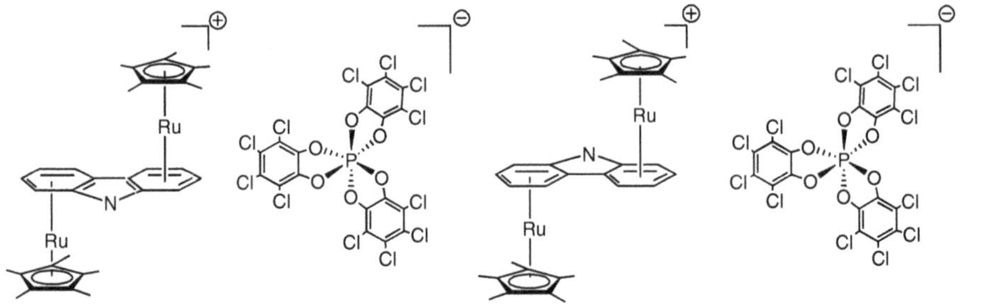

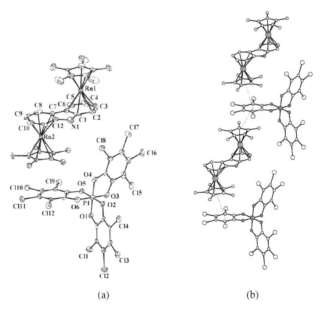

$(S_p, S_p: \Delta)$ (4.18a) $(R_p, R_p: \Delta)$ (4.18b)

Figure 4.9
Schematic drawing of the dinuclear ruthenium complexes of **4.18a** and **4.18b**.

is *Sp,Sp*. Further the CD curve of the analysed crystal **4.18a** (Figure 4.11) was recorded in CH_2Cl_2 in a micro cell (1 mm) allowing us to establish without ambiguity the identity of the major isomer in sample 1. Analysis of sample 2 by NMR spectroscopy and circular dichroism allowed us to establish the identity of the major isomer as **4.18b**.

The structure of **4.18a** shows that the two Cp*Ru units are bonded to the carbazolyl ligand and are disposed in a *trans* fashion. Most remarkable is the interaction between the TRISPHAT anion and the cationic metal complex. There are two π–π interactions between one of the tetrachloro-benzene rings of Δ-TRISPHAT and the two η^5-Cp*Ru units of two cationic metal complexes with ($d = 3.55$ (1) Å $\alpha = 19.24°$) for the C(13)-C(17) ring and with ($d = 3.67$ (2) Å $\alpha = 21.09°$) for the C(23)-C(27) ring. As a result the

(a)   (b)

Figure 4.10
(a) View of complex **4.18a** with atom numbering system; (b) 1D supramolecular chain formed through π–π contacts between the anion Δ-TRISPHAT and the cation *trans*-[(Sp,Sp)-bis(Cp*Ru) carbazolyl].

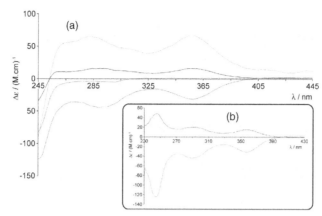

Figure 4.11

(a) CD curves for mounted crystal **4.18a** (*Sp,Sp*, Δ-TRISPHAT) (orange line), enriched **4.18a** (*Sp,Sp*, Δ-TRISPHAT) (blue line), enriched **4.18b** (*Rp,Rp*, Δ-TRISPHAT) (red line) and for Δ-TRISPHAT (green line) recorded in CH_2Cl_2 solution and at the same concentration (0.032 mM). (b) CD curves of only the two enantiomers (*Sp,Sp*) and (*Rp,Rp*) after substraction of the curve due to the Δ-TRISPHAT anion.

Δ-TRISPHAT anion intercalates between two cationic complexes providing a one-dimensional supramolecular chain. To our knowledge this example represented the first direct observation of chiral recognition between the Δ-TRISPHAT and any chiral organometallic species at least in the solid state. We also feel that this resolution originates from supramolecular control.[60]

The stereochemical relationship between **4.18a** (*Sp,Sp*, Δ) and **4.18b** (*Rp,Rp*, Δ) was assigned by circular dichroism (Figure 4.11), in which we note in both diastereomers the Δ-TRISPHAT shows a characteristic negative Cotton effect centred at 240 nm.

From Table 4.1, it is clearly shown that Δ-TRISPHAT is now a valuable tool which allowed many groups to differentiate and/or resolve a variety of organometallic and coordination complexes. More recently Lacour, Collin and Sauvage used this chiral anion to resolve dinuclear cyclometallated iridium compounds of the formula [(ppy)$_2$Ir(μ-L)Ir (ppy)$_2$]$^{2+}$ (**4.26**) {ppy = phenylpyridine; L = 3,8-dipyridyl-4,7-phenanthroline}.[59] The dinuclear species (**4.26**) exists as a mixture of meso form 'Δ, Λ' and a racemic form (enantiomeric pair 'Δ, Δ' and 'Λ, Λ'). Interestingly, Δ-TRISPHAT allowed the successful separation of the three stereoisomers. In the course of using the chiral counter-anion strategy to resolve cationic metal complexes other chiral anions were used and this is shown in the next section.

4.2.2   1,1′-binaphtyl-2,2′-diyl Phosphate Anion (BNP)

BNP anions possess axial chirality. Such anions have also been used to resolve cationic complexes. Sauvage and coworkers successfully reported the resolution of dinuclear trefoil not **4.27**.2OTf, M = Cu(I) and **4.28**.2OTf, M = Ag(I) using BNP as a chiral anion auxiliary (see Chapter 5, Section 5.1.2) (Figure 4.12).[61] Conversion of the racemic dinuclear trefoil knot of Cu(I) or Ag(I) occurred by anion metathesis to give the 1:1 diastereomers of **4.27**.2BNP and **4.28**.2BNP. The latter were separated by fractional crystallization. The BNP anions were then replaced back with either BF_4 or PF_6 to give

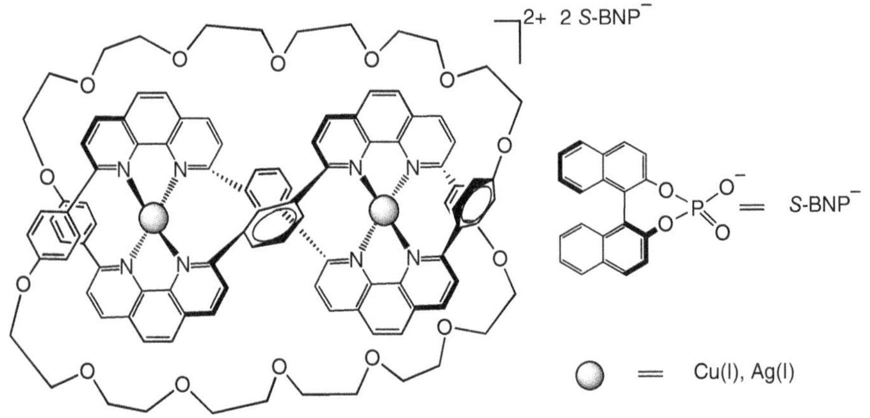

Figure 4.12
Schematic drawing of the trefoil knot of Cu(I) (**4.27**) and Ag(I) (**4.28**) with *S*-BNP as counter anion.

the optically pure dinuclear trefoil knot complex. In particular the X-ray molecular structure of the dextrorotatory dicopper trefoil knot (+)-**4.27**.2PF$_6$ was determined. The circular dichroism spectra of both enantiomers of the dicopper complex **4.27**.2PF$_6$ were also reported.

More recently Toste and coworkers reported the use of the *chiral counter-anion* strategy with a functionalized BNP (**4.29**), to carry out some asymmetric hydroxycylization reactions with gold (I) catalysts (see Chapter 3, Section 3.3.3) (Scheme 4.3).[62] The authors showed that the chiral counter-anion strategy has a profound effect on the enantioselectivity and generates products in 90–99% enantiomeric excess. These results hold promise for a golden future in this field.[63]

It is worth mentioning that BNP anions without a metal counterpart, were used as powerful organocatalysts to carry out a variety of asymmetric reactions such as epoxydation of enals reported by List *et al.*[64] The related chiral phosphoric acids HBNP, with bulky functional groups at the 3, 3' positions, are also powerful catalysts. Pioneered by Akiyama *et al.* and Terada and coworkers, they have recently been applied to a wide range of asymmetric organic transformations.[65–70] However, this area will not be discussed here since it is beyond the scope of this chapter but the reader may consult the cited references.[71–75]

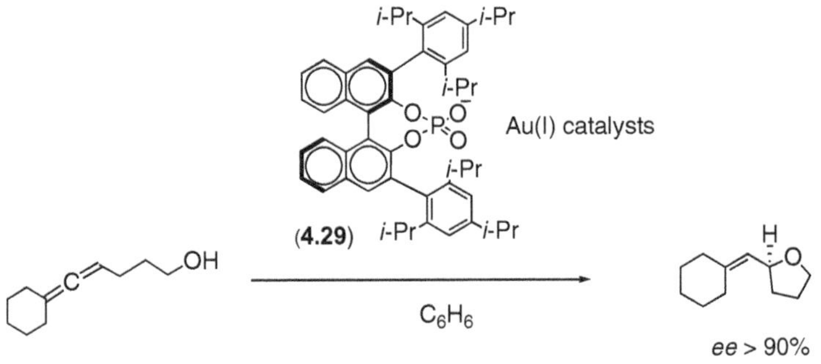

Scheme 4.3
Asymmetric hydoxycyclization reaction using the chiral counter-anion BNP (**4.29**) strategy.

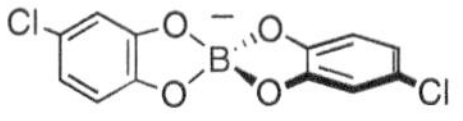

Figure 4.13
Some chiral borate anions.

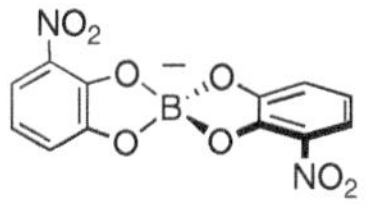

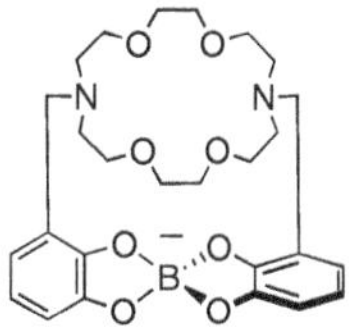

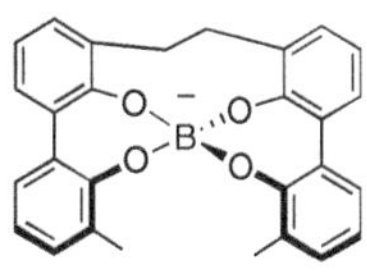

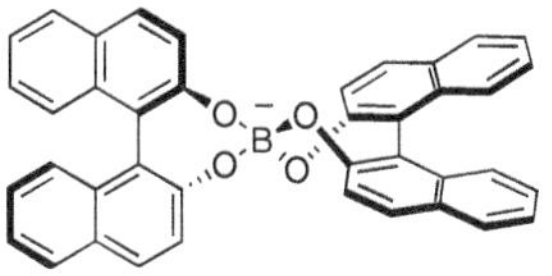

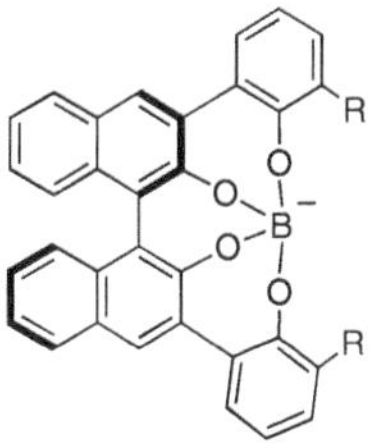

4.2.3 Bis(binaphthol) Borate Anion (BNB)

Only a few examples were reported where cationic transition metal complexes contained chiral borate anions. Chiral borate anions were first reported in 1925 when Böeseken and Mijs made the bis(catecholato) anions **4.30** and **4.31** (Figure 4.13) which were configurationally labile which limited their applications.[76] Later on, several borate anions **4.32**[77] and **4.33**[78] were prepared derived from the previous examples but most of these were configurationally labile.

Yamamoto and coworkers reported the synthesis of Brønsted acids of chiral borate anions **4.34** [79] and **4.35** [80] made or derived from enantiopure BINOL (1,1′ binaphthalenyl2,2′-diol).[81] The Bis(binaphthol)borate anion 'BNB' (**4.34**) is configurationally stable and was used as chiral counter anion with some Cu(I) complexes to promote enantioselective asymmetric aziridination of styrene by Andtsen and coworkers (Scheme 4.4).[82,83]

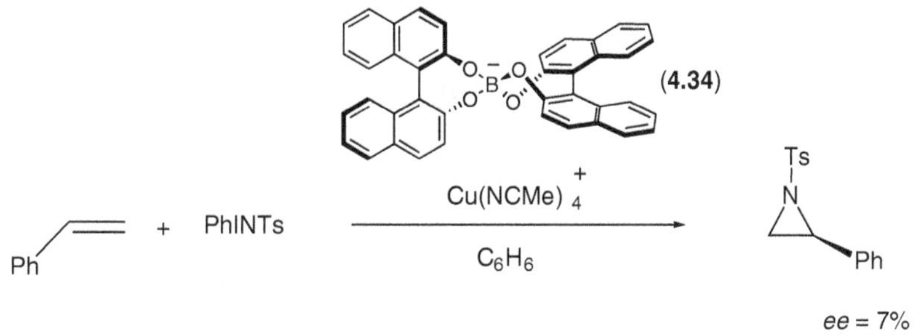

Scheme 4.4
Asymmetric aziridination of styrene using the chiral anion BNB Cu(I) complex.

4.3   Brief Introduction to DNA Discrimination by Octahedral Polypyridyl Metal Complexes

4.3.1   Introduction

A new book on '*chirality in transition metal chemistry*' which deals with chiral octahedral coordination compounds would be incomplete without some notes about their enantioselective interaction with double helix DNA. DNA 'deoxyribonucleic acid' is a chiral biological macromolecule whose structure was established in 1953 by J. D. Watson and F. H. C. Crick (Nobel Prize winners).[84,85] DNA has a double helical structure that involves two complementary strands linking together through hydrogen bonding and π–π stacking interactions. The basic components of DNA are nucleotides, that contain a nucleobase either adenine (A), thymine (T), cytosine (C) or guanine (G), attached to a sugar and a phosphate tail. The phosphate and sugar groups (the backbone of the DNA) form the outside of the structure, which is rather like a spiral staircase. The purine 'A, G' and pyrimidine 'C, T' bases are paired on the inside with guanine always opposite to cytosine (G≡C) and adenine always opposite to thymine(A=T). These specific base pairs are referred to as complementary bases.[86]

The three most common forms of DNA are (Figure 4.14):

(1) A-DNA (relatively frequent) has a right-handed double helix structure (Δ) and the base pairs are not parallel among themselves and are not perpendicular to the axis of the helix;
(2) B-DNA (natural DNA, most frequent) has a right-handed double helix structure (Δ) and the base pairs are parallel among themselves and are also perpendicular to the axis of the helix; and
(3) Z-DNA (rare) has a left-handed double helix structure (Λ).

The absence of a C_2-symmetry along the helicoidal axis between the two complementary strands generates a major and a minor groove.

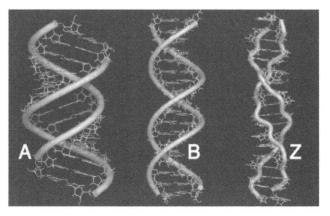

Figure 4.14
Views of the three most common forms of DNA.

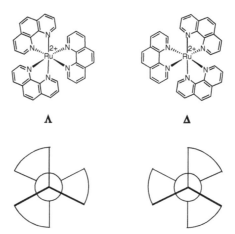

Figure 4.15
Simple model illustrating Λ- and Δ-configuration of $[Ru(phen)_3]^{2+}$ complex.

4.3.2 Background on DNA Binding with Chiral Octahedral Metal Complexes – the $[Ru(phen)_3]^{2+}$ Example

Transition metal complexes can interact with the DNA biomolecule either covalently, as with *cis*-platin, or noncovalently, when coordinatively saturated octahedral $[Ru(dimine)_3]^{2+}$ complexes or related are employed. The latter exists in two enantiomeric forms designated as the Δ and Λ optical isomers (Figure 4.15).[87] In solution at room temperature they are configurationally stable and kinetically inert to ligand substitution. Due to their geometry these compounds are ideally suited for DNA binding studies.

There are three kinds of noncovalent interactions:

(a) electrostatic binding of a positively charged metal complex to the negatively charged DNA backbone;
(b) surface binding, involving primarily hydrogen bonds in either the major or the minor groove of the DNA; and
(c) intercalative binding, where planar aromatic ligands insert themselves between stacked base pairs.

Only in (b) and (c) modes of binding would one expect enantioselective binding or chiral discrimination between the Δ- and Λ-enantiomer. It is believed that chiral discrimination is even greater in the intercalative binding mode.

DNA binding with octahedral metal complexes has been the focus of many studies of different groups;[18,88–90] however, Barton and coworkers were the first to report on *homochiral association* (Δ, Δ) between Δ-$[Ru(phen)_3]^{2+}$ with B-DNA (Δ-helix) on the basis of equilibrium dialysis data.[91,92] Although the enantiospecific interaction of the Δ- rather than the Λ-enantiomer of *rac*-$[Ru(phen)_3]^{2+}$ has been the subject of an intensive controversial debate,[93–97] Kane-Maguire showed that using the capillary electrophoresis (CE) technique on a sample of B-DNA in the presence of *rac*-$[Ru(phen)_3]^{2+}$, buffered at pH 5 or at pH 7, the *homo*-chiral association (Δ, Δ) was stronger.[98] For instance the electropherogram showed a base-to-base line separation of the Λ- and Δ-isomers of both complexes was achieved with Δ-[Ru(phen)$_3$]$^{2+}$ displaying the longer migration time. This result provides dramatic confirmation of earlier published studies involving equilibrium dialysis. It is important to mention that several papers and studies were devoted to determine whether the binding site was the major or the minor groove of the DNA, and whether the nature of the interaction is the intercalation or surface mode. We feel this kind of debate is beyond the scope of this book; however, it is highly important to note that the above studies do indeed show that a *homo*-chiral association (Δ, Δ) is favoured relative to a *hetero*-chiral one (Λ, Δ).

On the other hand, it has also been shown that interconversion between Δ- and Λ-enantiomers of [Fe(phen)$_3$]$^{2+}$ occurs easily and the Δ-isomer is enriched upon complexation of *rac*-[Fe(phen)$_3$]$^{2+}$ with a right-handed DNA double helix.[99] This behaviour, known as the Pfeiffer effect, was also observed when the same iron complex *rac*-[Fe(phen)$_3$]$^{2+}$, which is configurationally labile, was bound to the optically active species such as Δ-TRISPHAT (**4.7**) or the heptaanion modified cyclodextrin (**4.12**) (see Sections 4.1.3 and 4.2).

4.3.3 Molecular Light Switches for DNA

Barton and coworkers designed the octahedral chiral ruthenium complexes [Ru(L-L)$_2$(dppz)]$^{2+}$ with L-L = bpy (**4.36**), phen (**4.37**) and dppz is dipyrido[3,2-a:2′,3′-c] phenazine, a π-extended ligand (Figure 4.16) that is capable of intercalating between stacked base pairs of B-DNA. In aqueous solution the complex is nonluminescent while in the presence of double-stranded nucleic acids these complexes are strongly emissive and thus they function as molecular light switches for DNA.[100, 101] The absence of emission in aqueous media is attributed to excited state quenching by water that forms hydrogen bonding with the phenazine nitrogens. This is a remarkable example in the area of DNA binding studies and proves that the type of noncovalent interaction is the intercalation mode.[102] The binding of the Δ-enantiomer of the molecular light switches [Ru(phen)$_2$(dppz)]$^{2+}$ (**4.37**) and [Rh(phen)$_2$(phi)]$^{3+}$ (**4.38**) (phi = 9,10-phenanthrenequinonediimine) to B-DNA (Δ-helix) has been clearly proved by ^{1}H NMR, and fluorescence spectral studies.[103,104] The related rhodium intercalator Δ-[Rh(DPB)$_2$(phi)]$^{3+}$ (**4.39**) (DPB = 4,4′-diphenyl-2,2′-bipyridyl) also showed enantiospecific binding.[105]

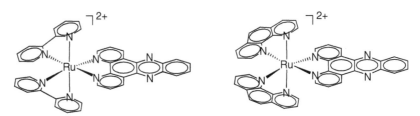

Figure 4.16
Schematic drawing of the ruthenium complexes [Ru(L-L)$_2$(dppz)]$^{2+}$ (**4.36–4.37**).

Summary

Throughout this chapter we have tried to present an overview of some recent examples in organometallic and coordination chemistry where a chiral recognition process occurs between a chiral complex and an optically active substrate or a receptor. Thus we have shown that a heterochiral association (Δ, Λ) occurs in metal complexes with helicoidal symmetry via π–π contacts between the extended π-ligands. In the presence of the optically pure Δ-TRISPHAT, a homochiral association (Δ, Δ; ion-pair) was observed with cationic octahedral $[M(\text{diimine})_3]^{2+}$ complexes with helicoidal-symmetry. The latter is known as the *chiral anion approach* and has been used successfully to resolve chiral organometallic complexes and also in asymmetric catalysis using Au(I) complexes associated with BNP chiral anions. We have also shown that chiral recognition occurs with metal complexes when a modified cylodextrin is employed. Finally an introduction to enantioselective DNA binding with cationic octahedral $[M(\text{diimine})_3]^{2+}$ complexes with helicoidal-symmetry was presented. This is one of the most active fields of modern bioinorganic chemistry and here also a chiral recognition phenomenon occurs between DNA – a chiral biomolecule – and an optically active metal complex. This is beautifully illustrated with the design of molecular light switches for DNA.

References

[1] G. Blaschke, H. P. Kraft, K. Fickentscher and F. Kohler, *Arzneimittel-Forschung/Drug Res.* **1979**, *29–2*, 1640.

[2] T. Nakanishi, N. Yamakawa, T. Asahi, N. Shibata, B. Ohtani and T. Osaka, *Chirality* **2004**, *16*, S36.

[3] L. Friedman and J. G. Miller, *Science* **1971**, *172*, 1044.

[4] S. C. Peacock, L. A. Domeier, F. C. A. Gaeta, R. C. Helgeson, J. M. Timko and D. J. Cram, *J. Am. Chem. Soc.* **1978**, *100*, 8190.

[5] M. Newcomb, J. L. Toner, R. C. Helgeson and D. J. Cram, *J. Am. Chem. Soc.* **1979**, *101*, 4941.

[6] J. M. Rivera, T. Martin and J. Rebek, Jr., *Science* **1998**, *279*, 1021.

[7] R. K. Castellano, C. Nuckolls and J. Rebek, Jr., *J. Am. Chem. Soc.* **1999**, *121*, 11156.

[8] J. M. Rivera and J. Rebek, Jr., *J. Am. Chem. Soc.* **2000**, *122*, 7811.

[9] J. M. Rivera, S. I. Craig, T. Martin and J. Rebek, Jr., *Angew. Chem. Int. Ed.* **2000**, *39*, 2130.

[10] A. R. Renslo and J. Rebek, Jr., *Angew. Chem. Int. Ed.* **2000**, *39*, 3281.

[11] A. Scarso, A. Shivanyuk, O. Hayashida and J. Rebek, Jr., *J. Am. Chem. Soc.* **2003**, *125*, 6239.

[12] J. Rebek, Jr., *Angew. Chem. Int. Ed.* **2005**, *44*, 2068.

[13] A. Scarso and J. Rebek, Jr., *Top. Curr. Chem.* **2006**, *265*, 1.

[14] Y. Masuda and H. Yamatera, *Bull. Chem. Soc. Jpn* **1984**, *57*, 58.

[15] D. Gut, A. Rudi, J. Kopilov, I. Goldberg and M. Kol, *J. Am. Chem. Soc.* **2002**, *124*, 5449.

[16] S. D. Bergman and M. Kol, *Inorg. Chem.* **2005**, *44*, 1647.

[17] F. R. Keene, *Coord. Chem. Rev.* **1997**, *166*, 121.

[18] K. E. Erkkila, D. T. Odom and J. K. Barton, *Chem. Rev.* **1999**, *99*, 2777.

[19] F. Favarger, C. Goujon-Ginglinger, D. Monchaud and J. Lacour, *J. Org. Chem.* **2004**, *69*, 8521.

[20] D. Monchaud, J. Lacour, C. Coudret and S. Fraysse, *J. Organomet. Chem.* **2001**, *624*, 388.

[21] J. Lacour and V. Hebbe-Viton, *Chem. Soc. Rev.* **2003**, *32*, 373.

[22] P. S. Pregosin, P. G. A. Kumar and I. Fernandez, *Chem. Rev.* **2005**, *105*, 2977.

[23] A. Macchioni, *Chem. Rev.* **2005**, *105*, 2039.

[24] O. Maury, J. Lacour and H. Le Bozec, *Eur. J. Inorg. Chem.* **2001**, 201.

[25] R. Caspar, H. Amouri, M. Gruselle, C. Cordier, B. Malezieux, R. Duval and H. Leveque, *Eur. J. Inorg. Chem.* **2003**, 499.

[26] I. Correia, H. Amouri and C. Cordier, *Organometallics* **2007**, *26*, 1150.

[27] K. Kano and H. Hasegawa, *Chem. Lett.* **2000**, 698.

[28] K. Kano and H. Hasegawa, *J. Am. Chem. Soc.* **2001**, *123*, 10616.

[29] K. Kano, *J. Phys. Org. Chem.* **1997**, *10*, 286.

[30] K. Kano, Y. Kato and M. Kodera, *J. Chem. Soc., Perkin Trans. 2* **1996**, 1211.

[31] K. Kano, H. Kamo, S. Negi, T. Kitae, R. Takaoka, M. Yamaguchi, H. Okubo and M. Hirama, *J. Chem. Soc., Perkin Trans 2* **1999**, 15.

[32] J. Szejtli, *Chem. Rev.* **1998**, *98*, 1743.

[33] P. Pfeiffer and K. Quehl, *Ber. Dtsh. Chem. Ges.* **1931**, *64*, 2667.

[34] P. Pfeiffer and K. Quehl, *Ber. Dtsh. Chem. Ges.* **1932**, *65*, 560.

[35] S. Kirshner, N. Ahmad and K. Magnell, *Coord. Chem. Rev.* **1968**, *3*, 201.

[36] P. Sun, A. Krishnan, A. Yadav, S. Singh, F. M. MacDonnell and D. W. Armstrong, *Inorg. Chem.* **2007**, *46*, 10312.

[37] I. V. Shevchenko, A. Fischer, P. G. Jones and R. Schmutzler, *Chem. Ber.* **1992**, *125*, 1325.

[38] J. Lacour and R. Frantz, *Org. Biomol. Chem.* **2005**, *3*, 15.

[39] J. Lacour, A. Londez, C. Goujon-Ginglinger, V. Buss and G. Bernardinelli, *Org. Lett.* **2000**, *2*, 4185.

[40] J. Lacour and A. Londez, *J. Organomet. Chem.* **2002**, *643–644*, 392.

[41] J. Lacour, A. Londez, D.-H. Tran, V. Desvergnes-Breuila, S. Constant and G. Bernardinelli, *Helv. Chim. Acta* **2002**, *85*, 1364.

[42] H. Ratni, J. J. Jodry, J. Lacour and E. P. Kuendig, *Organometallics* **2000**, *19*, 3997.

[43] H. Amouri, R. Thouvenot, M. Gruselle, B. Malezieux and J. Vaissermann, *Organometallics* **2001**, *20*, 1904.

[44] J. G. Planas, D. Prim, E. Rose, F. Rose-Munch, D. Monchaud and J. Lacour, *Organometallics* **2001**, *20*, 4107.

[45] A. Berger, J.-P. Djukic, M. Pfeffer, J. Lacour, L. Vial, A. De Cian and N. Kyritsakas-Gruber, *Organometallics* **2003**, *22*, 5243.

[46] M. Chavarot, S. Menage, O. Hamelin, F. Charnay, J. Pecaut and M. Fontecave, *Inorg. Chem.* **2003**, *42*, 4810.

[47] V. Desvergnes-Breuil, V. Hebbe, C. Dietrich-Buchecker, J.-P. Sauvage and J. Lacour, *Inorg. Chem.* **2003**, *42*, 255.

[48] H. Amouri, R. Thouvenot and M. Gruselle, *C. R. Chimie* **2002**, *5*, 257.

[49] J. G. Planas, D. Prim, F. Rose-Munch, E. Rose, R. Thouvenot and J. Vaissermann, *Organometallics* **2002**, *21*, 4385.

[50] A. Berger, J.-P. Djukic, M. Pfeffer, A. de Cian, N. Kyritsakas-Gruber, J. Lacour and L. Vial, *Chem. Comm.* **2003**, 658.

[51] H. Amouri, R. Caspar, M. Gruselle, C. Guyard-Duhayon, K. Boubekeur, D. A. Lev, L. S. B. Collins and D. B. Grotjahn, *Organometallics* **2004**, *23*, 4338.

[52] J. Lacour, J. J. Jodry, D. Monchaud, *Chem. Comm.* **2001**, 2302.

[53] G. Bruylants, C. Bresson, A. Boisdenghien, F. Pierard, A. Kirsch-De Mesmaeker, J. Lacour and K. Bartik, *New J. Chem.* **2003**, *27*, 748.

[54] M. Gruselle, R. Thouvenot, R. Caspar, K. Boubekeur, H. Amouri, M. Ivanov and K. Tonsuaadu, *Mendeleev Commun.* **2004**, 282.

[55] J. Lacour, J. J. Jodry, C. Ginglinger and S. Torche-Haldimann, *Angew. Chem. Int. Ed.* **1998**, *37*, 2379.

[56] R. Frantz, E. Guillamon, J. Lacour, R. Llusar, V. Polo and C. Vicent, *Inorg. Chem.* **2007**, *46*, 10717.

[57] L. Mimassi, C. Guyard-Duhayon, M. N. Rager and H. Amouri, *Inorg. Chem.* **2004**, *43*, 6644.

[58] L. Mimassi, C. Cordier, C. Guyard-Duhayon, B. E. Mann and H. Amouri, *Organometallics* **2007**, *26*, 860.

[59] A. Auffrant, A. Barbieri, F. Barigelletti, J. Lacour, P. Mobian, J. P. Collin, J. P. Sauvage and B. Ventura, *Inorg. Chem.* **2007**, *46*, 6911.

[60] L. Pérez-Garcia and D. B. Amabilino, *Chem. Soc. Rev.* **2002**, *31*, 342.

[61] C. Dietrich-Buchecker, G. Rapenne, J.-P. Sauvage, A. De Cian and J. Fischer, *Chem. Eur. J.* **1999**, *5*, 1432.

[62] G. L. Hamilton, E. J. Kang, M. Mba and F. D. Toste, *Science* **2007**, *317*, 496.

[63] N. Bongers and N. Krause, *Angew. Chem. Int. Ed.* **2008**, *47*, 2178.

[64] X. Wang and B. List, *Angew. Chem. Int. Ed.* **2008**, *47*, 1119.

[65] T. Akiyama, J. Itoh, K. Yokota and K. Fuchibe, *Angew. Chem. Int. Ed.* **2004**, *43*, 1566.

[66] T. Akiyama, H. Morita and K. Fuchibe, *J. Am. Chem. Soc.* **2006**, *128*, 13070.

[67] D. Uraguchi and M. Terada, *J. Am. Chem. Soc.* **2004**, *126*, 5356.

[68] D. Uraguchi, K. Sorimachi and M. Terada, *J. Am. Chem. Soc.* **2004**, *126*, 11804.

[69] S. Hoffmann, A. M. Seayad and B. List, *Angew. Chem. Int. Ed.* **2005**, *44*, 7424.

[70] Q.-S. Guo, D.-m. Du and J. Xu, *Angew. Chem. Int. Ed.* **2008**, *47*, 759.

[71] T. Akiyama, H. Morita, J. Itoh and K. Fuchibe, *Org. Lett.* **2005**, *7*, 2583.

[72] D. Uraguchi, K. Sorimachi and M. Terada, *J. Am. Chem. Soc.* **2005**, *127*, 9360.

[73] M. Terada and K. Sorimachi, *J. Am. Chem. Soc.* **2007**, *129*, 292.

[74] Y.-X. Jia, J. Zhong, S.-F. Zhu, C.-M. Zhang and Q.-L. Zhou, *Angew. Chem. Int. Ed.* **2007**, *46*, 5565.

[75] T. Akiyama, *Chem. Rev.* **2007**, *107*, 5744.

[76] J. Boeseken and J. A. Mijs, *Recl. Trav. Chim. Pays-Bas* **1925**, *44*, 758.

[77] E. Graf, R. Graff, M. W. Hosseini, C. Huguenard and F. Taulelle, *Chem. Commun.* **1997**, 1459.

[78] S. Green, A. Nelson, S. Warriner and B. Whittaker, *Perkin 1* **2000**, 4403.

[79] K. Ishihara, M. Miyata, K. Hattori, T. Tada and H. Yamamoto, *J. Am. Chem. Soc.* **1994**, *116*, 10520.

[80] K. Ishihara, H. Kurihara, M. Matsumoto and H. Yamamoto, *J. Am. Chem. Soc.* **1998**, *120*, 6920.

[81] M. Periasamy, L. Venkatraman, S. Sivakumar, N. Sampathkumar and C. R. Ramanathan, *J. Org. Chem.* **1999**, *64*, 7643.

[82] D. B. Llewellyn, D. Adamson and B. A. Arndtsen, *Org. Lett.* **2000**, *2*, 4165.

[83] D. B. Llewellyn and B. A. Arndtsen, *Organometallics* **2004**, *23*, 2838.

[84] J. D. Watson and F. H. C. Crick, *Nature* **1953**, *171*, 737.

[85] J. D. Watson and F. H. C. Crick, *Nature* **1953**, *171*, 964.

[86] C. R. Calladine and H. R. Drew, *Understanding DNA: The Molecule and How it Works*, Academic Press, New York, 2nd edn, **1997**.

[87] A. von Zelewsky, *Stereochemistry of Coordination Compounds*, John Wiley & Sons, Ltd, Chichester, **1996**.

[88] A. M. Pyle and J. K. Barton, in: *Progress in Inorganic Chemistry, Vol. 38* (S. J. Lippard, Ed.), John Wiley & Sons, Inc., New York, **1990**, pp. 413.

[89] P. U. Maheswari, V. Rajendiran, H. Stoeckli-Evans and M. Palaniandavar, *Inorg. Chem.* **2006**, *45*, 37.

[90] C. Moucheron and A. Kircsch-De Mesmaeker, *J. Phys. Org. Chem.* **1998**, *11*, 577.

[91] J. K. Barton, A. Danishefsky and J. Goldberg, *J. Am. Chem. Soc.* **1984**, *106*, 2172.

[92] J. K. Barton, L. A. Basile, A. Danishefsky and A. Alexandrescu, *Proc. Natl. Acad. Sci. USA* **1984**, *81*, 1961.

[93] M. T. Carter and A. J. Bard, *J. Am. Chem. Soc.* **1987**, *109*, 7528.

[94] M. Ericksson, M. Leijon, C. Hiort, B. Norden and A. Graslund, *J. Am. Chem. Soc.* **1992**, *114*, 4933.

[95] S. Satyanarayana, J. C. Dabrowiak and J. B. Chaires, *Biochemistry* **1992**, *31*, 9319.

[96] K. Naing, M. Takahashi, M. Taniguchi and A. Yamagishi, *J. Chem. Soc., Chem. Commun.* **1993**, 402.

[97] L. N. Ji, X. H. Zon and J. G. Ziu, *Coord. Chem. Rev.* **2001**, *216–217*, 513.

[98] N. A. P. Kane-Maguire and J. F. Wheeler, *Coord. Chem. Rev.* **2001**, *211*, 145.

[99] T. Härd and B. Norden, *Biopolymers* **1986**, *25*, 1209.

[100] A. E. Friedman, J. C. Chambron, J. P. Sauvage, N. J. Turro and J. K. Barton, *J. Am. Chem. Soc.* **1990**, *112*, 4960.

[101] R. M. Hartshorn and J. K. Barton, *J. Am. Chem. Soc.* **1992**, *114*, 5919.

[102] E. J. C. Olson, D. Hu, A. Hoermann, A. M. Jonkman, M. R. Arkin, E. D. A. Stemp, J. K. Barton and P. F. Barbara, *J. Am. Chem. Soc.* **1997**, *119*, 11458.

[103] C. M. Dupureur and J. K. Barton, *J. Am. Chem. Soc.* **1994**, *116*, 10286.

[104] C. M. Dupureur and J. K. Barton, *Inorg. Chem.* **1997**, *36*, 33.

[105] A. Sitlani, C. M. Dupureur and J. K. Barton, *J. Am. Chem. Soc.* **1993**, *115*, 12589.

5 Chirality in Supramolecular Coordination Compounds

Supramolecular chirality is widely manifested in nature; for example, the DNA double helix,[1] the protein single helices and, in humans, rhinovirus 14, a member of the major rhinovirus receptor class, possesses a protein capsid that is composed of 60 protomers arranged in an icosahedrally symmetric array.[2–4] As shown previously in this book, at the molecular level, chirality is very important in asymmetric catalysis for the creation of novel chiral molecules; however, more recently an increasing amount of attention has been drawn to chiral supramolecular assemblies. On the supramolecular level, chirality involves the nonsymmetric arrangement of molecular subunits in a noncovalent assembly via weak interactions such as hydrogen bonding, metal coordination and π–π interaction.[5–8]

In the first part of this chapter, we will present chiral supramolecular architectures formed from achiral building units that are linked via metal–ligand coordination, where chirality of the assembly results only from the asymmetric arrangement of the molecular components. In the second section, we will deal with chiral assemblies that result from resolved chiral bridging ligands, where chiral information can be transferred to the metal centre and hence a predetermination of the absolute configuration at the metal centre can be achieved and, eventually, the control of the chirality of the supramolecular architecture which can also be defined as '*stereoselective asymmetric supramolecular assembly*'.

Although the number of examples of chiral assemblies formed by the first method have been increasing steeply, and include helicates, trigonal antiprisms, tetrahedral, trigonal bipyramid and cuboctahedra, interest in stereoselective syntheses in coordination chemistry has only been developed in recent years and only a few examples have been reported.

5.1 Self-assembly of Chiral Polynuclear Complexes from Achiral Building Units

5.1.1 Helicates

Among the most eye-catching architectures resulting from the self-assembly process in supramolecular coordination chemistry are helicates. The term 'helicate' was introduced by Lehn and coworkers,[9] who made a great contribution to this area (see below). It is believed that the double helical structure of DNA represents a continuous inspiration in this area and chemists are tempted to produce such appealing motifs, namely by the use

Chirality in Transition Metal Chemistry: Molecules, Supramolecular Assemblies and Materials   H. Amouri and M. Gruselle
© 2008 John Wiley & Sons, Ltd

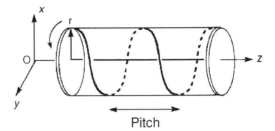

Figure 5.1
Helical screw about a defined axis.

of metal ions to template the organic threads into helical structures.[10–17] The basic
fundamental property of a helical complex is chirality, the latter being associated with the
sense of the screw about the helical axis. The pitch of the screw is defined as the distance
between one turn of the helix and the next (see Figure 5.1). To review this chemistry we
will try to cover the most important examples in a rational manner, thus starting with
simple single-stranded helicates up to the most complicated examples.

5.1.1.1 Single-stranded Helicates

A single-stranded diruthenium helical complex [Ru$_2$(*qpy*)(*tpy*)Cl][PF$_6$]$_3$ (**5.1**) was
reported by Constable and coworkers (Scheme 5.1).[18] Upon treatment of the quinquepyridine (*qpy*) with [Ru(*tpy*)Cl$_3$] (*tpy* = terpyridine) in the presence of *N*-ethylmorpho-
line, a red complex **5.1** was obtained. The X-ray crystal structure of the helical assembly
[Ru$_2$(*qpy*)(*tpy*)Cl][PF$_6$]$_3$ (**5.1**) was determined and this showed that the two six coordinate
ruthenium centres are bridged by the *qpy* ligand with one metal bonded to a *tpy* and a *tpy*
fragment of *qpy* ligand to give an N_6 environment similar to the [Ru(*bpy*)$_3$]$^{2+}$ cation. The
other metal centre is coordinated to a *tpy*, a *bpy* subunit of the pentadentate ligand **1**, and
to a chloride. The *qpy* is not planar and shows a helical twist of 74.9°, which results from
the torsion around the C−C bond between the two connected *bpy* and *tpy* subunits.

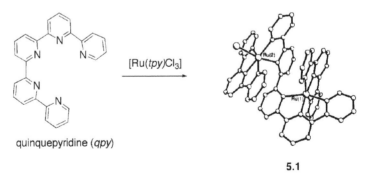

quinquepyridine (*qpy*)

5.1

Scheme 5.1
Synthesis of single-stranded helicate **5.1** (reprinted from reference[18], copyright 1990, The Royal
Society of Chemistry).

5.1.1.2 Double-stranded Helicates

The first double-stranded helicate [Cu$_3$(**2**)$_2$][CF$_3$COO]$_3$ (**5.3**) analysed by X-ray diffraction was reported by Lehn *et al.*[9] During the following decade, the helicate chemistry

was dominated by ligands with nitrogen donors such as oligopyrimidines. Later Raymond, Stack and Albrecht prepared a series of dicatechol ligands with different bridging groups that have been used for the preparation of triple-stranded heliates (see below). The trinuclear copper (I) complex **5.3** was assembled by treatment of an oligobipyridine ligand **2** containing three 2,2'bpy subunits separated by 2-oxapropylene bridges with $[Cu(CH_3CN)_4]^{+1}$ ions (Scheme 5.2). The X-ray molecular structure of the trifluoroacetate salt of the trimetallic complex was determined and confirmed the formation of a double-stranded helicate where the two oligobipyridine strands wrap around three tetrahedral copper ions. The pitch length is of $\approx 12\,\text{Å}$ with a radius of $\approx 6\,\text{Å}$. The stereochemistry of the double helix has been discussed from the viewpoint of its relation to the '*Coupe de Roi*' – when an apple is cut yielding two homochiral halves. The helices so far synthesized with this type of ligand seem to form homochiral helices and were obtained as racemates.[11]

Lehn and coworkers have also pointed out the presence of a 'positive cooperativity or self-organization' among the Cu(I) ions during the self-assembly of double-stranded di-, tri-, tetra- and pentanuclear homochiral helicates (**5.2–5.5**) using ligands **1–4**. This positive cooperativity is believed to occur such that, the configuration of one Cu(I) centre dictates that of its neighbour throughout the helicate to provide a homochiral motif. No doubt one explanation for such behaviour may arise from the high degree of conformational rigidity of the bpy subunits in the above bridging ligands **1–4**. Constable and coworkers reported the synthesis of a series of double helicates with various metals.[19–22]

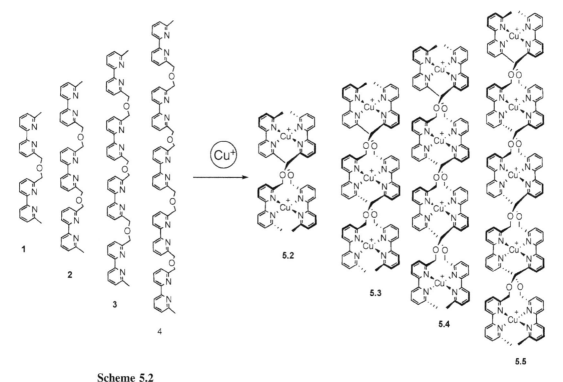

Scheme 5.2
Self-recognition in the self-assembly of double-stranded helicates from a mixture of oligobipyridine strands.

5.1.1.3 Triple-stranded helicates

Dinuclear triple-stranded helicates are formed generally via self-assembly processes from three ligand strands, each bearing two chelating units and two metal ions.[12,14] Thus triple helices require OC-6 building blocks and they may occur with a homochiral or heterochiral arrangement.[23] Raymond and coworkers isolated the first triple-stranded helicate [$Fe_2(RA)_3$] (H_2RA = rhodothorulic acid) that contains exclusively oxygen donors from hydroxamato groups.[24] The same group reported the formation of several triple helices [NMe_4]$_6$[$Ga_2(5)_3$], [NMe_4]$_6$[$Ga_2(6)_3$] and [NMe_4]$_6$[$Ga_2(7)_3$] (**5.6–5.8**) of Ga(III) with bis bidentate catecholate ligands H4[**5–7**] (Scheme 5.3).[13] In these systems a positive cooperativity or self-recognition occurs to give homochiral triple helices. Again, the nature of the strand, as well as the high degree of conformational rigidity of each donor subunit, provides a possible explanation for the self-recognition and formation of exclusively homo-triple helicates.

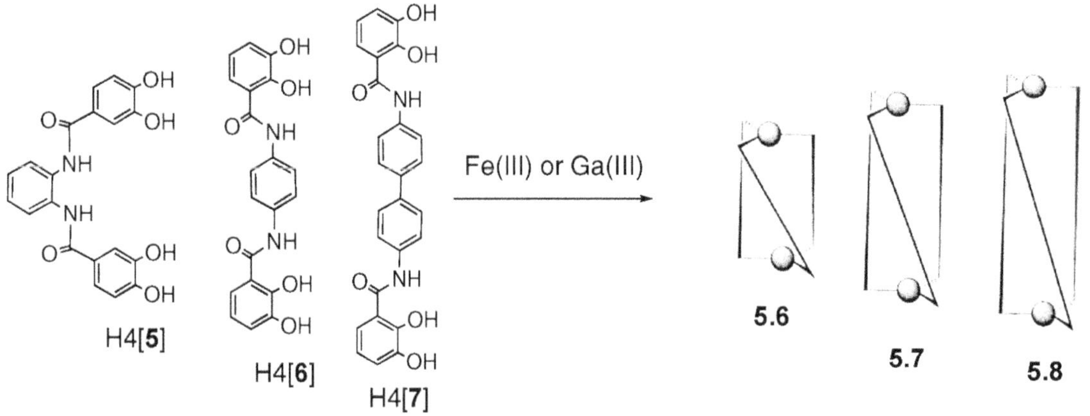

Scheme 5.3
Self-recognition and formation of triple-stranded helicates **5.6–5.8** (reprinted from reference[13]).

Albrecht and coworkers used alkyl-bridged bis(catecholate) ligands H4[**8-10**] and showed that the relative configuration of the two Ti(IV) centres in [M]$_3$[M⊂Ti$_2$(**8**)$_3$] (**5.9**), [M]$_3$[M⊂Ti$_2$(**9**)$_3$] (**5.10**) and [M]$_3$[M⊂Ti$_2$(**10**)$_3$] (**5.11**) (M = Li, Na) depends on the nature of the alkyl spacer (Scheme 5.4).[25] Ligands with an even number of methylene groups form a racemic mixture with either (Δ, Δ) or (Λ, Λ) configuration at the metal centres, whereas ligands with spacers of an odd number of methylene groups,

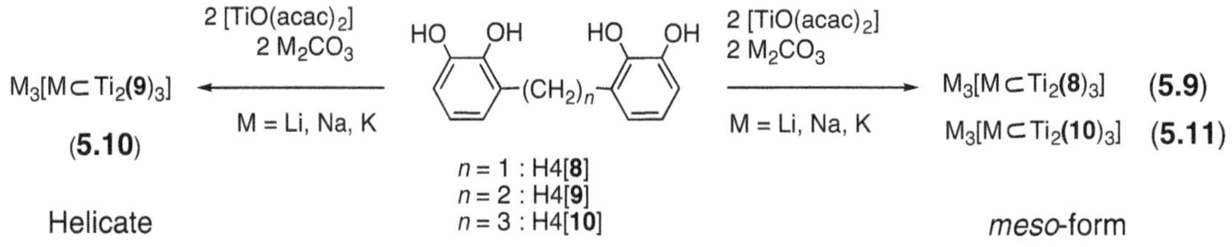

Scheme 5.4
Self-assembly of triple-stranded helicates **5.9–5.11** with even and odd spacers.

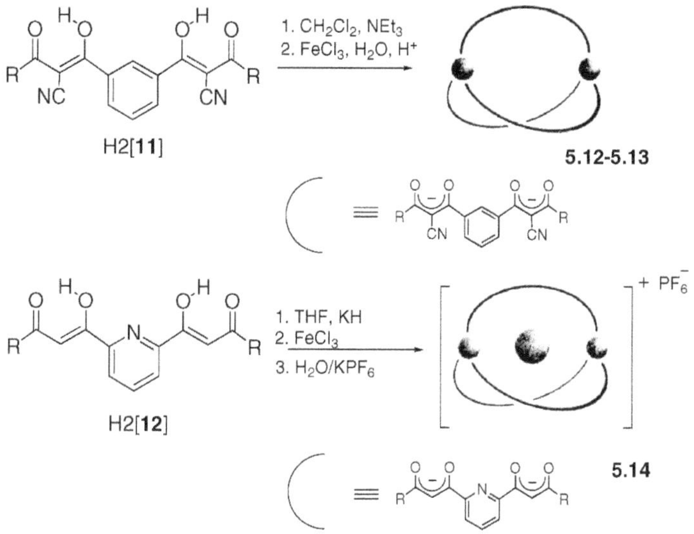

Scheme 5.5
Self-assembly of the triple helicates **5.12–5.14** (adapted from references[27–28]).

lead to achiral *meso*-triple helicates with heterochiral (Δ, Λ) configuration at the metal centres. An odd number of methylene units ($n = 1$, 3) leads to a conformation with an internal mirror plane. If this conformation is kept during the formation of a dinuclear complex, it consequently leads to the *meso*-form. In the case of an even number of methylene units, the preferred conformation possesses a C_2 axis, which upon complex formation will lead to the chiral helicate. In these helicates the guest cations are encapsulated in the cavity of the assemblies and act as templates to the formation of these helicates.[15]

More recently Hahn and coworkers reported the synthesis of several homochiral triple-stranded helicates of Ti(IV) using bis bidentate (benzene-*o*-dithiolate) ligands which are the thiol version of Raymond tetradentate ligands.[26]

Saalfrank and coworkers reported the synthesis of some triple helicates of Fe(III) using tailor-made tetradentate ligands with *m*-phenylene H2[**11**] and *m*-pyridylene H2[**12**] spacers (Scheme 5.5). The free-guest triple helicates [Fe$_2$(**11**)$_3$] (**5.12**) and [Fe$_2$(**12**)$_3$] (**5.13**) were obtained as racemate with (Δ, Δ) or (Λ, Λ) configurations at the metal centre and were defined as a *metallotopomer* of 2-cryptands. Interestingly the related helicates [K⊂(Fe$_2$(**12**)$_3$)][PF$_6$] (**5.14**), with a guest cation inside the cavity, exhibited achiral meso-configuration (Δ, Λ) at the metal centres. The latter were defined as *metallotopomer* of 2-cryptates.[27,28]

Williams and coworkers reported the synthesis of homochiral triple-stranded helicate of Co(II) (**5.15**) with a relatively rigid bis(bidentate) ligand **13** flanked by a methylene group (Scheme 5.6). The latter can be used as a 'probe group' in ^{1}H-NMR spectroscopy to determine whether a homochiral or a heterochiral arrangement has been obtained.[29–32] This is illustrated by the heterochiral triply helical dinuclear Fe(II) complex (**5.16**). Thus when the bis(bipyridine) ligand **14** with a methylene spacer was treated with the iron(II) complex, only one isomer was obtained and identified as a binuclear heterochiral triple-stranded helicate [Fe$_2$(**14**)$_3$]$^{4+}$ (**5.16**) (Scheme 5.7). Interestingly the ^{1}H-NMR

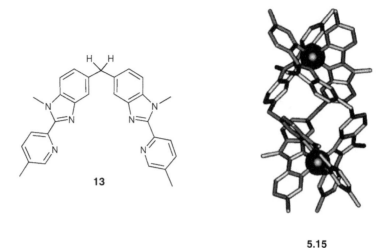

13

5.15

Scheme 5.6
Self-assembly of Co(II) triple-stranded helicate (adapted with permission from reference[30], copyright 1994, The Royal Society of Chemistry).

spectrum of **5.16** showed that the two protons of the –CH$_2$- bridge appear as an AB system reminiscent of a heterochiral system with a C_{3h} symmetry and thus the two –CH$_2$ protons lie on a mirror plane but they are unequivalent. In a homochiral arrangement with D_3 symmetry the two protons of the methylene bridge would be symmetry equivalent through a C_2 axis. When ligands analogous to **14** but with a larger spacer were used, both the homo- and the heterochiral isomers were formed.[33,34]

In very elegant studies, Piguet's group has reported several luminescent triple halicates with f-elements or f-block and d-block metals.[14,35,36] Efforts were devoted to the design of ligands such that some carried two tridentate units while others were mixed with a bidentate and tridentate ends. Thus the dinuclear helicate $[Ln_2(\mathbf{15})_3]^{6+}$ (**5.17**) was obtained by treatment of lanthanide(III) and ligand **15**. Each metal is bound to three chelating units, resulting in a nanocoordinated metal centre. X-ray molecular structural

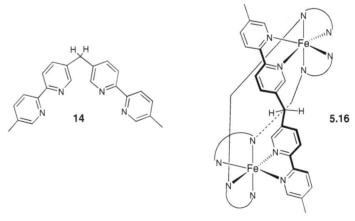

14

5.16

Scheme 5.7
Self-assembly of Fe(II)triple-stranded helicate.

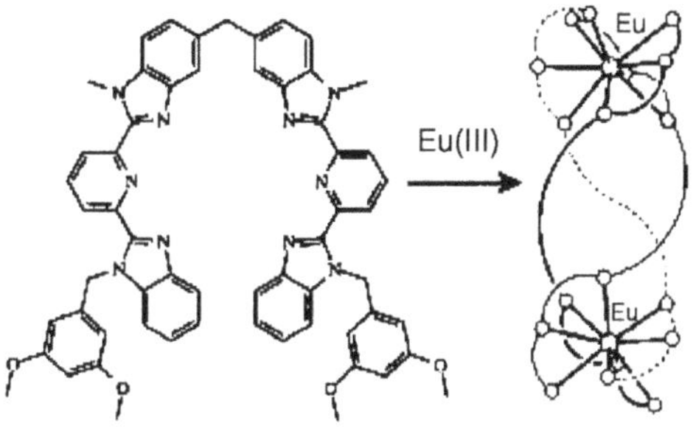

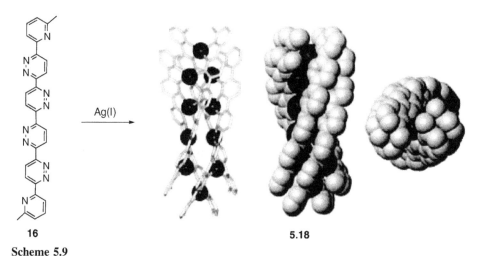

Scheme 5.8
Self-assembly of $Eu_2(\mathbf{15})_3^{6+}$ in acetonitrile (reproduced from reference[14]).

determination of $[Eu_2(\mathbf{15})_3]^{6+}$ shows the helical arrangement of the three ligands, which wrap around the two europium centres (Scheme 5.8).[37,38] Extension of his method to the preparation of triple-stranded heterodinuclear complexes has been accomplished with success.[39–44]

5.1.1.4 Quadruple-stranded Helicates

A rare example of a quadruple-stranded halicate $[Ag_{10}(\mathbf{16})_4][OTf]_{10}$ (**5.18**) was reported by Lehn and coworkers based on Ag(I) ions and the polydentate ligand **16**.[45] The latter was found to exist in dynamic equilibrium with a [4x 5]-$[Ag_{20}(\mathbf{16})_9][OTf]_{20}$ Grid (**5.19**). The quadruple strand can be viewed as two pairs of ligands **16** wrapped around the helix axis. The silver (I) ions are disposed into two parallel and nearly linear rows of four silver (I) ions each, while the top and the bottom are capped by the remaining silver (I) ions (Scheme 5.9). Each row of four silver ions is linked by two polydentate ligands **16** that

Scheme 5.9
Representation of decanuclear quadruple-stranded helicate **5.18** (reproduced with permission from reference[45]).

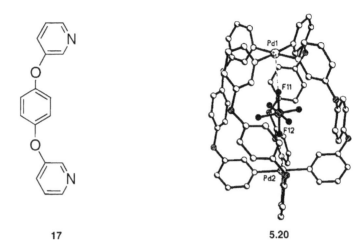

17   5.20

Figure 5.2
Structure of the quadruple-stranded helicate [PF$_6$⊂Pd$_2$(**17**)$_4$]$^{3+}$ encapsulating a PF$_6$ anion (reprinted with permission from reference[46]).

also interconnect to the top and bottom silver (I) ions. The ^{1}H NMR studies at VT showed that at low temperature the helicate structure **5.18** dominates while at 70°C the supramolecular [4x5]-[Ag$_{20}$(**16**)$_9$][OTf]$_{20}$ Grid (**5.19**) was more stable.

Another quadruple-stranded helicate [PF$_6$⊂Pd$_2$(**17**)$_4$][PF$_6$]$_3$ (**5.20**) based on Pd(II) ions and a bidentate ligand **17** was reported by Steel, McMorran and coworkers. The X-ray molecular structure of the helical cage was determined and shows that each Pd(II) centre adopts a square planar geometry coordinated to four bridging ligands **17**.[46] Furthermore a PF$_6$ anion is encapsulated inside the cage of the quadruple helix with two fluoride atoms of PF$_6$ weakly interacting with the palladium centres (Figure 5.2). The cage has an approximate D_4 symmetry and when viewed along the Pd—Pd axis the helical disposition of the four ligands is obvious. In solution the quadruple-stranded helicate is stable as confirmed by NMR spectroscopy performed on crystals of **5.20**. More recently Raymond and coworkers reported the X-ray molecular structure of a quadruple-stranded bis-bidentate helicate.[47]

5.1.1.5   Circular-stranded Helicates

Hannon and coworkers reported the self-assembly of a chiral ball {[Cu$_3$(**18**)$_3$]$_4$}$^{12+}$ (**5.21**) based on copper(I) ions and the tetradentate ligand **18**.[48] Remarkably, this supramolecular species **5.21** can be viewed as an assembly of four homochiral circular helicates [Cu$_3$(**18**)$_3$]$^{3+}$ that are maintained together through CH–π interactions (Figure 5.3). As expected, each copper (I) centre is coordinated to two imine ligands **18** and adopts a tetrahedral geometry. The circular trimeric helicate is stable in solution. It is worth mentioning that the chiral ball **5.21** is obtained from one pot reaction involving 48 achiral building blocks (24 aldehydes, 12 diamines and 12 copper(I) hexafluorophosphate units). This result illustrates the power of the self-assembly process in designing highly appealing architectures when a perfect match between the bridging ligand and the metal ions geometry exists.

More recently Ward and coworkers reported the self-assembly of a tetranuclear circular helicate [BF$_4$⊂Cu$_4$(**19**)$_4$][BF$_4$]$_3$ (**5.22**) in 1:1 metal to ligand ratio, from Cu(I)

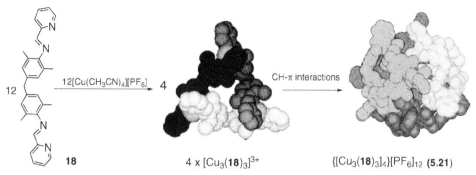

Figure 5.3
Formation of the chiral ball **5.21** (adapted with permission from reference[48]).

ions and the tetradentate ligand **19** made of a bis(pyrazolyl-pyridine) units connected to 1,8-naphthyl core via methylene spacers (Figure 5.4). All copper (I) ions are tetrahedral, the helical structure is a consequence of the four ligands having an over and under conformation around the complex. One of the tetrafluoroborate anions occupies the central cavity. The analogous Ag(I) complex gives a mononuclear species instead of a helical structure under the same experimental conditions.[49]

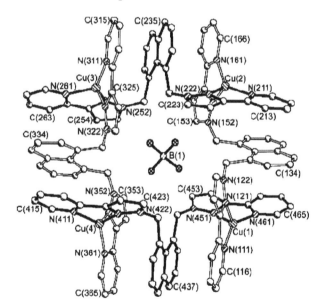

Figure 5.4
Formation of a tetranuclear circular helicate **5.22** (reprinted with permission from reference[49], copyright 2006, American Chemical Society).

5.1.2 Molecular Catenanes and Knots

A catenane is a molecule in which two or more rings are interlocked, in a manner resembling a daisy-chain. Chemists have speculated about the possibility of the synthesis of such motifs for a long time. The two arrangements of the interlinked cyclic molecules

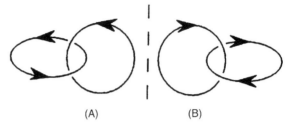

Figure 5.5
The two topological isomers A and B of interconnected rings (reproduced with permission from reference[50], copyright 1961, American Chemical Society).

(A) and (B) were defined as topological isomers by Frisch and Wassermann[50] (Figure 5.5). In this paper the authors link stereochemistry to topology.

In order to form a catenane rather than two separate, nonlinked rings, it is necessary to have an intermediate state in which one of the rings has formed and the second open-chain precursor is threaded through the ring. Dietrich-Buchecker and Sauvage made a great contribution to this area and developed an elegant synthetic strategy to prepare such interlinked molecular rings. For a review article, the reader may consult references[51,52] which also include the historical development of interlocked molecular ring chemistry.

Two synthetic strategies A and B (Scheme 5.10) were used to prepare [2]-catenand (two interlinked rings) which involves the use of copper (I) tetrahedral templates. In these templates the metal centre is coordinated to two bidentate ligands. It is worth noting that the tetrahedral copper (I) precursor is the key molecule to the success of either approach

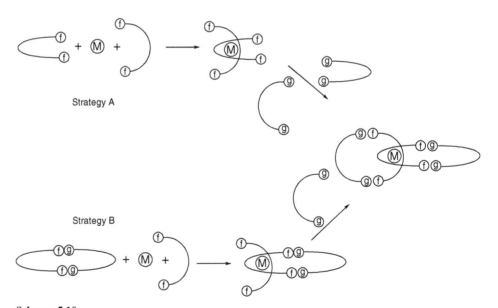

Scheme 5.10
Strategies for the synthesis of interconnected rings using a metal template; the groups f and g selectively bind to each other (from reference[51] p. 801).

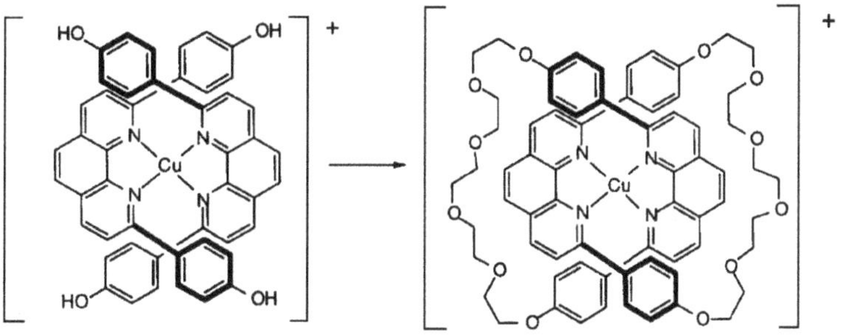

$[Cu(\textbf{20})_2]^+$ **5.23**

Scheme 5.11
Formation of the metallated two inner-linked rings (catenate) **5.23**.

(A) or (B). The interlocked metal-coordinated catenane thus formed is called a catenate. Upon demetallation with KCN the desired two interlinked macrocycles are formed.

An example to illustrate the above strategy consists of treatment of $[Cu(NCMe)_4]^+$ with 2,9-diphenol-1,10-phenanthroline (**20**), to give the tetrahedral copper complex $[Cu(\textbf{20})_2]^+$. To link the terminal phenol groups, complex $[Cu(\textbf{20})_2]^+$ was treated with $ICH_2(CH_2OCH_2)_4CH_2I$ to give the metallated two inner-linked rings (catenate) (**5.23**) (Scheme 5.11).[53] Upon treatment with CN^-, the two interlocked molecular rings (catenane) is thus obtained. When the correct combination of reagents is used, good yields of the catenane can be obtained.

Sauvage and coworkers pushed this chemistry further[51,54] and extended the above approach to prepare 3-catenanes and higher analogues.[55–57] In particular the preparation of the first molecular knot **5.25** about transition metal templates was performed in an elegant manner (Scheme 5.12) by treatment of the tetradentate ligand **21** with Cu(I) to give $[Cu_2(\textbf{21})_2]^{2+}$ (**5.24**); subsequent treatment with di(elecrophile) produced **5.25**. It is worth mentioning that the success in preparing **5.25** rather than any of the possible catenated or macrocyclic products is due to the correct orientation of the ligands about the two copper ions so that the reaction with $ICH_2(CH_2OCH_2)_4CH_2I$ leads to the target

Helical   **Trefoil knot**

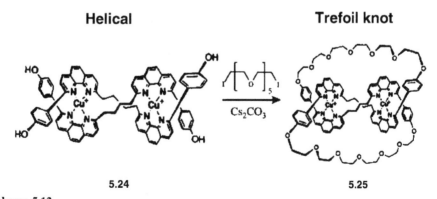

         5.24   **5.25**

Scheme 5.12
Formation of the metallated molecular knot **5.25** (reproduced from references[58–59], copyright 1992, Elsevier).

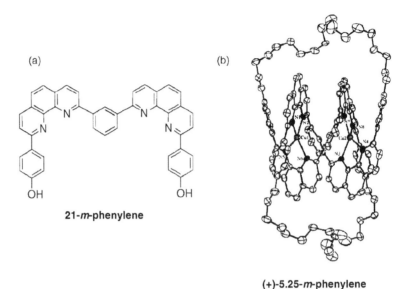

(a)

(b)

21-*m*-phenylene

(+)-5.25-*m*-phenylene

Figure 5.6
(a) Tetradentate ligand **21-*m*-pheynylene**; (b) crystal structure of the dextrorotatory dicopper trefoil knot (+)-**5.25-*m*-phenylene** (adapted with permission from reference[62]).

molecule **5.25**.[58,59] Moreover the same group has been able to extend this work to three metal centres, giving rise to the chiral double interlocked[2] catenate.[55,56,60]

The structure of the dicopper(I) knot complex **5.25** was confirmed by X-ray diffraction in 1990.[61] The metal-free trefoil knot **5.26** was obtained by treatment of **5.25** with cyanide anions. In 1999 Sauvage and coworkers prepared the related dicopper knot complex (**5.25-*m*-phenylene**) but using a tetradentate ligand **21-*m*-phenylene** where an *m*-phenylene linkage exists between the ligand subunits (Figure 5.6(a)). The resolution of this dicopper knot complex **5.25-*m*-phenylene** was successfully achieved using an enantiomerically pure chiral anion (*S*)-(+)-1,1' Binaphtyl-2,2'-diyl phosphate [(+)-BNP⁻¹]. The enantiomeric relationship was confirmed by circular dichroism spectroscopy and in particular the absolute configuration of the dextrorotatory knot corresponding to the left-handed knot was ascertained by X-ray diffraction study (Figure 5.6(b)).[62]

The knotted macrocycle is intrinsically chiral whether it is present as a metal complex or free metal molecules **5.26** and **5.26-*m*-phenylene**. The latter is defined as a *topologically chiral object*. The chirality nature of **5.26** was investigated by using 'Prickles reagent', an enantiomerically pure chiral shift reagent.[63,64] Analysis by ¹H-NMR spectroscopy confirmed the presence of two sets of proton signals reminiscent of the presence of two enantiomers that interact differently with the chiral shift reagent. As for the **5.26-*m*-phenylene**, both enantiomers were obtained separately by demetallation and their molar rotation was determined. A value of ±20 000 was found. Remetallation of the free knotted ligand **5.26-*m*-phenylene** gave the starting material **5.25-*m*-phenylene**.[62]

More recently von Zelewsky and coworkers were able to prepare the trefoil knot using optically pure tetradentate ligand with the α-pinene group. This interesting work is presented in the second section of this chapter.[65]

5.1.3 Chiral Tetrahedra

Supramolecular architectures with *T* symmetry are chiral because they have no improper symmetry operations, but possess a combination of two- and three-fold rotation axes. They represent the most investigated metal-based chiral assemblies.[66–69] There are three types of chiral tetrahedra with different metal-to-ligand stoichiometry, M_4L_6, M_6L_4 and M_4L_4 (see Figure 5.7).

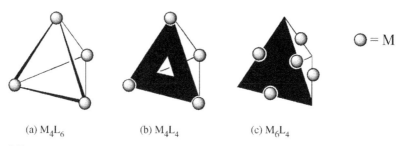

(a) M_4L_6 (b) M_4L_4 (c) M_6L_4

Figure 5.7
The three classes of molecular tetrahedra with different stoichiometries: M_4L_6, M_4L_4 and M_6L_4 (reprinted from reference[68]).

5.1.3.1 M_4L_6 Stoichiometry

The first example of chiral tetranuclear supramolecular assembly **5.27** was reported by Saalfrank and coworkers in 1988.[69] Thus the complex [(NH$_3$)$_4$∩(Mg (**22**)$_6$] (**5.27**) was obtained by serendipity upon treatment of the doubly bidentate ligand **22**, which is formed *in situ* from malonic ester and oxalyl chloride, with MeMgI in aqueous ammonium chloride solution (Scheme 5.13). Later an improved one-pot reaction[70] was used to prepare such tetrahedral complexes of magnesium, manganese, cobalt and nickel by treating the bis(chelate) ligand **22**, obtained *in situ*, with MeLi/MCl$_2$ followed

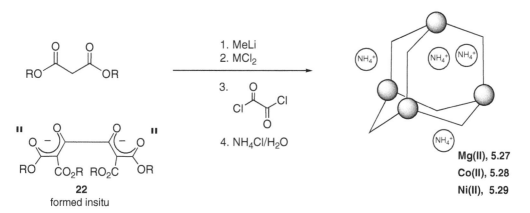

Scheme 5.13
The Saalfrank method to produce the first molecular tetrahedron.

by reaction work-up in aqueous ammonium chloride solution.[71–73] It is worth mentioning that ligand **22** was prepared *in situ* from the coupling of two dialkyl malonates with oxalyl chloride followed by deprotonation with MeLi; subsequent treatment with M(II) provided the chiral tetranuclear species **5.27–5.29**. This example gave an impetus to the development of the rational assembly of chiral tetranuclear complexes with a variety of bis(chelating) ligands.

The X-ray molecular structure of the cobalt complex **5.28** was determined and showed that the core of the complex is a tetrahedron composed of four Co(II) vertices which are linked by six doubly bidentate ligands acting as the tetrahedron edge.[74] Each cobalt centre has an octahedral geometry resulting from the coordination by three chelating oxygen centres of ligands **22**. Thus the cobalt centre is chiral with either Δ or Λ configuration. Complex **5.28** was obtained as a racemic mixture of homochiral ΔΔΔΔ and ΛΛΛΛ assemblies. Such an assembly possesses a *T*-symmetry (Figure 5.8). Mixed valence Fe(II)/Fe(III) tetrahedron assembly [Cs⊂Fe^{II}Fe$_3$^{III}(**22**)$_6$] (**5.30**) with endohedrally encapsulated Cs ions was also prepared and exhibited a similar *T*-symmetry.[75]

A neutral tetrahedron complex Fe$_4$(**23**)$_6$ (**5.31**) was obtained following the above procedure but using a metal centre with an oxidation state of three instead of two and a bis(bidentate) ligand **23** with a phenyl-spacer. Single crystal X-ray diffraction study on **5.31** revealed that the tetrahedral core possesses an S_4-symmetry, such that every two of the four iron centres have the same configuration ΔΔΛΛ.[76,77] Thus an achiral assembly was formed (*meso*-form). This result contrasts with those obtained with the preceding examples, highlighting the role and the nature of the ligand in determining the final superstructure. Related achiral tetrahedron assembly such as Ga$_4$L$_6$ with S_4-symmetry was reported by Raymond and coworkers where H2L = isophthal-di-*N*-(4-methylphenyl)hydroxamic acid.[78]

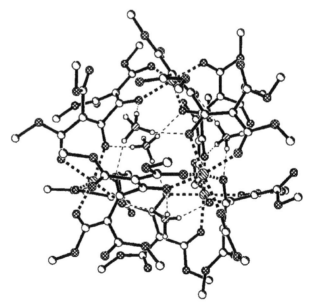

Figure 5.8
X-ray molecular structure of cobalt cage **5.28** (reprinted with permission from reference[74]).

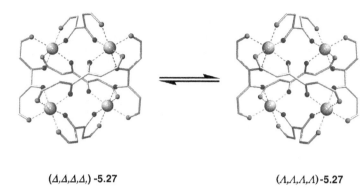

$(\Delta,\Delta,\Delta,\Delta,)$ -**5.27** $(\Lambda,\Lambda,\Lambda,\Lambda)$-**5.27**

Figure 5.9
Enantiomerization of ($\Delta\Delta\Delta\Delta$)–**5.27** to ($\Lambda\Lambda\Lambda\Lambda$)–**5.27** (courtesy of Professor Rolf W. Saalfrank, University Erlangen-Nürnberg, Germany, from reference[79]).

Interestingly Saalfrank and coworkers showed by NMR spectroscopy that the chiral tetranuclear magnesium species **5.27** undergoes a nondissociative enantiomerization process in solution converting one enantiomer to the other without metal–ligand decoordination. The proposed mechanism consists of four simultaneous Bailar twists at the four octahedrally coordinated Mg(II) centres and six atropenantiomerization processes around the C—C single bonds in the six ligands **22** (Figure 5.9).[79] The activation barrier for this enantiomerization process was found to be 14.3 kcal/mole from variable temperature ^{1}H-NMR spectroscopy.

More recently Saalfrank and coworkers extended this strategy to the preparation of enantiomerically pure Cu(II) cubanes using chiral bis-1,3-diketones as tetradentate ligands.[80]

Raymond *et al.* described the self-assembly of a series of chiral M_4L_6 coordination tetrahedra based on the bis-catecholate ligands **23**, **24** and **25** when combined with trivalent metal ions such as Al(III), Ga(III), In(III), Fe(III) and/or tetravalent ions Ti(IV), Sn(IV), Ge(IV).[67,81,82] For instance, in the tetrahedral assemblies $[Ga_4(\mathbf{23})_6]^{12-}$, $[Ga_4(\mathbf{24})_6]^{12-}$ and $[Ga_4(\mathbf{25})_6]^{12-}$ (**5.32–5.34**) each Ga(III) centre acquires an octahedral geometry and is chelated by three catecholate arms of the bis-tetradentate ligands that form the six edges of the tetrahedron (Scheme 5.14). Furthermore, the chirality established at one vertex is transmitted to the other three centres via the rigid backbones of the bridging ligands, resulting exclusively in a homochiral arrangement ($\Delta\Delta\Delta\Delta$ or $\Lambda\Lambda\Lambda\Lambda$). These coordination tetrahedra featured large cavities, appropriate to encapsulate tetraalkylammonium ions NR_4^+ (R = -Me, -Et, -Pr). It is worth mentioning that the formation of **5.32** occurred with or without a guest molecule in the cavity; however, **5.33** and **5.34** were only formed in the presence of NMe_4^+ and NEt_4^+ as thermodynamic templates to ensure formation. In the absence of the guest template, for example ligand **24** provides the triple-stranded helicate **5.35**.

Raymond and coworkers were able to resolve the racemic mixture of the supramolecular assembly **5.32** ($\Delta\Delta\Delta\Delta$ and $\Lambda\Lambda\Lambda\Lambda$) using the chiral guest (−)-(*S*)-N-methylnicotinium cation (**26**). Addition of **26** to a racemic mixture of **5.32** precipitated the diasteriomeric ion-pair ($\Delta\Delta\Delta\Delta$)-**5.32·26**, which was then separated by subsequent replacement of the chiral cation **26** by tetra-alkylammonium providing enaniomerically pure ($\Delta\Delta\Delta\Delta$)-**5.32** (Scheme 5.15).[83]

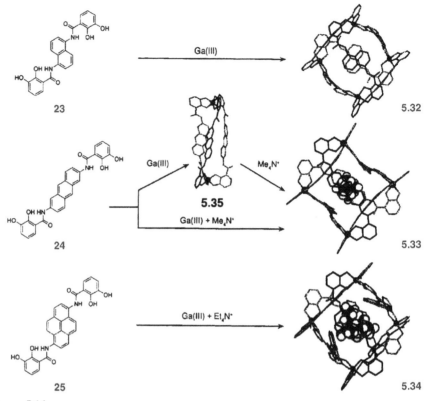

Scheme 5.14
Raymond method to a series of M_4L_6 tetrahedra (reprinted with permission from references[68,81–82], copyright 2001, American Chemical Society).

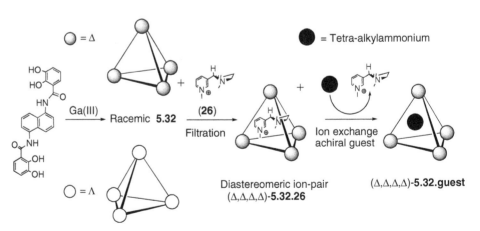

Scheme 5.15
Resolution of **5.32** using optically pure guest **26** (reproduced from reference[6], copyright 2006, Springer).

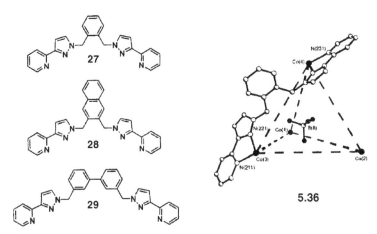

Figure 5.10
Tetranuclear assembly **5.36** with encapsulated BF$_4$ anion, and the bridging ligands **27–29** (reprinted with permission from references[66,87]).

The configurational stability of the resolved chiral tetrahedron **5.32** was investigated in solution by ^{1}H-NMR spectroscopy. Remarkably, the assembly retained its configuration for at least 8 months even in boiling alkaline solution. The authors attribute this high configurational stability of the assembly to the 'mechanical stiffness' which arises from the interconnection of the four metal centres through rigid bridging ligands to form a three-dimensional tetrahedral arrangement. This result prompted the authors to use such a supramolecular assembly as a nanovessel to carry out enantioselective reactions in supramolecular media (see below).[84]

Other M$_4$L$_6$ assemblies **5.36–5.38** but with anionic guests were reported by Ward and coworkers using the bis-(pyrazolyl-pyridine) ligands **27–29** with phenyl, naphthyl and biphenyl spacers and Co(II) metal ions.[66,85–87] In these assemblies the anion guest acts as a template. The X-ray molecular structure of [BF$_4$⊂Co$_4$(**27**)$_6$][BF$_4$]$_7$ (**5.36**) was determined and showed that each cobalt centre has an octahedral geometry and occupies the vertices of a tetrahedral assembly with six edges formed of the bis-(pyrazolyl-pyridine) ligands **27**. One of the BF$_4$ counter anions is encapsulated in the cage (Figure 5.10). VT^{19}F NMR spectra showed that it remains encapsulated in the cavity and does not exchange with the outside free anions; similar results were obtained with the achiral binuclear cobalt(II) cage [BF$_4$ ⊂ Co$_2$(L)$_4$][BF$_4$]$_3$ (**5.39**) (L = bis-benzimidazole).[88,89] In contrast the tetrahedral cage [BF$_4$ ⊂ Co$_4$(**29**)$_6$][BF$_4$]$_7$ (**5.38**) displaying ligand **29** with a larger spacer exchanged the encapsulated anion with the outside ones, the free energy of activation of this process was determined with $\Delta G = 15.4$ kcal/mole.

5.1.3.2 Catalysis in Supramolecular Media Using the Chiral M$_4$L$_6$ Capsule

As presented in the previous section, the chiral self-assembled M$_4$L$_6$ supramolecular tetrahedron can encapsulate a variety of tetra-alkyl ammonium guests and presents high configurational stability. Raymond, Bergman and coworkers, used the tetrahedral [Ga$_4$(**23**)$_6$]$^{12-}$ (**5.32**) (H4(**23**) = 1,5-bis(2,3-dihydroxybenzamido)naphthalene) (Figure 5.11) as a host to encapsulate reactive cationic organometallic complexes and studied some catalytic transformations within the chiral environment of the cavity of the supramolecular tetrahedron.[84]

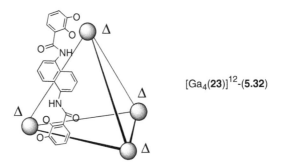

$[Ga_4(23)]^{12-}$**(5.32)**

Figure 5.11
The tetrahedral host $[Ga_4(23)_6]^{12-}$ (**5.32**) (adapted with permission from reference[84], copyright 2005, American Chemical Society).

The preparation of the desired host–guest complexes $[M \subset Ga_4L_6]^{11-}$ was quantitatively obtained, upon exchanging the weakly encapsulated NMe_4^+ in the starting material $[NMe_4 \subset Ga_4(23)_6]^{11-}$ by the cationic organometallic complex 'M'. The generation of the host–guest complexes $[M \subset Ga_4(23)_6]^{11-}$ was easily observed by ^1H-NMR spectroscopy. The resonances of the encapsulated guest were shifted significantly upfield due to shielding of the naphthalene moiety of the host assembly, illustrating the close contact between host and guest. For example, the guest resonances of $K_{11}[(CpRu(\eta^6-C_6H_6) \subset Ga_4(23)_6]$ (**5.40**) occurred at 2.60 ppm ($\eta^6-C_6H_6$) and 1.89 ppm (Cp), whereas the signals for the free ruthenium complex were at 6.04 and 5.29 ppm respectively (Figure 5.12).[90]

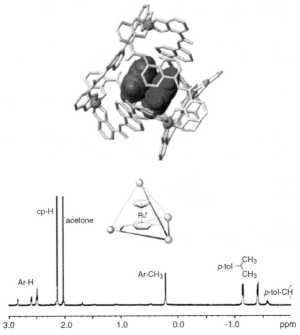

Figure 5.12
Cache model (top) of $[(CpRu(\eta^6-C_6H_6) \subset Ga_4(23)_6]^{11-}$ and ^1H NMR spectrum (bottom) of $K_{11}[(CpRu(p\text{-cymene}) \subset Ga_4(23)_6]$ (reprinted with permission from reference[84], copyright 2005, American Chemical Society).

Following the above procedure, a variety of organometallic guests such as Cp_2Fe^+, Cp_2Co^+, $[CpRu(p\text{-cymene})]^+$, $CpRu(diene)X$, $Cp^*Ru(diene)X$, $[Cp^*Ir(PMe_3)(Me)(C_2H_4)]^+$ and other related systems were encapsulated in the tetrahedral assembly $[Ga_4L_6]$.[12,91–92]

Upon encapsulation of chiral organometallic complexes such as $[Cp^*Ru(CH_2=CH-CHR^1=CHR^2)(H_2O)]^+$ in the chiral tetrahedral assembly $[Ga_4L_6]^{12-}$, four different host–guest stereoisomers are formed (Δ/R, Δ/S, Λ/R, Λ/S), that is two diastereomeric pairs of enantiomers. The ratio of the diastereomers varied from 52:48 up to 85:15 depending on the substituents R^1 and R^2 of the diene moiety.[91] Although the diastereomeric ratios are modest compared to those available by conventional reagents, the chiral recognition here relies only on weak van der Waals interactions. These results compared well with other diastereomeric ratios obtained in host–guest systems.[93]

Most interesting was the use of the chiral tetrahedral assembly $[Ga_4(\mathbf{23})_6]^{12-}$ (**5.32**) as a nanosized reaction vessel to carry out C-H activation reaction by encapsulated iridium complex. Combination of $Na_{12}[Ga_4(\mathbf{23})_6]$ with $Cp^*Ir(PMe_3)(Me)(CH_2=CH_2)$ resulted in the immediate and quantitative encapsulation to form $Na_{12}[Cp^*Ir(PMe_3)(Me)(CH_2=CH_2) \subset Ga_4(\mathbf{23})_6]$ (**5.41**). The corresponding host–guest assembly **5.41** displayed C-H bond activation at elevated temperatures in the presence of an aldehyde. The new products were identified as the host–guest assembly $[Cp^*(PMe_3)Ir(R)(CO) \subset Ga_4L_6]^{11-}$ (**5.42**), from the C-H activation of the aldehyde substrates used in the reaction.[92,94]

Interestingly the C-H activation results obtained with the host–guest assembly were different from those obtained from the same reaction but using the free iridium catalyst $Cp^*Ir(PMe_3)(Me)(CH_2=CH_2)$. For example, experiments conducted in the absence of $[Ga_4(\mathbf{23})_6]^{12-}$ (**5.32**) host demonstrated that the free catalyst $Cp^*Ir(PMe_3)(Me)(CH_2=CH_2)$ reacts with all of the aldehydes shown in Table 5.1 at similar rates; in contrast the host–guest catalyst **5.41** strictly regulates the ability of the aldehyde substrates to react with the iridium centre based on their size and shape. Large aldehydes such as benzaldehyde and pivalaldehyde are too bulky to enter the host cavity, while smaller aldehydes are readily activated since they are able to interact with the encapsulated catalyst $Cp^*Ir(PMe_3)(Me)(CH_2=CH_2)$. Because of this contrast between the free catalyst behaviour and the host–guest catalyst system it is concluded that the C-H bond activation reaction takes place within the host assembly (Table 5.1).[84]

More recently Raymond, Bergman and coworkers used the chiral tetrahedral host $[Ga_4(\mathbf{23})_6]^{12-}$ (**5.32**) as a catalyst by itself to carry out some catalytic reactions such as: 3-Aza-Cope rearrangement of Enammonium cations to give γ, δ-unsaturated aldehyde products, via formation of imminium cations first and subsequent hydrolysis[95] (Scheme 5.16), and acid hydrolysis of orthoformates in basic solution (Scheme 5.17).[96] It is worth noting that in both reactions the catalytic rate is enhanced up to 1000-fold for the former and up to 890-fold for the latter. It is believed that due to reaction space restriction inside the capsule **5.32**, as well as preorganization of the substrates into reactive conformations, the catalytic reactions described above are accelerated.

On the other hand, we would like to mention that the use of the restricted area or confined cavity to mimic enzyme-like reaction was first reported in the late 1990s by Rebek and coworker using the free metal hydrogen assemblies to carry out Diels Alder reactions.[97–99] Since this book is dedicated to 'chirality in transition metal chemistry' this fascinating work will not be reviewed here. We should also mention that Fujita *et al.* have used achiral hexanuclear Pd(II) assemblies to carry out stoichiometric $[2 + 2]$ cycloadditions[100] which is also beyond the scope of this book.

Table 5.1 C–H bond activation of aldehydes using $Na_{12}[Cp*Ir(PMe_3)(Me)(CH_2=CH_2)$ $\subset Ga_4(23)_6]$ (**5.41**) (reproduced from reference[84]).

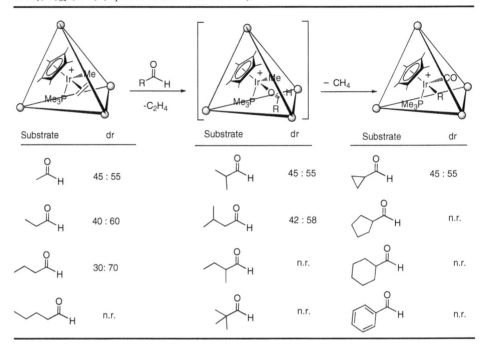

Substrate	dr	Substrate	dr	Substrate	dr
(acetaldehyde)	45 : 55	(isobutyraldehyde)	45 : 55	(cyclopropanecarbaldehyde)	45 : 55
(propionaldehyde)	40 : 60	(isovaleraldehyde)	42 : 58	(cyclopentanecarbaldehyde)	n.r.
(butyraldehyde)	30 : 70	(2-methylbutyraldehyde)	n.r.	(cyclohexanecarbaldehyde)	n.r.
(valeraldehyde)	n.r.	(pivaldehyde)	n.r.	(benzaldehyde)	n.r.

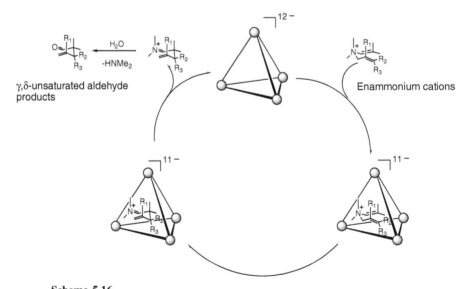

Scheme 5.16
Catalytic cycle of the Aza-Cope rearrangement using **5.32** (adapted with permission from reference[95]).

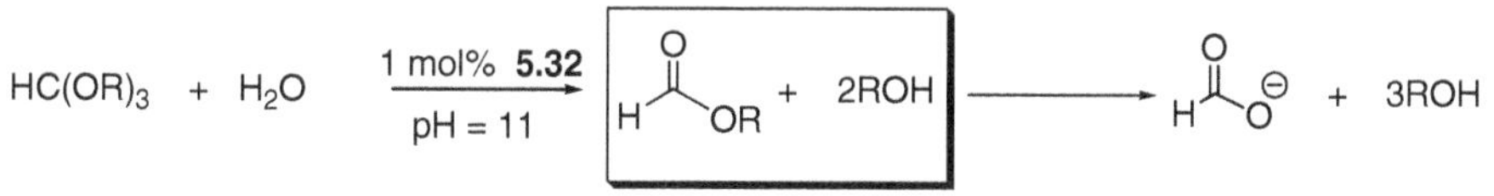

$$HC(OR)_3 \; + \; H_2O \xrightarrow[\text{pH} = 11]{\text{1 mol\% } \mathbf{5.32}} \boxed{\underset{H}{\overset{O}{\|}}_{OR} \; + \; 2ROH} \longrightarrow \underset{H}{\overset{O}{\|}}_{O}^{\ominus} \; + \; 3ROH$$

Scheme 5.17
Acid hydrolysis of orthoformates in basic solution using **5.32**.

5.1.3.3 M₆L₄ Stoichiometry

Chiral molecular tetrahedra with inverted stoichiometry M_6L_4 are made from tridentate ligands L which form the four faces of the tetrahedron while the metal ions are located on the edges. The chirality arises generally from the screw orientation of the tridentate ligand that spans the face of the tetrahedral assembly. Although this exact stoichiometry has not been realized so far, related examples were reported by Robson and Muller with the stoichiometry $M_{12}L_4$.[68,101–103] The authors used guanidine-based ligands **30** and **31** that form polyhedra assemblies of $Cd_{12}(\mathbf{30})_4$ (**5.43**) and $Cd_{12}(\mathbf{31})_4$ (**5.44**) respectively (Scheme 5.18).[68] In these assemblies the four tris-chelating ligands form the faces of a tetrahedron and the 12 Cd(II) metal ions are disposed close to the six edges of the tetrahedron. Further, the faces of the tetrahedron are linked by a square $[CdO]_2$ coordination motif. Each Cd centre adopts a square pyramidal coordination environment with the Cl^- ligand occupying the apical position. Other related high-symmetry coordination assemblies were reported by the same groups using these tridentate ligands but with Pd(II) metal ion centres.

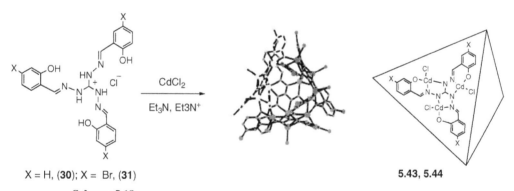

X = H, (**30**); X = Br, (**31**)   **5.43, 5.44**

Scheme 5.18
Formation of **5.43** and **5.44** (adapted with permission from reference[68]).

5.1.3.4 M₄L₄ Stoichiometry

Another alternative synthetic procedure to the formation of chiral tetrahedral architectures with M_4L_4 stoichiometry, is the use of trigonally symmetric tris-bidentate ligands that occupy the faces of the tetrahedron with four pseudo-octahedral metals at the vertices. The metal centre is thus coordinated to three bidentate ligands and adopts an octahedral geometry with either a Δ or Λ configuration. Such an assembly may possess overall T ($\Delta\Delta\Delta\Delta$ or $\Lambda\Lambda\Lambda\Lambda$), C_3 ($\Delta\Delta\Delta\Lambda$ or $\Delta\Lambda\Lambda\Lambda$) or S_4 ($\Delta\Delta\Lambda\Lambda$ or $\Lambda\Lambda\Delta\Delta$) symmetry.

In 1995 McCleverty and Ward reported the synthesis of the cationic tetrahedral assembly $[Mn_4(\mathbf{32})_4][PF_6]_4$, (**5.45**) by treatment of $Mn(CH_3COO)_2 \cdot 4H_2O$ with one

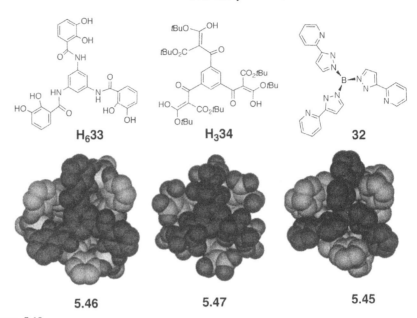

Scheme 5.19
Formation of tetrahedral M_4L_4 **5.45–5.47** (reprinted with permission from references[104–106]).

equivalent of the tris(bidentate) pyrazyl-pyridyl ligand **32** (Scheme 5.19).[104] The X-ray structure of complex **5.45** was determined and shows that the four crystallographyically independent metal centres are disposed on the vertices of an octahedron with a homochiral configuration, and obtained as a racemic mixture. Each ligand **32** spans the face of the tetrahedral and is chelating three metal centres on the vertices of that face.

Raymond and coworkers described a rational approach to such a M_4L_4 family of tetrahedral assembly using a tris-bidentate ligand with three-fold symmetry based on a central benzene ring connected via amide linkage to three catecholate units in the 1, 3 and 5 positions.[105] Thus treatment of the tris(catecholatemide) $H_6[33]$ ligand with one equivalent of $Ti(OBu)_4$ in methanol followed by reaction work-up provided the racemic homochiral anionic assembly $[HNEt_3]_8[Ti_4(33)_4]$ (**5.46**) (Figure 5.13). A variety of

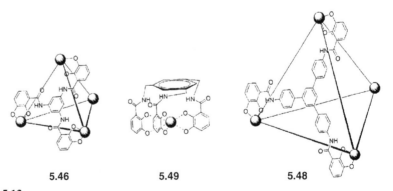

Figure 5.13
Three different cages of **5.46**, **5.48** and **5.49** (reproduced with permission from references[105,107], copyright 2005, American Chemical Society).

anionic assemblies were obtained with metal ions such as Al(III), Ga(III), Sn(IV), Fe(III) and Ti(IV). Interestingly the neutral $Fe_4(34)_4$ (**5.47**) was reported by Saalfrank and Raymond using the tris (bidentate) ligand $H_3(34)$.[106] All the previous assemblies displayed small cavities and no guest molecules or ions were encapsulated.

To increase the cavity dimension for guest encapsulation purposes, Raymond *et al.* followed the above procedure but used a larger tris(catecholate) ligand $H_6[35]$ with a phenyl extension relative to $H_6[33]$. Thus anionic tetrahedral assemblies of the type $[M_4(35)_4]^{n-}$ (**5.48**) (Figure 5.13) were formed with a variety of metal ions with M = Al(III), Ga(III), In(III) and Ti(IV).[107] The rigid phenyl extension appears necessary to ensure a tetrahedron of the type **5.48**, since the more flexible methylene linkers **36** result in the formation of mononuclear $M_1(36)_1$ complexes of the type **5.49**.[108]

The tetrahedron cages **5.48** were able to encapsulate guest cations such as NMe_4^+, NEt_4^+ and PEt_4^+ inside the huge cavity. This host–guest interaction was thoroughly investigated by the authors.[107]

5.1.4   Chiral Anti(prism)

Fujita and coworkers reported the chiral assembly [(guest)⊂M_3L_2] with a structure that conforms to a trigonal prism. Thus, upon treatment of the planar Pd(II) complex with the tridentate ligand **37**, in the presence of guest molecules, a dynamic equilibrium between chiral cage **5.50** and achiral cage **5.51** was observed depending on the nature of the encapsulated guest molecule (Scheme 5.20).[109,110]

Upon encapsulation of a flat guest such as 1,3,5-benzene tricarboxylic acid a chiral assembly was observed by ^1H NMR technique; the chirality arises from the asymmetric orientation of the two ligands in the cage framework (Figure 5.14). On the other hand the cage becomes symmetric and achiral when a larger spheric guest such as $CBrCl_3$ is encapsulated. Again loss of chirality arises from the symmetric coordination modes of two tridentate ligands in the cage framework. Competition studies in the presence of both guests showed a slight preference for the chiral assembly with the encapsulated flat guest.

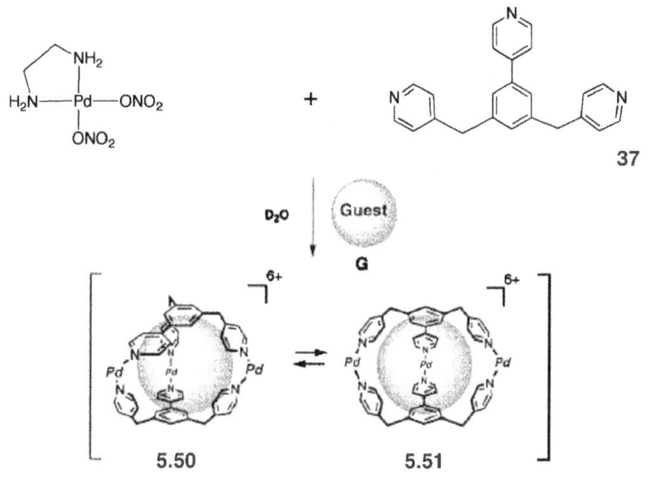

Scheme 5.20
Formation of trigonal prismatic **5.50** and **5.51** (reproduced with permission from reference[109], copyright 1999, American Chemical Society).

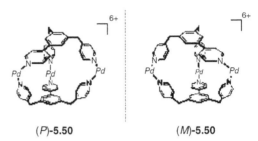

(*P*)-**5.50**   (*M*)-**5.50**

Figure 5.14
Illustration of right- (*P*) and left- (*M*) handed chirality of cage **5.50** (from reference[109]).

Two diastereomeric complexes were observed in solution when (*R*)-mandelic acid was added, suggesting that racemization of one diastereomer to the other, which would involve cleavage of two Pd-N bonds, does not take place on the NMR time scale. When racemic mandelic acid was added, diastereomers were not observed. This suggests that the (*R*) and (*S*) forms of the guest exchange quickly from one host enantiomer to the other, showing the host enantiomers to be spectroscopically identical.

Another example of chiral trigonal prismatic assembly of the formula [(guest)⊂M$_6$L$_3$] was reported by Fujita and coworkers. In this assembly, the tetratopic ligand **38** is obtained from attaching 4 Py units to a central Zn(II)-porphyrin scaffold. Upon combining three ligands of **38** with six metal ions of (en)Pd(II) a trigonal prism of the formula [Pd$_6$(**38**)$_3$]$^{12+}$ (**5.52**) was formed.[111] In this achiral assembly the Pd(II) ions occupy the corners of the prism and the three ligands form the faces. The whole structure acquires a D_{3h} symmetry (see Figure 5.15). Treatment of [Pd$_6$(**38**)$_3$]$^{12+}$ (**5.52**) with pyrene at 80°C resulted in the desymmetrization of the prismatic structure due to pyrene encapsulation and formation of [(pyrene)⊂Pd$_6$(**38**)$_3$]$^{12+}$ (**5.53**). The structure of **5.53** was confirmed by ^{1}H-NMR and cold-spray ionization mass spectroscopy. The authors suggest that the novel

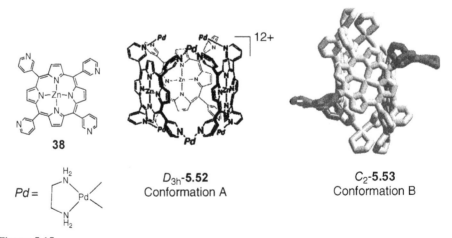

38

$Pd =$

D_{3h}-**5.52**
Conformation A

C_2-**5.53**
Conformation B

Figure 5.15
Illustration of the conformational change in cage **5.52** upon guest encapsulation (adapted with permission from reference[111]).

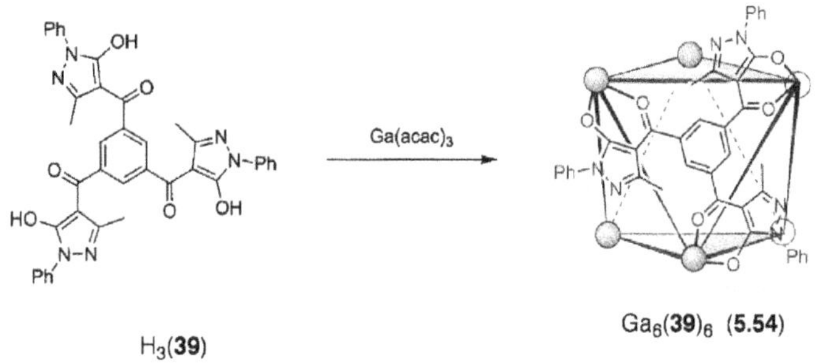

Scheme 5.21

Chiral antiprismatic species **5.54** (reprinted with permission from reference[112]).

assembly [(pyrene)⊂Pd$_6$(**38**)$_3$]$^{12+}$ (**5.53**) is chiral and exhibits a C_2 symmetry, which is triggered by guest encapsulation, during which two apical Py-Pd-Py hinges are flipped to diagonal positions to provide enough space for the guest molecule in the new assembly **5.53** (see Figure 5.15).

Raymond and Saalfrank reported the assembly of a chiral antiprismatic species Ga$_6$(**39**)$_6$ (**5.54**) (Scheme 5.21). The threefold symmetric tris diketone ligand H$_3$(**39**) was designed. Treatment of the tris bidentate ligand H$_3$(**39**) with Ga(acac)$_3$ in DMSO gave a precipitate **5.54**. Single X-ray diffraction allowed the identification of this assembly as Ga$_6$(**39**)$_6$ (**5.54**).[112] The X-ray structure confirms the formation of a distorted trigonal antiprism, in which six ligands of **39** form the equatorial faces with the Ga centres at the corners of the antiprism structure. The assembly exists as a racemic mixture of homochiral **5.54** (ΔΔΔΔΔΔ or ΛΛΛΛΛΛ) in the solid state and in solution. The configurational stability of the supramolecular structure was examined by VT NMR in the temperature range −40°C to 120°C. The results show no interconversion to any of the possible 64 isomers has occurred.

Another trigonal antiprism assembly [Pd(PPh$_3$)$_2$]$_3$[**40**]$_2$ (**5.55**) was reported by Ikeda, Shinkai and coworkers but unlike the previous example the chirality here originates from the twist of ligand **40**. The tridentate ligand was designed by attaching three Py units at the *meta*-position to a bowl-shape homo-oxacalix[3]arene scaffold. Due to the *meta*-substitution, and upon coordination of the two bowl-shape tridentate ligands to the metal centre, the resulting capsule **5.55** is helically twisted (Figure 5.16).[113,114] The ^{1}H NMR

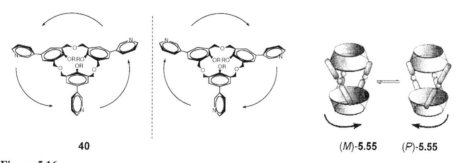

Figure 5.16

Formation of the dimeric capsule **5.55** based on calixarene ligand **40** (reprinted with permission from reference[114], copyright 2001, American Chemical Society).

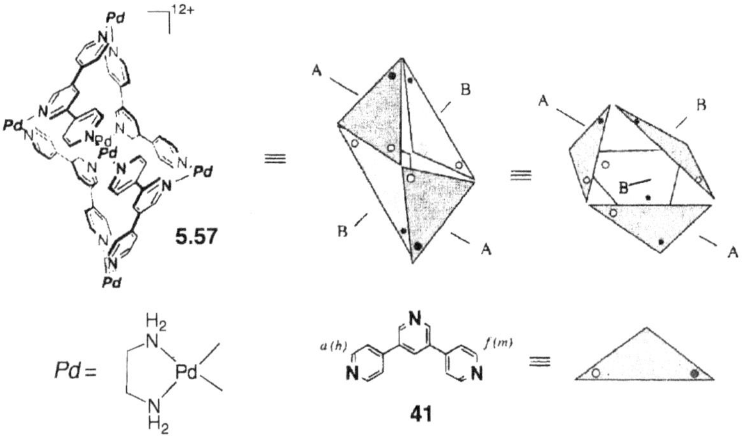

Figure 5.17
Assembly of molecular octahedron **5.57** from six end-capped Pd(II) and ligand **41** (reprinted with permission from reference[115], copyright 2002, The Chemical Society of Japan).

spectroscopy and CD studies revealed that the ratio of the enantiomers can be controlled by the inclusion of chiral guests {e.g. the (*S*)-2-methylbutylamonium ion}; however, the helical orientation of the diastereomers aggregates was not established. A similar chiral M_3L_2 assembly involving a pyridyl based cyclotriveratrylene and *cis*-protected Pd units has also been reported by the same authors.

5.1.5 Chiral Octahedra and Cuboctahedra

Fujita and coworkers reported the self-assembly of chiral octahedron cage complexes $[(L_2Pd)_6(41)_4]^{12+}$ (**5.56–5.57**), by mixing the tridentate based pyridyl ligand **41** with *cis* end-capped $L_2Pd(NO_3)_2$ building blocks {L_2 = 2-2′-bipyridine (bpy), ethylenediamine (en)} in the presence of *p*-dichlorobenzene which acts as a template guest molecule. The structure of the octahedron assembly **5.57** was determined by X-ray crystal diffraction studies and showed that the assembly has a C_2 symmetry where two ligands are linked together by their terminal pyridine rings to form an S-shaped helical motif, whereas the other two ligands are linked by their central pyridine units, resulting in a mesomeric X-shaped motif (Figure 5.17).[115] When using resolved building blocks $L_2^*Pd(NO_3)_2$ {$L_2^* = (R,R)$-1,2-diaminocyclohexane or (*S,S*)-1,2-diaminocyclohexane} in the above self-assembly process, modest chiral induction was observed, the absolute configuration of the major isomer was not reported because ΔG is estimated to be only 0.16 kcal/mol.

The only example of a chiral cuboctahedral assembly **5.58** was reported by Kimura *et al.* The cuboctahedron was self-assembled from trimeric Zn(II)-cyclen building blocks and the tridentate ligand trianionic trithiocyanurate 'TCA^{3-}' (**42**) in 4:4 ratio during guest encapsulation and was characterized in the solid state and in solution (Figure 5.18).[116]

The X-ray molecular structure of **5.58** was determined and showed that each tridentate ligand **42** links to three Zn(II)-cyclen moieties through Apical C=S–Zn coordination forming a triangular face of the cuboctahedron. Depending on the rotational sense, in which the C=S–Zn linkages are bent, the exterior of the cage twists right or left and therefore the assembly **5.58** becomes chiral. The interior of the cage is described as a

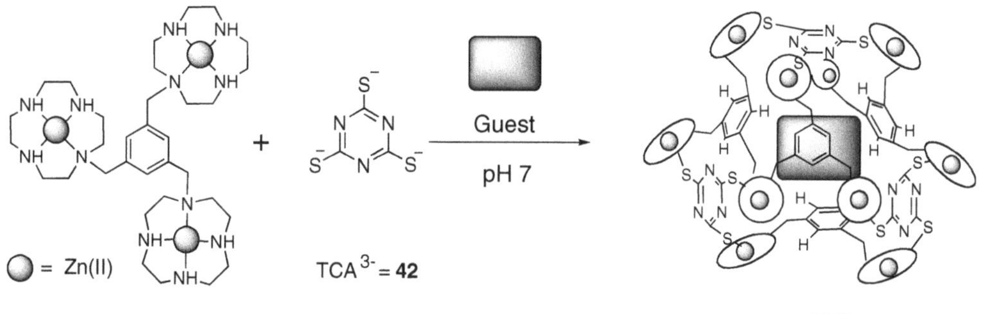

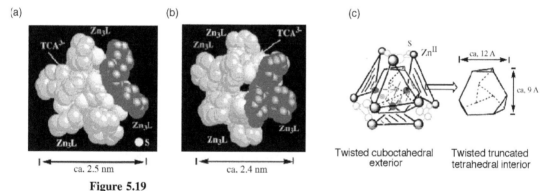

Figure 5.18

Self-assembly of the cuboctahedron **5.58** from 1:1 ratio of Zn(II)-cyclen and the tridentate ligand TCA^{3-} (reproduced with permission from reference[116]).

truncated tetrahedron and is screwed clockwise or counterclockwise (Figure 5.19). It is noteworthy that crystals were spontaneously resolved into a conglomerate mixture.

The ability of this chiral cage **5.58** to encapsulte chiral guest molecules such as (1S,2R,4R)-(−)-camphorsultam was investigated by ^{1}H NMR spectroscopy, indicating that an encapsulated chiral guest can control the exterior chirality of the assembly **5.58**.

Figure 5.19

Cuboctahedron **5.58** space filling viewed along C_3-symmetry axis. The four units of *tris*(Zn^{II}-cyclen) are shown in light green, green, yellow and blue (reprinted with permission from reference[116]).

5.1.6 Chiral Metallo-macrocycles with Organometallic Half-sandwich Complexes

We showed in Chapter 3 that half-sandwich complexes with three-legged piano stool geometry with different substituents are archetypal examples of optically active chiral-at-metal centre. Such species are ubiquitous in organometallic synthesis and catalysis; however, more recently it has become clear that they can be very interesting building blocks for supramolecular chemistry (Figure 5.20).[117–120]

The organic π-ligands of (arene)Ru and Cp*M (M = Rh, Ir) complexes are relatively inert to substitution reactions, hence they act as spectator ligands; further, the Cp* moiety increases the solubility of the resulting supramolecular complexes compared to those with Cp or simple arene ligands. The three facial coordination sites opposite to the π-ligand can be used to coordinate various ligands with N-, O-, S- or P-donor groups

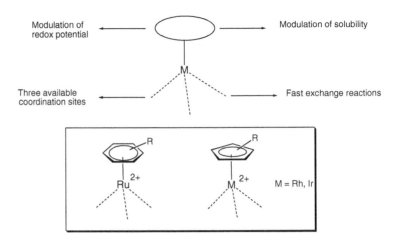

Figure 5.20
Organometallic half-sandwich piano stool compounds Cp*M and/or (arene)M as building blocks for supramolecular chemistry (reprinted with permission from reference[117]).

providing thermodynamically stable compounds. The lability of the ligands bound opposite to the π-ligand is illustrated by the aqua complex $[Cp^*Rh(H_2O)_3]^{2+}$ for which a water exchange rate constant of $k = 1.6 \times 10^5 \text{ s}^{-1}$ has been determined.[121] This value is much higher than that observed for the homoleptic complex $[Rh(H_2O)_6]^{2+}$ $k = 2.2 \times 10^{-9} \text{ s}^{-1}$.[122] This implies that solvent molecules in $[Cp^*M(solvent)_3]^{2+}$ (M = Rh, Ir) are loosely bound.and can be easily substituted by assembling ligands of different shapes and structures to form a variety of supramolecular structures. In this part we will only review the chiral metallo-macrocycles formed from half-sandwich metallated bricks.

5.1.6.1 Self-assembly of the Chiral Metallo-macrocycles

Chiral metallo-macrocycles can be formed by treating the solvated half-sandwich complexes $[(\pi\text{-arene})M(solvent)_3]^{2+}$ (π-arene = η-C_6H_6, η-Cp, η-Cp*) with trifunctional ligands.[117,123–130] In these metallo-macrocycles the ligand chelates one metal centre by two coordinating atoms and the remaining donor centre binds to the second metal centre. Using this approach tri-, tetra- and hexanuclear matallo-macrocycles have been reported.[119,131] In this section, we will only focus on the chiral trimetallic macrocycles, since this is the most investigated class of supramolecular macrocycles in this area.[132] These macrocycles were reported as racemates, through homochiral self-assembly (all metal centres have the same configuration) $R_M R_M R_M$ and $S_M S_M S_M$ with a 1:1 ratio of both diastereomers (Scheme 5.22).

Several groups have investigated the synthesis of such compounds, for instance the cationic metallo-macrocycles $[(\eta\text{-ring})M(Aa)]_3[BF_4]_3$ (Aa = amino acidate) act as catalysts for hydrogen transfer to ketones[129] while the neutral metallo-macrocycles $[(\eta\text{-ring})ML]_3$ are efficient hosts for alkali metals (Li and Na).[133] However, the chiral aspect of these species in solution was not investigated and only Yamanari and coworkers[127] have reported diastereomeric separations, but no NMR studies have been made to differentiate these chiral macrocycles by the use of a chiral auxiliary. Recently the chemistry of these macrocycles was reviewed by Severin, who contributed largely to the development of this field.[117]

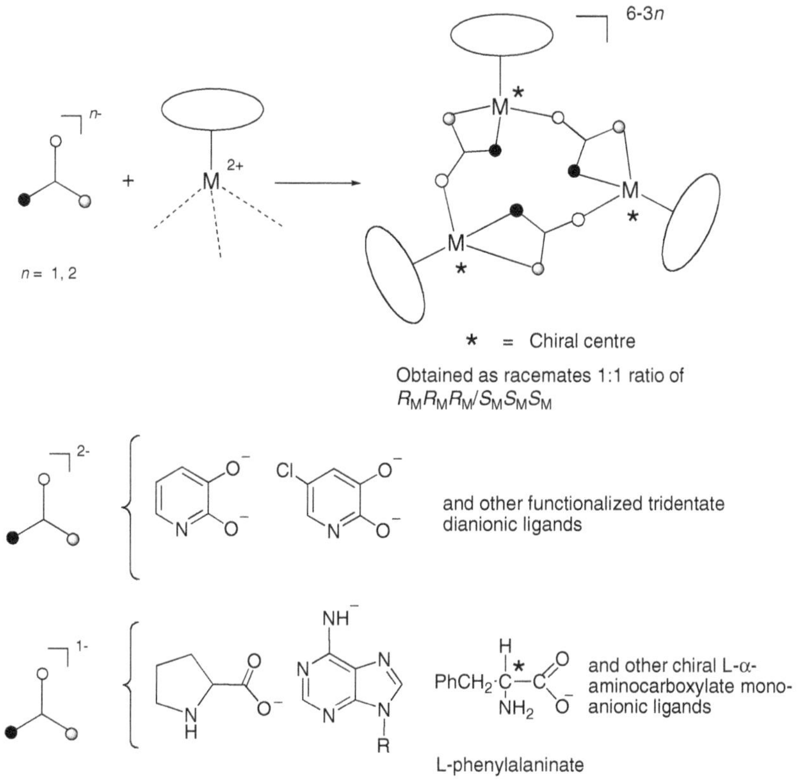

* = Chiral centre
Obtained as racemates 1:1 ratio of
$R_M R_M R_M / S_M S_M S_M$

and other functionalized tridentate
dianionic ligands

and other chiral L-α-
aminocarboxylate mono-
anionic ligands

L-phenylalaninate

Scheme 5.22
Self-assembly of chiral metallo-macrocycles using trifunctional assembling mono-anionic and dianionic ligands.

5.1.6.2 Diastereomeric Separation of the Rhodium Triangular Hosts Using a Chiral Auxiliary

Amouri and coworkers reported the synthesis of a family of chiral neutral macrocycles of the general formula $[(\eta\text{-ring})ML]_3$ following the procedure described by Severin *et al.* {with $(\eta\text{-ring})M = Cp^*Rh$ (**5.59**), Cp^*Ir (**5.60**) and $(p\text{-cymene})Ru$ (**5.61**); $L = $ 5-chloro-2,3-dioxopyridine, a trifunctional ligand} which were obtained as racemic mixture (Scheme 5.23).[134,135] The aim of the work was to investigate the homochiral self-assembly in this class of supramolecular triangular hosts as well as the epimerization process ($R_M R_M R_M \rightleftarrows S_M S_M S_M$) using an external optically pure anion such as Δ-Trisphat {Trisphat = tris[tetrachlorobenzene-1,2-bis(olato)]phosphate} reported by Lacour and coworkers (Figure 5.21).[136] Such anion forms a strong ion pairing with the cationic organometallic and coordination compounds.[137,138]

Conversion of the racemic complexes **5.59–5.61** to a mixture of diastereomeric salts began with the encapsulation of a lithium cation to give **5.62–5.63**. Subsequent replacement of the chloride anions by triflates produced the racemic precursors $[Li \subset \{(\eta\text{-ring})ML\}_3]$ [OTf] (**5.64–5.65**). The key step in this synthesis was the metathesis of the triflate anion of the racemic precursors $[Li \subset \{(\eta\text{-ring})ML\}_3][OTf]$ by the chiral anion Δ-Trisphat (Scheme 5.23) to give the target 1:1 mixture of the two diastereomeric complexes $[Li \subset (R,R,R)-\{(\eta\text{-ring})ML\}_3][\Delta\text{-Trisphat}]$ and $[Li \subset (S,S,S)-\{(\eta\text{-ring})ML\}_3][\Delta\text{-Trisphat}]$. Among

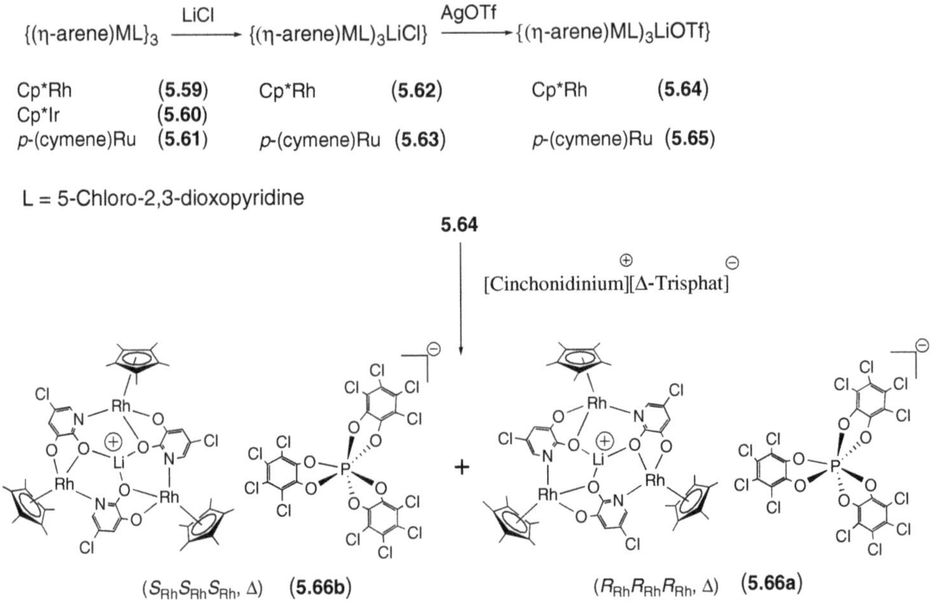

$$\{(\eta\text{-arene})ML\}_3 \xrightarrow{\text{LiCl}} \{(\eta\text{-arene})ML\}_3LiCl\} \xrightarrow{\text{AgOTf}} \{(\eta\text{-arene})ML\}_3LiOTf\}$$

Cp*Rh	(5.59)	Cp*Rh	(5.62)	Cp*Rh	(5.64)
Cp*Ir	(5.60)				
p-(cymene)Ru	(5.61)	p-(cymene)Ru	(5.63)	p-(cymene)Ru	(5.65)

L = 5-Chloro-2,3-dioxopyridine

5.64

[Cinchonidinium][Δ-Trisphat]

$(S_{Rh}S_{Rh}S_{Rh}, \Delta)$ **(5.66b)** $(R_{Rh}R_{Rh}R_{Rh}, \Delta)$ **(5.66a)**

Scheme 5.23
Formation of the diastereomers [Li⊂(R,R,R)-{Cp*Rh(**L**)}₃][Δ-Trisphat] (**5.66a**) and [Li⊂(S,S,S)-{Cp*Rh(**L**)}₃][Δ-Trisphat] (**5.66b**).

the chiral triangular hosts investigated, only the rhodium metallo-macrocycle underwent a smooth anion metathesis reaction providing a 1:1 mixture of the two diastereomers of triangular metallo-macrocycles [Li⊂(R,R,R)-{Cp*Rh(**L**)}₃][Δ-Trisphat] (**5.66a**) and [Li⊂(S,S,S)-{Cp*Rh(**L**)}₃][Δ-Trisphat] (**5.66b**). The latter were separated by fractional crystallization.

The X-ray molecular structure of one of the diastereomers (**5.66a**) was determined and provided valuable information about the mean of chiral recognition between the optically pure anion Δ-Trisphat and the single enantiomer cation (R_{Rh}, R_{Rh}, R_{Rh}). This is manifested by π–π interaction ($d = 3.6$ Å, $\alpha = 20°$) between the π-deficient tetrachlorocatecholate ring and π-donor 'Cp*Rh^{2+}' moiety (Figure 5.22).[134]

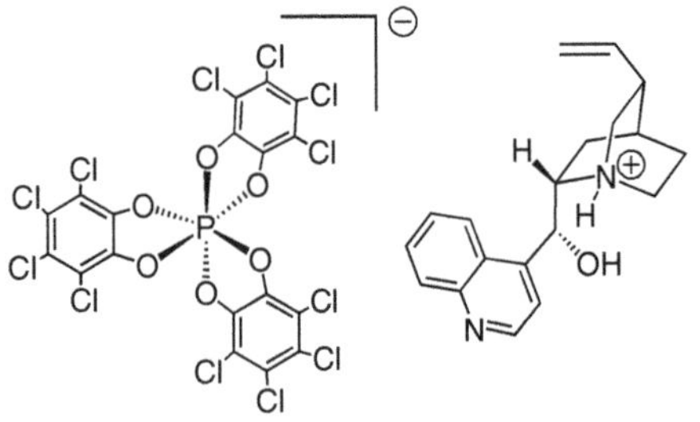

Figure 5.21
Schematic drawing of [Cinchonidinium][Δ-Trisphat].

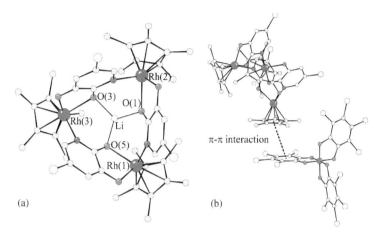

Figure 5.22
(a) Frontal projection of the cationic part of **5.66a** showing Li^+ encapsulation with the atom numbering system, and (b) sidewise projection of **5.66a** showing the π–π interaction between Δ-Trisphat and triangular host (reprinted from reference[134]).

The stereochemical relationship between **5.66a** ((R,R,R), Δ) and **5.66b** ((S,S,S), Δ) was assigned by circular dichroism (Figure 5.23). In both diastereomers the Δ-Trisphat shows a characteristic negative Cotton effect centred at 240 nm.

The solution behaviour of these chiral triangular hosts was monitored by ^{1}H-NMR studies (Figure 5.24). Thus when **5.66b** (95% de) was left in solution a slow epimerization process occurred at room temperature which after 90 days produced approximately

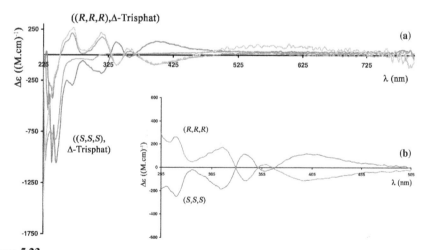

Figure 5.23
(a) CD curves for analysed crystal **5.66a** ((R,R,R), Δ-Trisphat) (orange line), for sample 2, enriched **5.66a** ((R,R,R), Δ-Trisphat) (red line), sample 1 enriched **5.66b** ((S,S,S), Δ-Trisphat) (blue line) and for [n-Bu$_4$N][Δ-Trisphat] (green line) recorded in CH$_2$Cl$_2$ solution and at same concentration (0.03 mM). (b) CD curves of only the two enantiomers (R,R,R) and (S,S,S) after substraction of the curve due to the Δ-Trisphat anion (reference[134]).

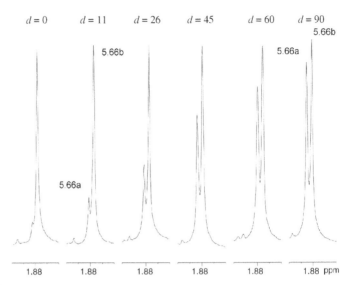

Figure 5.24
Partial 500 MHz ^{1}H NMR spectrum of **5.66b** in $CD_2Cl_2/CD_3C_6D_5$ at 290 K as a function of time showing the C_5Me_5 signal (d = days) (reference[135]).

a 1:1 mixture of **5.66ab**, suggesting that the configurational stability is highly enhanced in the trimer host compared to analogous chiral mononuclear species of half-sandwich Cp*M compounds with three-legged piano stool geometry.

It is probable that the first step of the racemization of the cationic part of **5.66** involves bond dissociation at one rhodium to give a 16-electron intermediate. In theory, this could produce $[(S_{Rh}, R_{Rh}, R_{Rh}), \Delta]$ and $[(S_{Rh}, S_{Rh}, R_{Rh}), \Delta]$ intermediates, but using molecular models, such species appear to be under high levels of strain and are therefore unlikely. It therefore seems reasonable to propose a mechanism for **5.66**, which involves complete decoordination of the Li^+ followed by dissociation into monomers. Subsequent self-assembly produces the other diastereomer (Figure 5.25).

Previous studies of the solution behaviour of chiral trimers $[(\eta\text{-ring})M(Aa)]_3[BF_4]_3$, where Aa is an optically pure amino acidate with either L or D configuration, have been

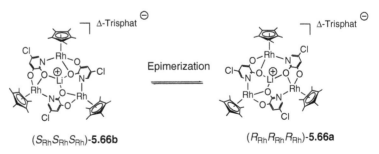

Figure 5.25
Racemization of the rhodium supramolecular host **5.66**.

reported. In fact the metal centres in these trimers adopted a preferential configuration imposed by the optically pure ligand.[129] For instance, in the solid state $R_M R_M R_M$ diastereomers were observed for rhodium and iridium centres with various amino acidate ligands; however, for prolinate compounds the $S_M S_M S_M$ diastereomers were favoured. In the above example the chiral auxiliary Δ-Trisphat is a free anion and not covalently linked to the trimer structure as in the previous examples.[134,135] Thus it acts as a chiral probe for a detailed analysis of the cation racemization process ($R_M R_M R_M \rightleftarrows S_M S_M S_M$).

In this part of Chapter 5, we have tried to review the self-assembly of chiral supramolecular assemblies obtained from achiral components. A variety of examples with different structural forms have been presented which attest to the rapid growth of this subject where a complete coverage of all examples remains a difficult task. The second section of this chapter will deal with the self-assembly of chiral objects but starting with optically pure constituents. As one may expect there are fewer examples due to difficulties in preparing enantiomericcally pure bridging ligands and/or auxiliaries.

5.2 Chirality Transfer in Polynuclear Complexes: Enantioselective Synthesis

A classical method for the preparation of enantiopure compounds is the resolution of racemate. However, it is much more effective to use the selective synthesis of the desired enantiopure substance via *enantioselective approach*. Stereoselective methods of synthesis have been widely developed in organic chemistry. The method of asymmetric synthesis has been known since the nineteenth century[139] and asymmetric catalysis has witnessed an enormous amount of development in recent decades as shown in Chapter 3. In contrast, the asymmetric synthesis of coordination compounds has only recently become a subject of systematic investigation. This is no doubt related to the fact that the chirality of coordination compounds is a much more complex phenomenon than that of organic compounds, because of higher coordination and the multitude of possible central atoms. Furthermore, while in organic chemistry the chiral tetrahedral carbon centres can be prepared without racemization, in contrast *T-4* metal centres are very often labile. In fact it is even difficult to prepare compounds with a metal centre coordinated to four different monodentate ligands, and thus the possibility of obtaining one enantiomer is excluded in most cases.

Asymmetric synthesis of coordination compounds was started by Smirnoff in 1920.[140] Shortly after Werner's death, a report published by Smirnoff, represented a breakthrough in the field. Smirnoff studied the coordination of Pt(IV) complexes with (S)- and (R)-1,2-diaminopropane. Several complexes were prepared with one, two or three optically active diamine ligands. Smirnoff expected that the molar rotations of these compounds would be proportional to the number of optically active ligands in the complex,that is in a ratio 1:2:3. However, the measurements showed a significantly high value for the trisubstituted $[Pt(L^*)_3]^{4+}$ complex. In his publication of 1920, Smirnoff deduced the following observation: '*The excess of rotation must be due to the asymmetry of the complex itself.*'[140] The author concluded that the optically pure ligands should be bound to the metal in only one of the two possible forms of the mirror isomers. This example clearly illustrates the concept of predetermination of chirality in coordination compounds.

In this part of Chapter 5 we will present the most recent examples of chirality transfer in coordination chemistry, but only polynuclear complexes will be reviewed. There are

4, 5-pinene pyridine (**43**) 5, 6-pinene pyridine (**44**)

Figure 5.26
Schematic drawings of 4,5- and 5,6-pinene pyridine **43** and **44**.

two approaches known in the literature. The first method consists of using optically pure ligands to coordinate a transition metal complex so that the chirality is transferred from the ligand to the metal centre. In the second approach, a resolved metallo-brick is used which, upon self-assembly with a bridging ligand, propagates the predetermined chirality throughout the whole polynuclear system.

5.2.1 Chirality Transfer via Resolved Bridging Ligands

5.2.1.1 Polynuclear Complexes with Chiralized Pyridine and Polypyridine Ligands

von Zelewsky and coworkers reported a rare and modern approach for chiral determination in coordination synthesis.[141,142] The idea consisted of the *chiralization* of pyridine and polypyridine through annellation with terpenes especially with α-pinene. The first example was reported in 1992 when 4,5-pinene pyridine (**43**) and the 5,6 pinene pyridine (**44**) were reported via Kröhnke type synthesis[143] (Figure 5.26). Since then about 200 different *chiralized* ligands of this family have been made. Such *chiralized* ligands coordinate to a transition metal centre (OC-6, T-4, SP-4) and impose a defined configuration at the metal centre. As one might expect the *chiralized* ligand coordinates the metal through only one possible configuration and hence a predetermination of the chirality occurs at the metal centre. Using this approach von Zelewsky reported a variety of supramolecular coordination assemblies where chirality at the metal centre in a polynuclear assembly could be controlled. We will review this work by the nature and denticity of the ligand used.

(a) Bidentate ligands (pinene-thienylpyridine and pinene-phenylpyridine). Thienylpyridine is an achiral ligand and upon reaction with $RhCl_3.3H_2O$, dinuclear rhodium (III) complexes are formed where two of the bidentate ligands are coordinated to Rh (III) as cyclometallated ligands. The two rhodium centres are bridged by two chloride ligands. Due to the mode of coordination of the ligand, only N/N-*trans* and C/C-*cis* isomers are formed but the reaction is not at all enantioselective (Scheme 5.24). In fact two enantiomers with ΔΔ and ΛΛ and the meso-form ΔΛ are produced in the statistical ratio 1:1:2.[144] If the same reaction is repeated but using 4,5-(*R*,*R*)-pinene-thienylpyridine **H[45]**, only two diastereomers ΔΔ and ΛΛ are formed in a ratio of 9:1. (Scheme 5.24) The latter can be separated by chromatographic methods.

Cleavage of the resolved dimer by chloride abstraction provides a mononuclear complex where the configuration at the metal is preserved. Such compounds, with either Δ or Λ configuration, can be used as building blocks for further stereoselective coordination synthesis.

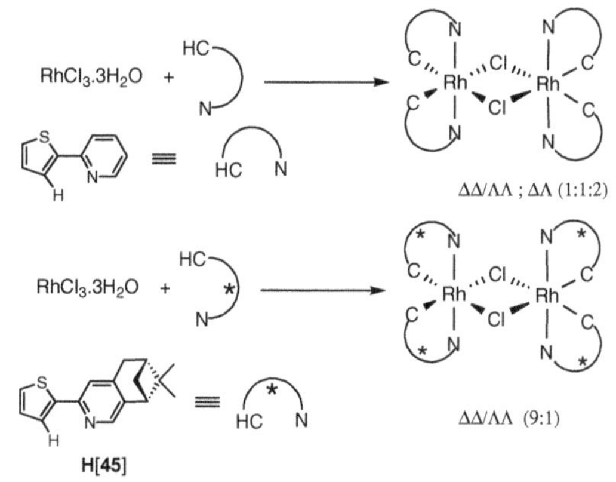

Scheme 5.24
Synthesis of the dinuclear rhodium complexes with thienylpyridine and with the related pinene version.

In a similar way the pinene-phenylpyridine **H[46]** was prepared. Thus upon treatment of 4,5-(*R*,*R*)-pinene-2-phenylpyridine **H[46]** with $IrCl_3.3H_2O$, the HEXOL-type tetranuclear iridium complexes $[Ir\{IrCl_2(46)_2\}_3]$ were obtained as two isomers (**5.67ab**) in a ratio of 5:3 (Scheme 5.25). Each complex can be described as an inner core of HEXOL-type $[Ir(IrCl_2)_3]^{6+}$ units and a surface of six chiral bidentate cyclometallated ligands (**46**).[145]

Since each metal centre can have either a Δ- or a Λ-configuration, in theory there should be eight possible stereoisomers of the tetranuclear HEXOL-type complex (Figure 5. 27). An NMR analysis and X-ray structure determination were carried out on both isomers which showed that complexes **5.67ab** acquire C_2-symmetric configuration. Furthermore, in the two isomers **5.67a** and **5.67b** the central iridium atom has a Δ-configuration while the peripheral iridium atoms Ir2, Ir3 and Ir4 have Δ-, Λ-, Λ for **5.67a** and Δ-, Λ-, Δ-configurations for **5.67b**.

The authors suggest that the stereoselectivity is complete for the central iridium atom since it acquires a Δ-configuration in both isomers of the tetranuclear Ir-HEXOL type complex. However, mixed configurations were obtained for the

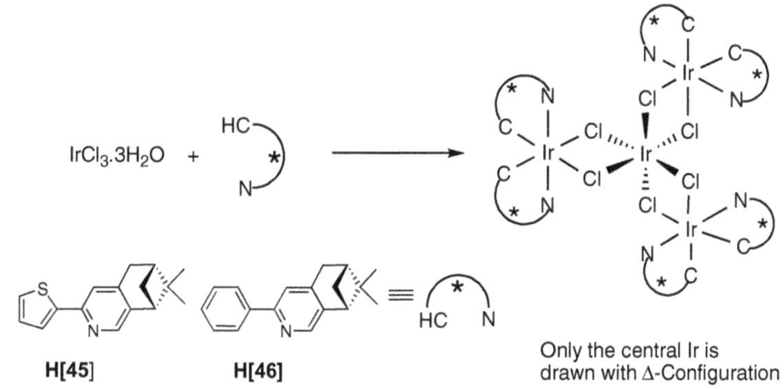

Only the central Ir is drawn with Δ-Configuration

Scheme 5.25
Synthesis of the tetranuclear Ir-HEXOL-type complexes.

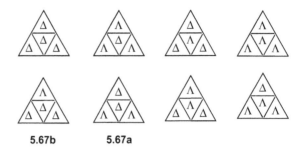

5.67b   **5.67a**

Figure 5.27
Eight possible stereoisomers of the tetranuclear Ir-HEXOL-type complexes (reference[145]).

peripheral iridium centres. When the reaction was repeated with the optically active pinene-thienylpyridine **H[45]** the two analogous isomers were obtained but in a 1:5 ratio. It is noteworthy that the achiral phenylpyridine ligand reacts with iridium trichloride hydrate to give the well-known cyclometallated dimer [Ir(ppy)$_2$Cl]$_2$, hence the formation of these HEXOL-type complexes are caused by the pinene group annellated to the pyridine ring. However, the most striking results reported by von Zelewsky were obtained with the CHIRAGEN ligands presented below.

(b) *Tetradentate CHIRAGEN type of ligands (CHIRAGEN = chiral generator)*. A remarkable property of the pinene annellated pyridine ligands is the acidity of the $-CH_2-$ group that is adjacent to the pyridine ring. The two diastereotopic protons can be stereoselectively deprotonated by strong bases. This type of reactivity opens the field for the synthesis of the so-called CHIRAGEN type of ligands. Figure 5.28 illustrates several CHIRAGEN ligands (**47–49**) obtained with different spacers.

CHIRAGEN-type ligands were used to produce self-assembled supramolecular coordination species with predetermined chirality. For instance 4,5-CHIRAGEN[*m*-xylyl] (**47**) reacted with octahedral metal ions to form homochiral-helicates of M$_2$(**47**)$_3$ stoichiometry.[146] The 5,6-CHIRAGEN-type ligands do not form any well-defined assemblies with metals that coordinate in octahedral geometry. This is related to the steric effect caused by the pinene moiety annellated to the pyridine ring. In contrast, they showed interesting coordination chemistry with tetrahedral metal ions. Thus when (−)5,6-CHIRAGEN[*p*-xylyl] (**48**) was treated with either Cu(I) or Ag(I) a circular helicate of six metal ions was formed [M$_6$(**48**)$_6$]$^{6+}$ {M = Cu(I) (**5.68**); M = Ag(I) (**5.69**)}. The X-ray

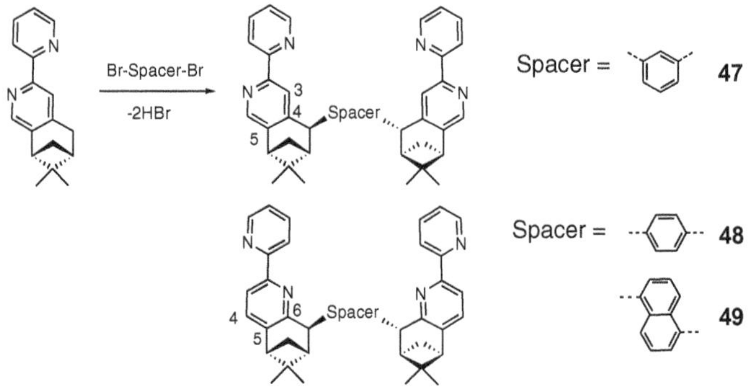

Figure 5.28
CHIRAGEN type ligands **47–49**.

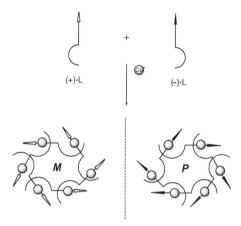

Scheme 5.26
Chiral recognition between CHIRAGEN ligands and the related assembled structures (from references[142,148]).

structures of **5.68** and **5.69** were determined and they are iso-structural and confirmed the formation of a circular, hexameric, single-stranded helicate.[147] The symmetry in the crystal is C_6, that is the sixfold rotation axis is polar. The outside diameter of the hexagon structure is about 3 nm and has a diameter of 0.84 nm. The thickness of the hexagonal disc is about 1.4 nm. Solution behaviour of these species has been investigated by [109]Ag-NMR and ^{1}H-NMR. The results suggest an equilibrium exists between the hexameric and a tetrameric species. The hexamer is largely dominant at ambient temperature and pressure. Remarkably a complete mutual recognition of the configuration of the ligands within the supramolecular assembly has occurred. For instance, starting with CHIRAGEN ligands with $(-)$-α-pinene, indicated that the handedness of the helix is *P* (Scheme 5.26).[142,148]

When the *p*-xylyl spacer in **48** was modified by 1,5-dimethylnaphtalene a new CHIRA-GEN type ligand **49** was made. This tetradentate ligand prodived a completely different supramolecular assembly (**5.70**) with Ag(I) when compared to that obtained with **48**. The solid state crystal structure of **5.70** was determined and showed the formation of an infinite double helix.[149] The helical pitch is 3.417 nm and contains five Ag(**49**) coordination units. Remarkably the chirality of the double helix is *P*. The *P* helicity combined with local Λ configuration at each metal centre was also observed in the previous example. Once again the predetermined chirality of the supramolecular assembly originates from the chirality of the enantiopure CHIRAGEN-type ligands. In solution it is believed that a hexameric circular species is formed. The authors suggest that upon crystallization a double helix is formed and the driving force behind this assembly is the π–π stacking. A mechanical model is proposed to explain the mechanism of the helix formation (see Figure 5.29).[142]

Finally the use of CHIRAGEN-type of ligands to construct stereoselectively molecular knots was achieved with success (Figure 5.30). Thus treatment of the CHIRAGEN type ligand **50** with Cu(I) forms a dinuclear double helix **5.71** with predetermined configuration Λ at the individual copper centres. A ring-closing metathesis (RCM) reaction is then applied and subsequent hydrogenation leads to the chiral dicopper knotted complex **5.72**. The CD spectrum of **5.72** is very similar to that of the double helix **5.71**.[65]

Demetallation of the dicopper knotted complex **5.72** occurs by treatment with KCN to give the trefoil **5.73** which acquires a topological chirality. The CD curve of this species was recorded and shows a pronounced signal in the range 280–340 nm. Unfortunately no X-ray structures could be determined for any of three compounds **5.71**–**5.73**.

Figure 5.29
Mechanical model showing the formation of a double helix from a hexamer upon crystallization (reproduced from reference[142], copyright 2003, Elsevier).

The use of chiral resolved ligands to produce supramolecular assemblies with predetermined configuration was also investigated by Lin and coworkers (see the following paragraph).

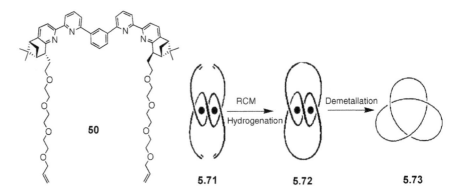

Figure 5.30
Strategy for the construction of the metallated knot **5.72** and the trefoil knot **5.73** using the CHIRAGEN type ligand **50** (adapted with permission from reference[65]).

5.2.1.2 Chiral Molecular Polygons

The use of chiral bridging ligands to construct chiral polygons was thoroughly investigated by Lin and coworkers.[150,151] Their approach consisted of using enantiopure *rigid atropisomeric* bridging ligands and appropriate metal connectors as the building blocks. Further, by taking advantage of the kinetic inertness of the robust Pt-alkynyl linkage, they were able to describe the efficient one-pot self-assembly of chiral metallo-macrocycles ranging from triangles to octagons by using *trans*-Pt(PEt$_3$)$_2$ as a linear metal linker. The enantiopure ligand is 6, 6'-diacetyl-1, 1'binaphthyl-6,6'-bis(ethyne) H$_2$(**51**), which can be prepared in several steps from the commercially available 1,1'-bi-2-naphthol (BINOL).

Upon treatment of the (*R*)-ligand H$_2$(**51**) with one equivalent of *trans*-Pt(PEt$_3$)$_2$Cl$_2$, in the presence of a catalytic amount of CuCl in CH$_2$Cl$_2$ and HNEt$_2$ at room temperature, the related (*R*)-polygons [(*R*)-*trans*-Pt(Et)$_2$(**51**)]$_n$ (*n* = 3–8, **5.74–5.79**) of different sizes were formed (Scheme 5.27). These metallo-macrocycles were separated through

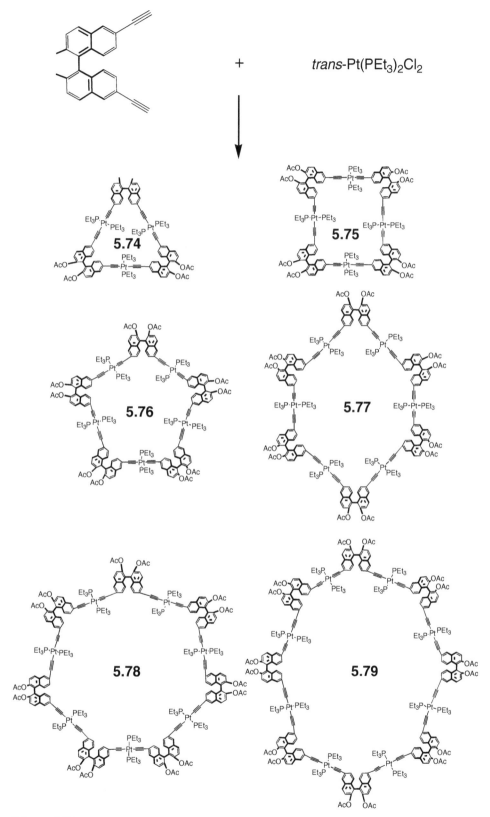

Scheme 5.27
Self-assembly of chiral molecular polygons (reproduced with permission from reference[150], copyright 2003, American Chemical Society).

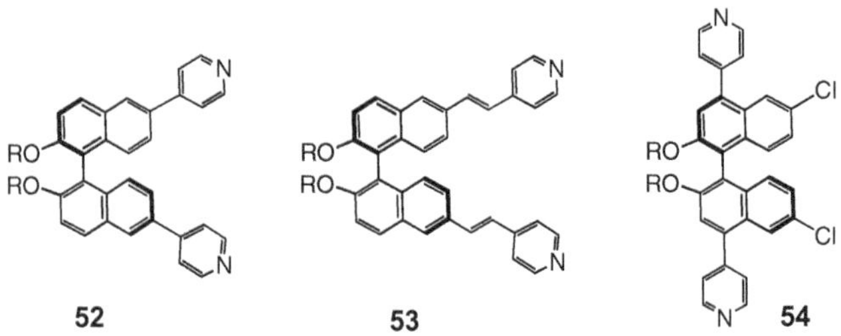

Figure 5.31
Angular bipyridine ligands (**52–54**) based on a 1,1' binaphthyl framework.

column chromatography but the yields were low and varied from 5–18% depending on the size of the macrocycle. The X-ray molecular structure of one of these chiral tetrametallic-macrocycles **5.75** was determined and confirmed the formation of a single isomer.

The chirality of the macrocycles **5.74–5.79** was also confirmed by recording their CD spectra. Interestingly the CD signals of the assembly increased steadily as the size of the metallo-cycle increased. The authors believe that the limited flexibility of the bridging ligand is the key to the facile one-pot self-assembly of the chiral molecular polygons. On the other hand, Lin and coworkers took advantage of the chiral cavity displayed by such assemblies to carry out some enantioselective sensing and asymmetric catalysis.[152,153]

The authors extended their method to design other chiral bridging ligands. Thus several angular bipyridine ligands **52–54** were prepared derived from the 1,1' binaphtyl framework (Figure 5.31). These ligands were successfully used with metallo-corners such as [M(dppe)]$^{2+}$ [M = Pd, Pt; dppe = bis(diphenylphophino)ethane)] or 'ClRe (CO)$_3$' to form chiral metallo-macrocycles.[152,154]

It is worth noting that the chiral metallo-macrocycles assembled with these angular bipyridine ligands were less robust than the previous ones formed from the strong Pt-alkynyl linkage. This is due to the lability of the M-pyridyl bonds.

More recently Lin and coworkers extended this methodology to the preparation of chiral mesoscopic metallo-cycles with controllable size. In this way large homochiral metallocycles with up 38 bridging ligands and 38 *trans*-Pt(Et$_3$)$_2$ centres, and with cavities as large as 22 nm in diameter, were obtained. The novelty of their approach consisted in extending the lengths of the oligomeric building blocks such as L$_m$[Pt(Et$_2$)]$_{m+1}$Cl$_2$ ($m = 1, 2, 3, 5, 7, 11, 19$ and 31) and [Pt(Et$_2$)]$_n$L$_{n+1}$Cl$_2$ ($n = 1, 2, 3, 5, 6, 10, 18$ and 30) (Scheme 5. 28). Subsequent cyclization of the different lengths of such building blocks provided the nanoscopic and mesoscopic chiral molecular polygons (Scheme 5.29).[155] These large macrocycles were characterized by spectroscopic techniques ^{1}H{^{31}P}, ^{31}P{^{1}H} and ^{13}C{^{1}H} NMR and MALDI-TOF MS characterization. Unlike the oligomeric starting material, the NMR spectra of the chiral molecular polygons exhibited a single ligand environment consistent with the formation of cyclic species of D_n symmetry. Furthermore, the X-ray molecular structure the *meso*-metallocycle (*RSRS*)-[*trans*-(PEt$_3$)$_2$Pt (L-OEt)]$_4$ was determined.

Lin and coworkers extended their method to the synthesis of chiral coordination polymers based on the use of rigid bridged ligands derived from the 1,1' binaphtyl framework. Thus homochiral porous coordination networks were prepared; however this work will be reviewed in the next chapter of this book.[156]

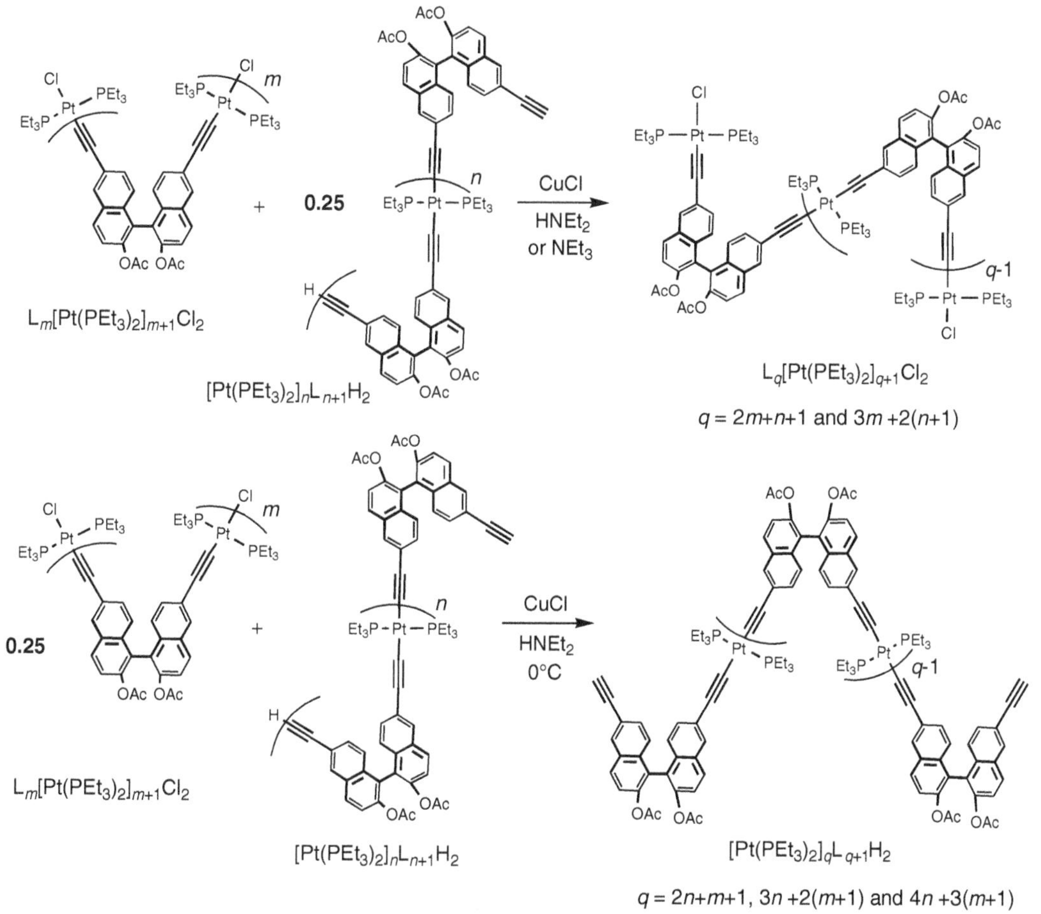

Scheme 5.28
Preparation of the chiral oligomeric building blocks (reprinted from reference[155], copyright 2003, American Chemical Society).

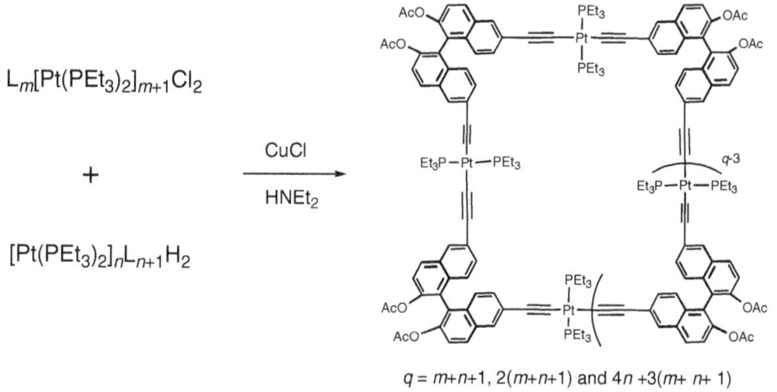

Scheme 5.29
Self-assembly of the large homochiral polygons (from reference[155]).

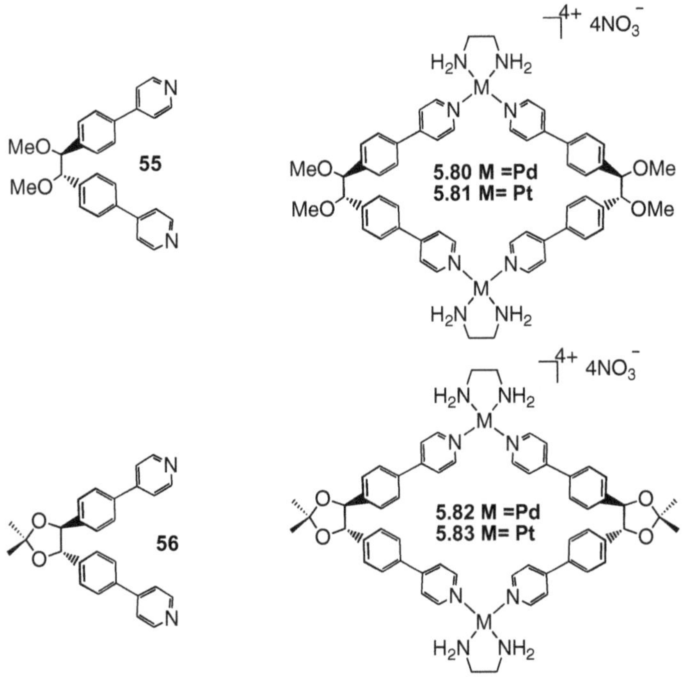

Figure 5.32
Angular bipyridine ligands **55** and **56** with chiral backbones and the related enantiopure metallo-rhombs **5.80–5.81** and **5.82–5.83**.

On the other hand, use of the angular bridging bipyridine ligands with chiral back-bones **55** and **56** was also employed by other groups to prepare chiral supramolecular rhombs **5.80** and **5.81**. Based on the Fujita procedure,[157] the bipyridine bridging ligand – either **55** or **56** was treated with (en)M(NO$_3$)$_2$ (M = Pd, Pt) to give the related supramolecular rhombs **5.80–5.81** and **5.82–5.83** (Figure 5.32). Starting with the (S,S)-bipyridine ligand the related (S,S,S,S)-metallo-rhombs were obtained.[158] No X-ray molecular structure was reported. The identity of the supramolecular species was ascertained by ESI-FT-ICR mass spectra combined with ^{1}H-NMR spectroscopy. It is believed that in solution these metallo-rhombs exist in equilibrium between a major [2:2] macrocycle with a minor species of < 10% [3:3] of macrocycle. The CD spectra showed opposite curves for (S) and the (R) enantiomers. Interestingly upon deposition of these enantiomeric rhombs on a Cu(100) electrode, a chirality on the global level was observed as confirmed by STM images. This trend originated from the flat orientation of the adsorbed enantiopure macrocycles.

To achieve discrete chiral supramolecular assemblies another approach was used based on chiral auxiliaries. This approach was investigated thoroughly by Stang and coworkers, as presented in the following section.

5.2.2 Chirality Transfer via a Resolved Chiral Auxiliary Coordinated to a Metal or the Use of Resolved Metallo-bricks

Stang has proposed five ways of creating chiral supramolecular species via spontaneous self-assembly: (1) use of chiral auxiliary coordinated to a metal, (2) use of an inherently

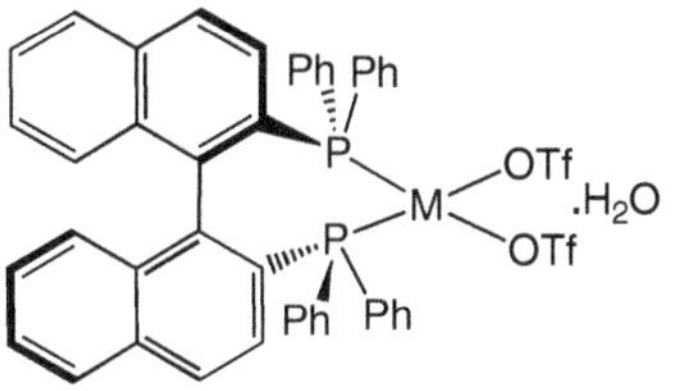

Figure 5.33
(From references[160,161]).

chiral octahedral metal complex, (3) use of an optically active atropisomeric diaza-bisheterocycle (substituted 4′,4′-bipyridines, bisquinolines, etc.) as linker ligands, (4) helicity or twist due to the use of diaza ligands which lack a rotation symmetry about their linkage axis, and (5) a combination of the above methods.[159] Stang's group used the first approach which consisted of a first step of reacting the achiral metallo-corners [L$_2$M][OTf]$_2$ M = Pd, Pt with optically pure auxiliaries such as *R*-(+)-BINAP which leads to the formation of chiral metal containing building units [M(*R*-(+)-BINAP)(H$_2$O)] [OTf]$_2$ M = Pd (**5.84**), Pt (**5.85**) (Figure 5.33). Upon reacting these chiral building blocks with various bridging ligands a variety of chiral supramolecular assemblies were successfully prepared.[160,161]

5.2.2.1 Chiral Squares and Polyhedra

When the reactive chiral building unit [M(*R*-(+)-BINAP)(H$_2$O)][OTf]$_2$ with M = Pd (**5.84**) or M = Pt (**5.85**) was treated with the bridging ligand bis [4-(4′-pyridyl)phenyl] iodonium triflate (**57**), the related chiral hybrid supramolecular squares were formed (**5.86**–**5.87**) (hybrid = presence of iodonium and transition metal corners in the assembly). These hybrid molecular squares are optically active due to the chiral transition metal auxiliary (BINAP) in the assembly. Interestingly chiral induction was observed in these metallo-macrocycles since the preferred diastereomer in each preparation was formed in excess compared with the other isomers (Figure 5.34).[162]

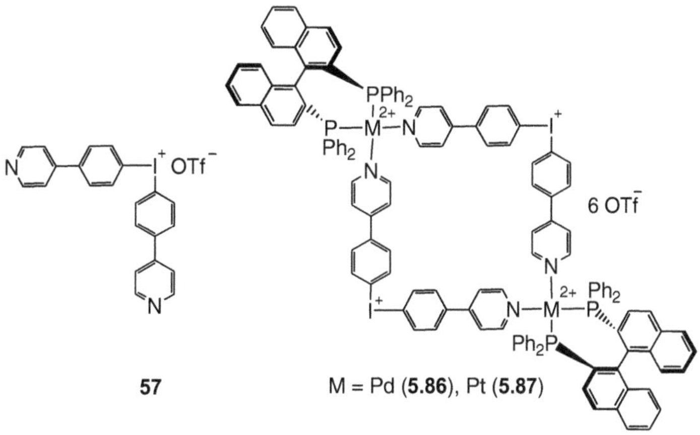

Figure 5.34
Formation of the chiral hybrid supramolecular squares (from reference[162]).

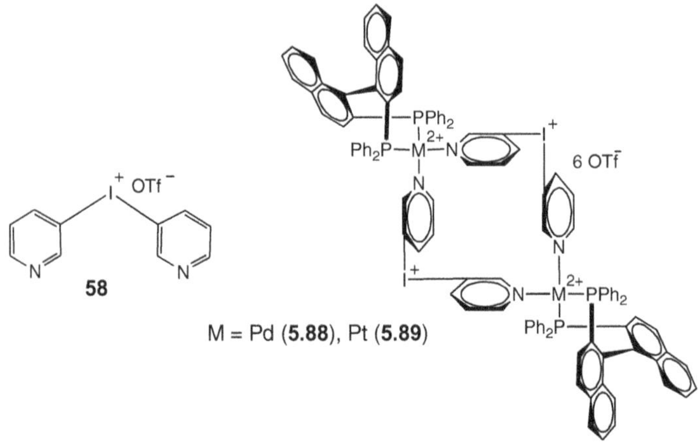

Figure 5.35
The hybrid chiral molecular squares **5.88** and **5.89** obtained with bis(3-pyridyliodonium triflate (**58**) (reproduced from reference[162], copyright 1996, American Chemical Society).

By changing the symmetry and nature of the bridging iodonium ligand novel hybrid supramolecular squares were prepared with elements of helicity (twist). For instance the bis(3-pyridyl)iodonium triflate (**58**) lacks the rotation symmetry about its linkages, that is the nitrogen lone pair and the carbon iodine bond. Reaction of the bisphosphine complexes **5.84–5.85** with the iodonium bridging ligand (**58**) afforded the expected metallo-macrocycles with the formation of the preferred diastereomers of [Pd(R)-(+)-BINAP)(**58**)$_2$][OTf]$_6$ (**5.88**) and [Pt(R)-(+)-BINAP)(**58**)$_2$][OTf]$_6$ (**5.89**) as confirmed from the ^{31}P{^{1}H} NMR analysis (Figure 5.35). In the case of the Pt-metallo-macrocyle (**5.89**), a significant amount of the other diastereomers were formed as well. The authors attributed this result to the relative conformational flexibility of the iodonium moiety. In this example, only a modest degree of asymmetric induction is achieved.

It is worth mentioning that the reaction of **58** with the achiral diphosphine ligands [Pd(*cis*-Et$_3$P)$_2$][OTf]$_2$ and [Pt(*cis*-Et$_3$P)$_2$][OTf]$_2$ under the above self-assembly conditions provided a diastereomeric mixture of products. In theory, six possible isomers could be formed (Figure 5.36). In this example the chirality of the metallo-squares arises from the chiral conformation of the iodonium moiety in the assembly and the restricted rotation of the pyridine ligands around the M–N bond.

At this stage a comment on the origin of the chirality of the hybrid squares obtained with achiral phosphine is required. Structures I and II are chiral but enantiomers, squares III and IV are achiral while V and VI are chiral but also enantiomers of each other. The chirality in I, II, V and VI arises only from the *chiral conformation* of the ligands in the assemblies. Hence one might also expect to form chiral assemblies just as a result of a *chiral conformation* of the bridging ligand, which could be considered as a sixth approach to the five possibilities presented above by Stang for creating chiral supramolecular species.

Stang extended this approach to prepare chiral supramolecular squares featuring only chiral metal corners. Thus the C_{2h}-symmetrical diaza bridging ligands, 2,6-diazaanthracene (DAA, **59**) and 2,6-diazaanthracene-9,10-dione (DAAD, **60**) were used (Figure 5.37).[162]

Upon treatment of [M(R)-(+)-BINAP)(H$_2$O)][OTf]$_2$, M = Pd (**5.84**) or M = Pt (**5.85**) with DAA (**59**), a single diastereomer of each square of [Pd(R)-(+)-BINAP)(DAA)]$_4$[OTf]$_8$

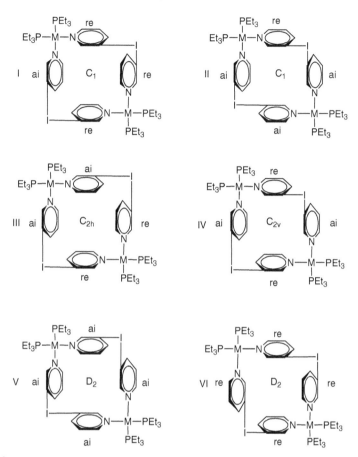

Figure 5.36
The six possible isomers self-assembled from bis(3-pyridyliodonium triflate (**58**) and achiral bisphosphine [M(*cis*-Et$_3$P)$_2$][OTf]$_2$, M = Pd, Pt (from reference[162]).

(**5.90**) or of [[Pt(*R*)-(+)-BINAP)(DAA)]$_4$[OTf]$_8$ (**5.91**) was formed as confirmed by ^{31}P NMR analysis. The X-ray molecular structure of this isomer was determined and its absolute configuration was assigned a D_4 symmetry (Figure 5.38).[163] In contrast, when the DAAD was used as the assembling ligand, a major diastereomer was formed along with minor amounts of other diastereomers as suggested by ^{31}P-NMR spectra of [Pd(*R*)-(+)-BINAP)(DAAD)]$_4$[OTf]$_8$ (**5.92**) and [[Pt(*R*)-(+)-BINAP)(DAAD)]$_4$ [OTf]$_8$ (**5.93**).

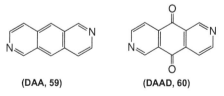

(DAA, 59)   (DAAD, 60)

Figure 5.37
The C_{2h}-symmetrical diaza bridging ligands DAA and DAAD.

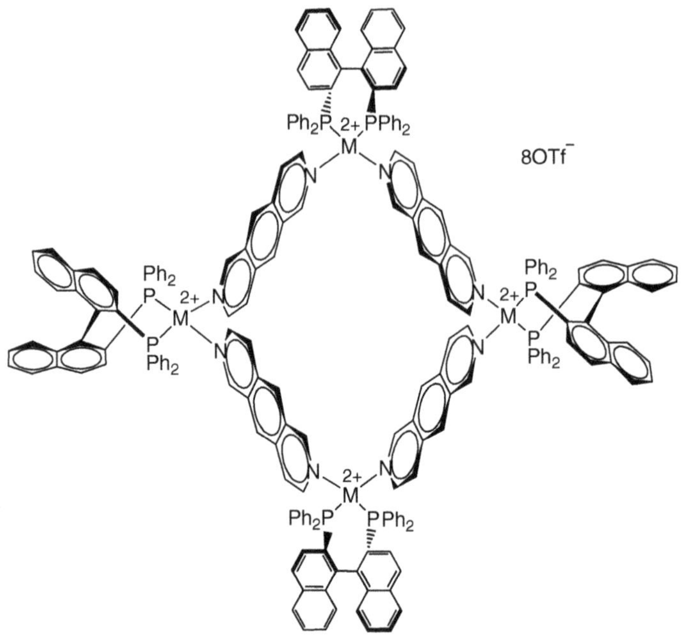

Figure 5.38
Chiral squares of [M(*R*)-(+)-BINAP)(DAA)]₄[OTf]₈ (M = Pd, **5.90**; M = Pt, **5.91**) obtained with the DAA connector (**59**) (reproduced with permission from reference[163]).

In comparison to the hybrid squares [Pd(*R*)-(+)-BINAP)(**58**)₂][OTf]₆ (**5.88**) and [Pt(*R*)-(+)-BINAP)(**58**)₂][OTf]₆ (**5.89**), the all metal squares were formed in significantly greater diastereomeric excesses and were conformationally much more rigid. This was confirmed by VT ¹H-NMR studies carried on these chiral squares with only metal corners. In brief the C_{2h}-symmetrical connector ligands are locked into a chiral conformation in the square assemblies and must be restricted in rotation due to the presence of two – *not* just one – metal corners. As expected the use of the opposite chiral metallo-corners based on the (*S*)-(−)-BINAP ligand allowed the preparation of the opposite enantiomers of chiral molecular squares. The CD spectra of these chiral species confirmed the expected enantiomeric relationship.

Stang and coworkers extended their approach in the preparation of chiral supramolecular polygons to that of three-dimensional chiral polyhedra. To this end the tridentate connector ligand tris(1,3,5-(4′-pyridyl)ethynyl)benzene (**61**) was prepared. Upon treatment of **61** with the chiral metal bistriflates [M(*R*)-(+)-BINAP)(H₂O)][OTf]₂, M = Pd (**5.84**) or M = Pt (**5.85**) a highly symmetrical entity with a stoichiometry of 6:4 was formed (Figure 5.39). Analysis and spectroscopic data, as well as force-field simulations, suggested that the supramolecular assembly obtained is a macrocyclic polyhedron with six metal vertices {[M(*R*)-(+)-BINAP)]₆(**61**)₄}[OTf]₁₂ M = Pd (**5.94**) and M = Pt (**5.95**).[159] Such chiral assemblies are rare examples of molecules belonging to the T-symmetrical point group which are found only for few organic molecules.[164,165] It is worth mentioning that similar achiral assemblies were reported by Fujita and coworkers using 2,4, 6 tris-(4-pyridyl)1-3,5-triazene) ligand and Pd(en)(NO₃)₂ as metal corners. The structure of the latter was ascertained by X-ray analysis.[166]

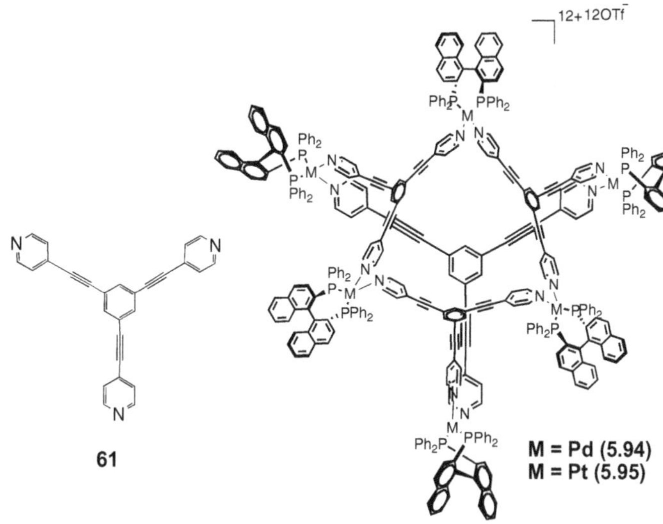

Figure 5.39

Formation of the chiral three-dimensional polyhedrons $\{[M(R\text{-}(+)\text{-}BINAP)]_6(\mathbf{61})_4\}[OTf]_{12}$ M = Pd (**5.94**) and M = Pt (**5.95**) using the tridentate connector **61** (from reference[159]).

The examples presented in this section illustrate an elegant approach by Stang and coworkers to the design of chiral supramolecular polygons (squares, cages) and also chiral three-dimensional polyhedrons. The combination of tailored ligand connectors with the metal corners derived from the BINAP framework should allow the preparation of a variety of different supramolecular chiral objects.[160]

Another approach which uses chiral metal ions for the preparation of chiral supramolecular species is the use of chiral octahedral tris-chelated metal centre with achiral multidentate ligands as connectors; this research approach is presented in the next section.

5.2.2.2 Polynuclear Systems with Inherently Chiral Octahedral (Tris-chelated) Metal Subunits

Octahedral hexacoordinated complexes of the type $[M(L\text{-}L)_3]$ M = Ru(II), Os(II) and L-L = diimine ligand (Figure 5.40) have received considerable attention because of (a) their potential use as chiral probes for biological molecules (polynucleotides, DNA),[167,168] and (b) their photoluminescence and electroluminescence properties especially ruthenium, making them promising light emitting devices.[169,170] More recently efforts were devoted to use these complexes as chiral building blocks for stereoselective synthesis of polynuclear supramolecular assemblies. Multimetallic centre complexes prepared from racemic material often give a mixture of products with complex NMR spectra, and hence rendering their characterization a very difficult task. In contrast, polynuclear complexes prepared from resolved mononuclear constituents, often give rise to simple NMR signals. Furthermore, the simple synthetic procedures allow access to enantiomerically pure mononuclear building blocks such as Δ– /Λ– [Ru(bpy)$_2$(py)$_2$]$^{2+}$ [171–175] (**5.96**) or Δ–/Λ– [Ru(Phen)$_2$(py)$_2$]$^{2+}$ [176–179] (**5.97**) and Δ–/Λ– [Ru(Mebpy)$_2$(CO)$_2$]$^{2+}$ [180,181] (**5.98**) making them attractive starting material for the preparation of chiral supramolecular assemblies. Moreover, due to configurational

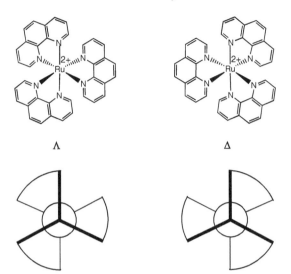

Figure 5.40
Representation of the Δ- and Λ-configuration of an octahedral $[Ru(L-L)_3]^{2+}$ cation with D_3 symmetry; L-L = phenanthroline.

stability of the metal centre in these mononuclear building blocks, it is expected that only a single diastereomer will be formed.

One approach to the formation of enantiomerically pure assemblies consists in functionalizing the ligand periphery of the chiral building blocks which ultimately provide a way to bridge the monomers. For instance Lehn and Torr have shown halogen and/or ethynyl functional groups on chiral $[Ru(phen)_3]^{2+}$ core can form optically pure assemblies with phenylene and acetylene bridges (Figure 5.41).[182,183]

Furthermore, Lehn's group used a convenient method to form rigid polydentate ligands with extended aromatic rings via condensation of *o*-diamines and *o*-quinones functions on the periphery of the monomer starting materials. These ligands reacted with enantiopure $[Ru(Phen)_2(phendione)]^{2+}$ to give stereoisomerically controlled dinuclear and polynuclear ruthenium architectures (Figure 5.42).[184,185]

In a similar approach MacDonnell and coworkers developed the peripheral functionalization of phenanthroline ligands, 1,10-phenanthroline-5,6-dione (phendione) and 1,10-phenanthroline-5,6-diamine (phendiamine) which can undergo a condensation reaction while coordinated to form a tpphz bridge (tpphz = tetrapyrido[3,2-*a*:2′3′-*c*:3″,

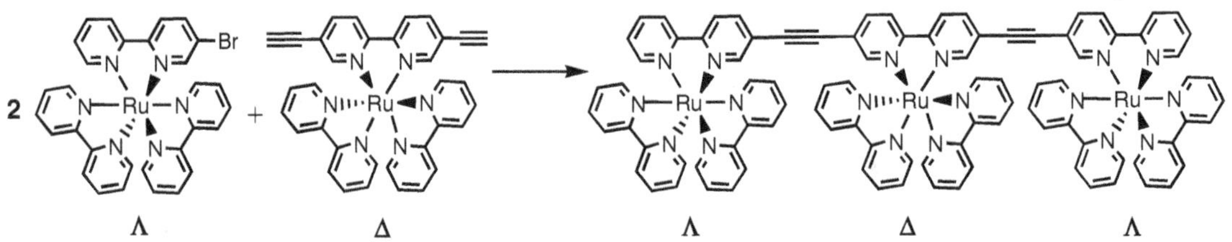

Figure 5.41
Formation of a supramolecular assembly using peripheral acetylene bridges (from reference[182]).

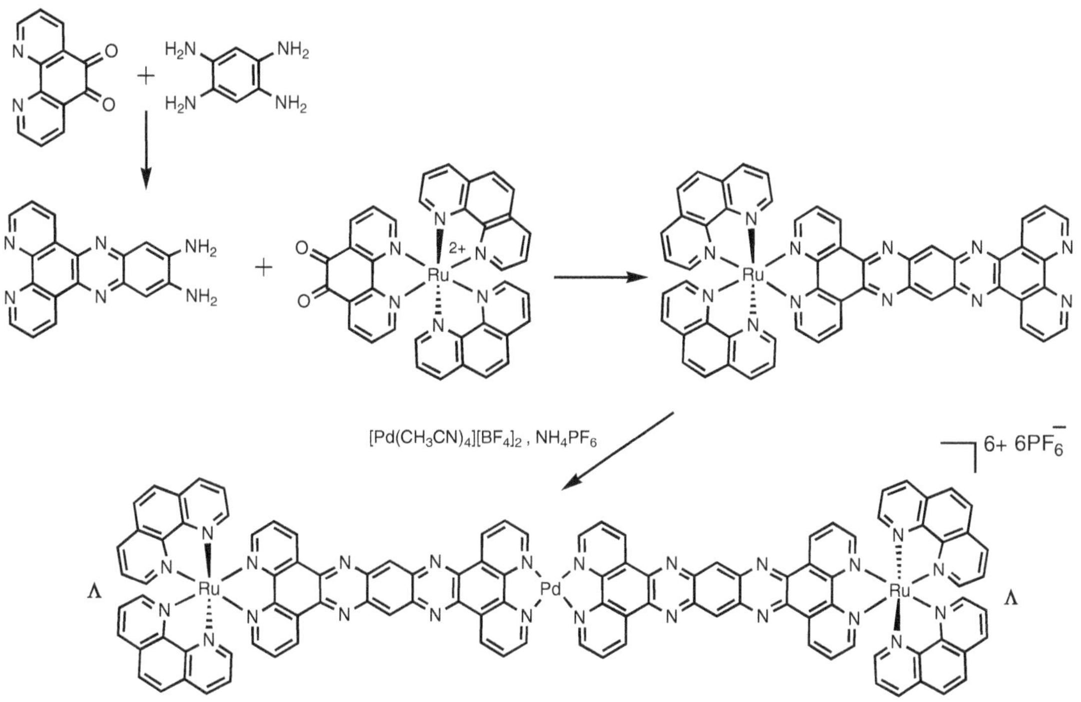

Figure 5.42
Formation of stereoisomerically controlled architectures (from reference[184]).

2″-*h*:2″,3″-*j*]phenazine) between the monomers. Following this method, they were able to prepare several chiral ruthenium dimers with controlled configuration at the metal centre.[186] The chiral nature of the enantiomers ΔΔ and ΛΛ was confirmed by CD spectra.[187] In particular the formation of the tetranuclear ruthenium assembly ΔΛΛΛ-Ru₄ (**5.101**) was successfully made from three Λ components of [(Phen)₂Ru (phendiamine)]²⁺ (**5.99**) and one Δ component of [Ru(phendione)₃]²⁺ (**5.100**) via a condensation reaction between the two complementary building blocks (Figure 5.43). Similarly, starting with one Λ component of **5.100** and three Δ of **5.99**, the other enantiomer ΛΔΔΔRu₄ (**5.101**) was formed. The CD spectra of both assemblies were determined and confirmed the enantiomeric relationship.[173]

The second more common approach to the preparation of enantiopure assemblies consists of using the Δ– or Λ- *cis*-[Ru(diimine)₂L₂] (**5.96–5.98**) as chiral building blocks. Then, when the latter is treated with multidentate ligands and subsequent displacement of monodentate labile ligands L occurs, the related supramolecualr assemblies are formed. It should be pointed out that such reactions often occur at relatively high temperatures in order to displace the monodentate ligands; in some cases this results in low yields of the polynuclear products.

Keene and coworkers made the chiral trinuclear ruthenium complex [(Ru(Phen)₃(**62**)] (**5.102**) by treatment of the hexadentate bridging ligand HAT (**62**) (HAT = 1,4,5,8,9,12-Hexaazatriphenylene) with three equivalents of the Δ–/Λ–[Ru(Phen)₂(CO)₂]²⁺ building block (Figure 5.44). Keene and coworkers investigated this approach thoroughly and reported the synthesis of various chiral supramolecular assemblies.[188–190]

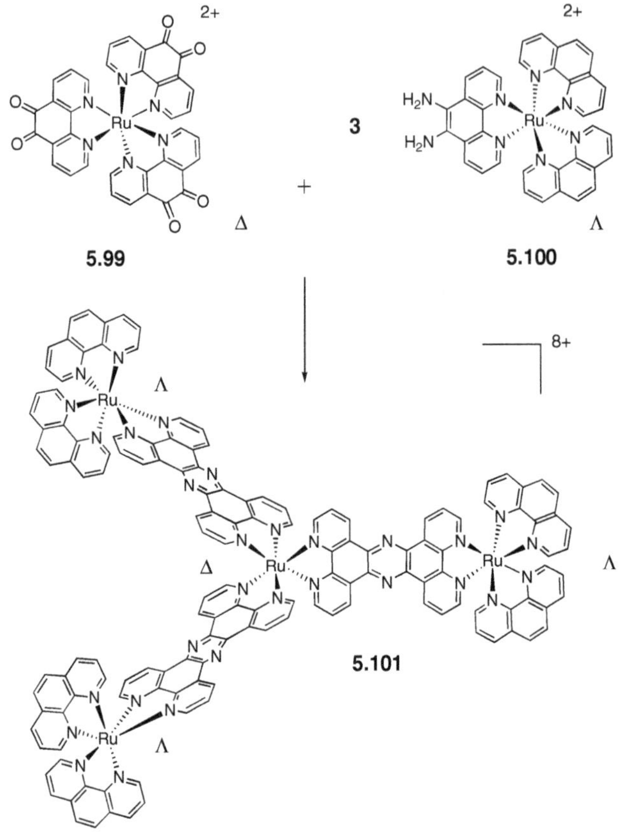

Figure 5.43
Formation of the tetranuclear assembly $\Delta\Lambda\Lambda\Lambda$-Ru$_4$ (**5.101**) with tpphz bridge (from reference[173]).

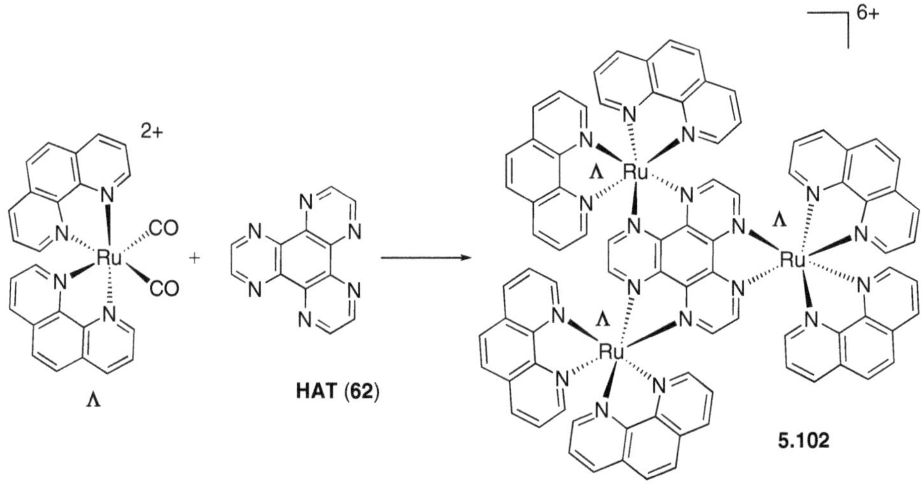

Figure 5.44
Self-assembly of the trinuclear ruthenium complex $\Lambda\Lambda\Lambda$Ru$_3$ (**5.102**) with HAT bridging ligand **62** (from reference[189]).

More recently and following the above procedure Elsenbaumer and coworkers reported the synthesis of homochiral multinuclear ruthenium complexes from oligomeric benzimidazoles as bridging ligands and enantiopure mononuclear $[Ru(bpy)_2(py)_2]^{2+}$ (**5.96**) as a building block.[191]

Reaction of the enantiopure building block Λ-[Ru(bpy)$_2$(py)$_2$][(−)-O,O'-dibenzoyl-L-tartrate].12H$_2$O[171] (**5.96**) with the oligomeric benzimidazole bridging ligand (**63**) and subsequent anion metathesis with NH$_4$PF$_6$ allowed the preparation of homochiral multinuclear ruthenium complexes with retention of configuration. The X-ray structure of the tetra-ruthenium complex ($\Lambda\Lambda\Lambda\Lambda$)-[(Ru(bpy)$_2$)$_4$(bis(BiBzIm))][PF$_6$]$_4$ (**5.103**) was determined and revealed that all four metal centres had Λ-configuration (Figure 5.45).

It is expected that this approach will continue to be explored and more appealing chiral nonracemic supramolecular assemblies will be prepared, due to the availability of enantiopure mononuclear building blocks, which are configurationally stable, and the possibility of designing different geometrical bridging ligands that are only limited by the boundless imagination of a synthetic chemist.

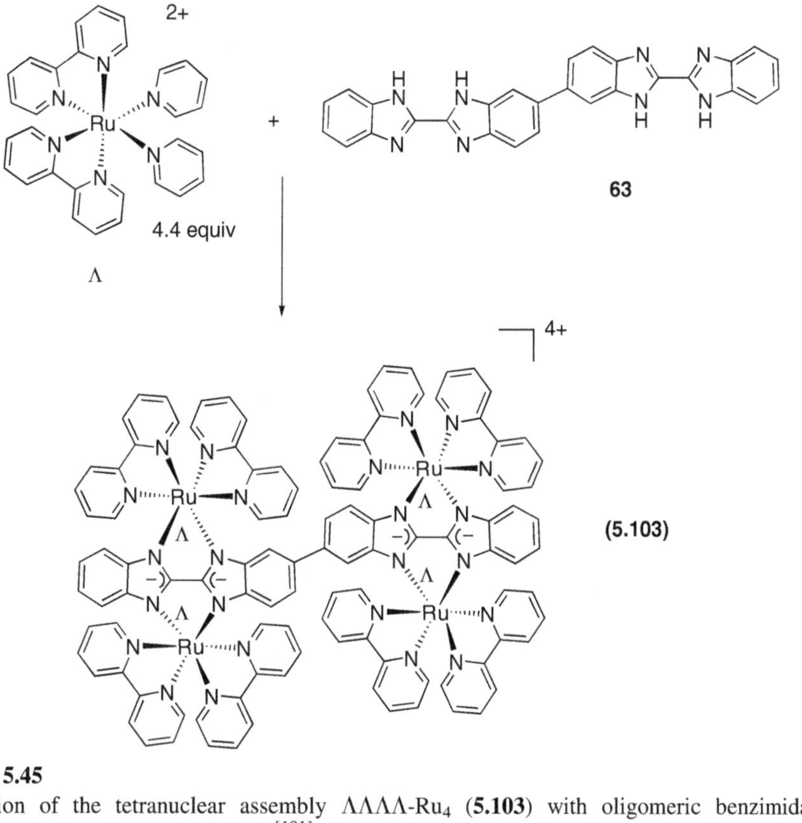

Figure 5.45
Formation of the tetranuclear assembly $\Lambda\Lambda\Lambda\Lambda$-Ru$_4$ (**5.103**) with oligomeric benzimidazole bridging ligand (**63**) (from reference[191]).

Summary

Throughout this chapter we have tried to cover the different methods described in the literature to prepare chiral supramolecular assemblies. In the first part of the chapter we described the use of achiral bridging ligands of the right match that bind to the transition metal components of different geometry to give chiral supramolecular architectures where the chirality arises only from the asymmetric disposition of these achiral constituents in three-dimensional space. Many examples were reported and this is attributed to the simple access of the reactants where there is no need to prepare beforehand the optically pure constituents.

In the second part of the chapter we presented two different methods based on the use of enantiopure constituents for the assembly of chiral nonracemic supramolecular assemblies; one could also define it as 'enantioselective supramolecular assembly'. The first method consisted of the preparation of resolved bridging ligands which, upon coordination to the metal centre, imposed a certain configuration and ultimately a single supramolecular diastereomer was expected to be formed. In the second approach, the chirality is introduced at the metal centre either by using a chiral auxiliary (and hence the transition metal building blocks becomes chiral nonracemic) or by using inherently chiral metal building blocks, which up to the present time is restricted to octahedral metal complexes with chelating ligands. Upon combing the chiral metal constituents with different bridging ligands, this should often lead to the formation of enantiopure supramolecular architectures. It is not surprising that there are not many examples of supramolecular assemblies obtained from resolved constituents in the literature due to the difficulty in finding and resolving the constituents first before using them in the assembly process.

References

[1] C. R. Calladine and H. R. Drew, *Understanding DNA: The Molecule and How it Works*, Academic Press, New York, 2nd edn, **1997**.

[2] M. G. Rossmann, E. Arnold, J. W. Erickson, E. A. Frankenberger, J. P. Griffith, H. J. Hecht, J. E. Johnson, G. Kamer, M. Luo, A. G. Mosser, R. R. Rueckert, B. Sherry and G. Vriend, *Nature* **1985**, *317*, 145.

[3] E. Arnold and M. G. Rothmann, *Acta Crystallog., Sec. A* **1988**, *44*, 270.

[4] J. Badger, I. Minor, M. J. Kremer, M. A. Oliveira, T. J. Smith, J. P. Griffith, D. M. Guerin, S. Krishnaswamy, M. Luo, M. G. Rossmann, M. A. McKinlay, G. D. Diana, F. J. Dutko, M. Fancher, R. R. Rueckert and B. A. Heinz, *Proc. Natl. Acad. Sci. USA* **1988**, *85*, 3304.

[5] A. von Zelewsky, *Stereochemistry of Coordination Compounds*, John Wiley & Sons, Ltd, Chichester, **1996**.

[6] G. Seeber, B. E. F. Tiedemann, and K. N. Raymond, *Top. Curr. Chem.* **2006**, *265*, 147.

[7] T. D. Hamilton and L. R. MacGillivray, *Cryst. Growth & Design* **2004**, *4*, 419.

[8] J. W. Steed and J. L. Atwood, *Supramolecular Chemistry*, John Wiley & Sons, Ltd, Chichester, **2000**.

[9] J. M. Lehn, A. Rigault, J. Siegel, J. Harrowfield, B. Chevrier and D. Moras, *Proc. Natl. Acad. Sci. USA* **1987**, *84*, 2565.

[10] R. Kramer, J. M. Lehn and A. Marquis-Rigault, *Proc. Natl. Acad. Sci. USA* **1993**, *90*, 5394.

[11] J. M. Lehn, *Supramolecular Chemistry*, Wiley-VCH Verlag GmbH, Weinheim, **1995**.

[12] E. C. Constable, *Comp. Supramol. Chem.* **1996**, *9*, 213.

[13] D. L. Caulder and K. N. Raymond, *Angew. Chem. Int. Ed.* **1997**, *36*, 1440.

[14] C. Piguet, G. Bernardinelli and G. Hopfgartner, *Chem. Rev.* **1997**, *97*, 2005.

[15] M. Albrecht, *Chem. Rev.* **2001**, *101*, 3457.

[16] J. A. Johnson, J. W. Kampf and V. L. Pecoraro, *Angew. Chem. Int. Ed.* **2003**, *42*, 546.

[17] A. D. Cutland-Van Noord, J. W. Kampf and V. L. Pecoraro, *Angew. Chem. Int. Ed.* **2002**, *41*, 4667.

[18] C. J. Cathey, E. C. Constable, M. J. Hannon, D. A. Tocher and M. D. Ward, *J. Chem. Soc., Chem. Commun.* **1990**, 621.

[19] E. C. Constable, M. D. Ward and D. A. Tocher, *J. Am. Chem. Soc.* **1990**, *112*, 1256.

[20] E. C. Constable, A. J. Edwards, M. J. Hannon and P. R. Raithby, *J. Chem. Soc., Chem. Commun.* **1994**, 1991.

[21] E. C. Constable and R. Chotalia, *J. Chem. Soc., Chem. Commun.* **1992**, 64.

[22] E. C. Constable, M. J. Hannon, A. Martin, P. R. Raithby and D. A. Tocher, *Polyhedron* **1992**, *11*, 2967.

[23] R. C. Scarrow, D. L. White and K. N. Raymond, *J. Am. Chem. Soc.* **1985**, *107*, 6540.

[24] C. J. Carrano and K. N. Raymond, *J. Am. Chem. Soc.* **1978**, *100*, 5371.

[25] M. Albrecht and S. Kotila, *Chem. Commun.* **1996**, 2309.

[26] T. Kreickmann, C. Diedrich, T. Pape, H. V. Huynh, S. Grimme and F. E. Hahn, *J. Am. Chem. Soc.* **2006**, *128*, 11808.

[27] R. W. Saalfrank, A. Dresel, V. Seitz, S. Trummer, F. Hampel, M. Teichert, D. Stalke, C. Stadler, J. Daub, V. Schunemann and A. X. Trautwein, *Chem. Eur. J.* **1997**, *3*, 2058.

[28] R. W. Saalfrank, V. Seitz, D. L. Caulder, K. N. Raymond, M. Teichert and D. Stalke, *Eur. J. Inorg. Chem.* **1998**, 1313.

[29] C. Piguet, G. Bernardinelli, B. Bocquet, O. Schaad and A. F. Williams, *Inorg. Chem.* **1994**, *33*, 4112.

[30] L. J. Charbonniere, G. Bernardinelli, C. Piguet, A. M. Sargeson and A. F. Williams, *J. Chem. Soc., Chem. Commun.* **1994**, 1419.

[31] L. J. Charbonniere, A. F. Williams, C. Piguet, G. Bernardinelli and E. Rivara-Minten, *Chem. Eur. J.* **1998**, *4*, 485.

[32] C. Provent and A. F. Williams, *Persp. Supramol. Chem.* **1999**, *5*, 135.

[33] A. von Zelewsky and J. Nachbaur, *personal communication*.

[34] A. von Zelewsky, *Stereochemistry of Coordination Compounds*, John Wiley & Sons, Ltd, Chichester, **1996**, p. 193.

[35] C. Piguet and J.-C. Bunzli, *Chimia* **1998**, *52*, 579.

[36] C. Piguet, C. Edder, S. Rigault, G. Bernardinelli, J.-C. Bunzli and G. Hopfgartner, *J. Chem. Soc., Dalton Trans.* **2000**, 3999.

[37] G. Bernardinelli, C. Piguet and A. F. Williams, *Angew. Chem. Int. Ed.* **1992**, *31*, 1622.

[38] C. Piguet, *Chimia* **1996**, *50*, 144.

[39] C. Edder, C. Piguet, J.-C. G. Bunzli and G. Hopfgartner, *J. Chem. Soc., Dalton Trans.* **1997**, 4657.

[40] S. Rigault, C. Piguet, G. Bernardinelli, and G. Hopfgartner, *Angew. Chem. Int. Ed.* **1998**, *37*, 169.

[41] C. Edder, C. Piguet, G. Bernardinelli, J. Mareda, C. G. Bochet, J. C. Bunzli and G. Hopfgartner, *Inorg. Chem.* **2000**, *39*, 5059.

[42] C. Piguet, J.-C. G. Buenzli, G. Bernardinelli, G. Hopfgartner, S. Petoud and O. Schaad, *J. Am. Chem. Soc.* **1996**, *118*, 6681.

[43] C. Piguet, E. Rivara-Minten, G. Bernardinelli, J.-C. G. Buenzli and G. Hopfgartner, *J. Chem. Soc., Dalton Trans.* **1997**, 421.

[44] S. Rigault, C. Piguet and J.-C. G. Bunzli, *Dalton Trans.* **2000**, 2045.

[45] P. N. W. Baxter, J.-M. Lehn, G. Baum and D. Fenske, *Chem. Eur. J.* **2000**, *6*, 4510.

[46] D. A. McMorran and P. J. Steel, *Angew. Chem. Int. Ed.* **1998**, *37*, 3295.

[47] J. Xu and K. N. Raymond, *Angew. Chem. Int. Ed.* **2006**, *45*, 6480.

[48] L. J. Childs, N. W. Alcock and M. J. Hannon, *Angew. Chem. Int. Ed.* **2002**, *41*, 4244.

[49] S. P. Argent, H. Adams, T. Riis-Johannessen, J. C. Jeffery, L. P. Harding, O. Mamula and M. D. Ward, *Inorg. Chem.* **2006**, *45*, 3905.

[50] H. L. Frisch and E. Wasserman, *J. Am. Chem. Soc.* **1961**, *83*, 3789.

[51] C. O. Dietrich-Buchecker and J.-P. Sauvage, *Chem. Rev.* **1987**, *87*, 795.

[52] J.-P. Sauvage, *Acc. Chem. Res.* **1998**, *31*, 611.

[53] C. O. Dietrich-Buchecker, J.-P. Sauvage and J. M. Kern, *J. Am. Chem. Soc.* **1984**, *106*, 3043.

[54] J.-P. Sauvage, *Acc. Chem. Res.* **1990**, *23*, 319.

[55] J. F. Nierengarten, C. O. Dietrich-Buchecker and J.-P. Sauvage, *J. Am. Chem. Soc.* **1994**, *116*, 375.

[56] J.-F. Nierengarten, C. O. Dietrich-Buchecker and J.-P. Sauvage, *New J. Chem.* **1996**, *20*, 685.

[57] R. F. Carina, C. Dietrich-Buchecker and J.-P. Sauvage, *J. Am. Chem. Soc.* **1996**, *118*, 9110.

[58] C. O. Dietrich-Buchecker, J.-P. Sauvage, J. P. Kintzinger, P. Maltese, C. Pascard and J. Guilhem, *New J. Chem.* **1992**, *16*, 931.

[59] C. O. Dietrich-Buchecker, J. F. Nierengarten, J. P. Sauvage, N. Armaroli, V. Balzani and L. De Cola, *J. Am. Chem. Soc.* **1993**, *115*, 11237.

[60] C. Dietrich-Buchecker and J.-P. Sauvage, *Chem. Commun.* **1999**, 615.

[61] C. O. Dietrich-Buchecker, J. Guilhem, C. Pascard and J.-P. Sauvage, *Angew. Chem. Int. Ed.* **1990**, *29*, 1154.

[62] C. Dietrich-Buchecker, G. Rapenne, J.-P. Sauvage, A. De Cian and J. Fischer, *Chem. Eur. J.* **1999**, *5*, 1432.

[63] W. H. Prickle and M. S. Hoekstra, *J. Am. Chem. Soc.* **1976**, *98*, 1832.

[64] W. H. Prickle and D. J. Hoover, *Top. Stereochem.* **1982**, *13*, 263.

[65] L.-E. Perret-Aebi, A. von Zelewsky, C. Dietrich-Buchecker and J.-P. Sauvage, *Angew. Chem. Int. Ed.* **2004**, *43*, 4482.

[66] J. S. Fleming, K. L. V. Mann, C.-A. Carraz, E. Psillakis, J. C. Jeffery, J. A. McCleverty and M. D. Ward, *Angew. Chem. Int. Ed.* **1998**, *37*, 1279.

[67] D. L. Caulder, R. E. Powers, T. N. Parac and K. N. Raymond, *Angew. Chem. Int. Ed.* **1998**, *37*, 1840.

[68] I. M. Müller, D. Möller and C. A. Schalley, *Angew. Chem. Int. Ed.* **2005**, *44*, 480.

[69] R. W. Saalfrank, A. Stark, K. Peters and H. G. von Schnering, *Angew. Chem. Int. Ed.* **1988**, *27*, 851.

[70] R. W. Saalfrank, B. Demleitner, H. Glaser, H. Maid, S. Reihs, W. Bauer, M. Maluenga, F. Hampel, M. Teichert and H. Krautscheid, *Eur. J. Inorg. Chem.* **2003**, 822.

[71] R. W. Saalfrank, A. Stark and R. Burak, *Inorg. Synth.* **1992**, *29*, 275.

[72] R. W. Saalfrank and B. Demleitner, *Persp. Suramol. Chem.* **1999**, *5*, 1.

[73] R. W. Saalfrank, E. Uller, B. Demleitner and I. Bernt, *Struct. Bond. (Berlin)* **2000**, *96*, 149.

[74] R. W. Saalfrank, R. Burak, S. Reihs, N. Loew, F. Hampel, H.-D. Stachel, J. Lantmaier, K. Peters, E.-M. Peters *et al.*, *Angew. Chem. Int. Ed.* **1995**, *34*, 993.

[75] R. W. Saalfrank, R. Burak, A. Breit, D. Stalke, R. Herbst-Irmer, J. Daub, M. Porsch, E. Bill, M. Müther and A. X. Trautwein, *Angew. Chem. Int. Ed.* **1994**, *33*, 1621.

[76] R. W. Saalfrank, B. Hörner, D. Stalke and J. Salbeck, *Angew. Chem. Int. Ed.* **1993**, *32*, 1179.

[77] R. W. Saalfrank, R. Harbig, J. Nachtrab, W. Bauer, K.-P. Zeller, D. Stalke and M. Teichert, *Chem. Eur. J.* **1996**, *2*, 1363.

[78] T. Beissel, R. E. Powers and K. N. Raymond, *Angew. Chem. Int. Ed.* **1996**, *35*, 1084.

[79] R. W. Saalfrank, B. Demleitner, H. Glaser, H. Maid, D. Bathelt, F. Hampel, W. Bauer and M. Teichert, *Chem. Eur. J.* **2002**, *8*, 2679.

[80] R. W. Saalfrank, C. Schmidt, H. Maid, F. Hampel, W. Bauer and A. Scheurer, *Angew. Chem. Int. Ed.* **2006**, *45*, 315.

[81] M. Scherer, D. L. Caulder, D. W. Johnson and K. N. Raymond, *Angew. Chem. Int. Ed.* **1999**, *38*, 1588.

[82] D. W. Johnson and K. N. Raymond, *Inorg. Chem.* **2001**, *40*, 5157.

[83] A. J. Terpin, M. Ziegler, D. W. Johnson and K. N. Raymond, *Angew. Chem. Int. Ed.* **2001**, *40*, 157.

[84] D. Fiedler, D. H. Leung, R. G. Bergman and K. N. Raymond, *Acc. Chem. Res.* **2005**, *38*, 349.

[85] R. L. Paul, Z. R. Bell, J. S. Fleming, J. C. Jeffery, J. A. McCleverty and M. D. Ward, *Heteroatom. Chem.* **2002**, *13*, 567.

[86] R. L. Paul, Z. R. Bell, J. C. Jeffery, J. A. McCleverty and M. D. Ward, *Proc. Natl. Acad. Sci. USA* **2002**, *99*, 4883.

[87] R. L. Paul, S. P. Argent, J. C. Jeffery, L. P. Harding, J. M. Lynam and M. D. Ward, *Dalton Trans.* **2004**, *21*, 3453.

[88] H. Amouri, L. Mimassi, M. N. Rager, B. E. Mann, C. Guyard-Duhayon and L. Raehm, *Angew. Chem. Int. Ed.* **2005**, *44*, 4543.

[89] H. Amouri, C. Desmarets, A. Bettoschi, M. N. Rager, K. Boubekeur, P. Rabu and M. Drillon, *Chem. Eur. J.* **2007**, *13*, 5401.

[90] D. Fiedler, D. Pagliero, J. L. Brumaghim, R. G. Bergman and K. N. Raymond, *Inorg. Chem.* **2004**, *43*, 846.

[91] D. Fiedler, D. H. Leung, R. G. Bergman and K. N. Raymond, *J. Am. Chem. Soc.* **2004**, *126*, 3674.

[92] D. H. Leung, D. Fiedler, R. G. Bergman and K. N. Raymond, *Angew. Chem. Int. Ed.* **2004**, *43*, 963.

[93] C. Nuckolls, F. Hof, T. Martin and J. Rebek, Jr., *J. Am. Chem. Soc.* **1999**, *121*, 10281.

[94] D. H. Leung, R. G. Bergman and K. N. Raymond, *J. Am. Chem. Soc.* **2006**, *128*, 9781.

[95] D. Fiedler, R. G. Bergman and K. N. Raymond, *Angew. Chem. Int. Ed.* **2004**, *43*, 6748.

[96] M. D. Pluth, R. G. Bergman and K. N. Raymond, *Science* **2007**, *316*, 85.

[97] J. Rebek, Jr., *Chem. Soc. Rev.* **1996**, 255.

[98] J. Kang and J. Rebek, Jr., *Nature* **1997**, *385*, 50.

[99] J. Kang, G. Hilmersson, J. Santamaria and J. Rebek, Jr., *J. Am. Chem. Soc.* **1998**, *120*, 3650.

[100] M. Fujita, K. Umemoto, M. Yoshizawa, N. Fujita, T. Kusukawa and K. Biradha, *Chem. Commun.* **2001**, 509.

[101] I. M. Müller, R. Robson and F. Separovic, *Angew. Chem. Int. Ed.* **2001**, *40*, 4385.

[102] I. M. Müller, S. Spillmann, H. Franck and R. Pietschnig, *Chem. Eur. J.* **2004**, *10*, 2207.

[103] I. M. Müller and D. Möller, *Angew. Chem. Int. Ed.* **2005**, *44*, 2969.

[104] A. J. Amoroso, J. C. Jeffery, P. L. Jones, J. A. McCleverty, P. Thornton and M. D. Ward, *Angew. Chem. Int. Ed.* **1995**, *34*, 1443.

[105] C. Bruckner, R. E. Powers and K. N. Raymond, *Angew. Chem. Int. Ed.* **1998**, *37*, 1837.

[106] R. W. Saalfrank, H. Glaser, B. Demleitner, F. Hampel, M. M. Chowdhry, V. Schunemann, A. X. Trautwein, G. B. M. Vaughan, R. Yeh, A. V. Davis and K. N. Raymond, *Chem. Eur. J.* **2002**, *8*, 493.

[107] R. M. Yeh, J. Xu, G. Seeber and K. N. Raymond, *Inorg. Chem.* **2005**, *44*, 6228.

[108] F. L. Weitl and K. N. Raymond, *J. Am. Chem. Soc.* **1979**, *101*, 2728.

[109] S. Hiraoka and M. Fujita, *J. Am. Chem. Soc.* **1999**, *121*, 10239.

[110] S. Hiraoka, Y. Kubota and M. Fujita, *Chem. Commun.* **2000**, 1509.

[111] N. Fujita, K. Biradha, M. Fujita, S. Sakamoto and K. Yamaguchi, *Angew. Chem. Int. Ed.* **2001**, *40*, 1718.

[112] D. W. Johnson, J. Xu, R. W. Saalfrank and K. N. Raymond, *Angew. Chem. Int. Ed.* **1999**, *38*, 2882.

[113] Z. Zhong, A. Ikeda, S. Shinkai, S. Sakamoto and K. Yamaguchi, *Org. Lett.* **2001**, *3*, 1085.

[114] A. Ikeda, H. Udzu, Z. Zhong, S. Shinkai, S. Sakamoto and K. Yamaguchi, *J. Am. Chem. Soc.* **2001**, *123*, 3872.

[115] Y. Kubota, K. Biradha, M. Fujita, S. Sakamoto and K. Yamaguchi, *Bull. Chem. Soc. Jpn* **2002**, *75*, 559.

[116] S. Aoki, M. Shiro and E. Kimura, *Chem. Eur. J.* **2002**, *8*, 929.

[117] K. Severin, *Chem. Commun.* **2006**, 3859.

[118] H. Amouri, M. N. Rager, F. Cagnol and J. Vaissermann, *Angew. Chem. Int. Ed.* **2001**, *40*, 3636.

[119] K. Yamanari, S. Yamamoto, R. Ito, Y. Kushi, A. Fuyuhiro, N. Kubota, T. Fukuo and R. Arakawa, *Angew. Chem. Int. Ed.* **2001**, *40*, 2268.

[120] H. Amouri, C. Guyard-Duhayon, J. Vaissermann and M. N. Rager, *Inorg. Chem.* **2002**, *41*, 1397.

[121] L. Dadci, H. Elias, U. Frey, A. Hörning, U. Koelle, A. E. Merbach, H. Paulus and J. S. Schneider, *Inorg. Chem.* **1995**, *34*, 306.

[122] G. Laurenczy, U. Frey, D. T. Richens and A. E. Merbach, *Magn. Reson. Chem.* **1991**, *29*, S45.

[123] K. Severin, R. Bergs and W. Beck, *Angew. Chem. Int. Ed.* **1998**, *37*, 1634.

[124] H. Chen, M. F. Maestre and R. H. Fish, *J. Am. Chem. Soc.* **1995**, *117*, 3631.

[125] S. Ogo, b. H. Chen, M. M. Olmstead and R. H. Fish, *Organometallics* **1996**, *15*, 2009.

[126] H. Chen, S. Ogo and R. H. Fish, *J. Am. Chem. Soc.* **1996**, *118*, 4993.

[127] K. Yamanari, R. Ito, S. Yamamoto and A. Fuyuhiro, *Chem. Commun.* **2001**, 1414.

[128] T. Poth, H. Paulus, H. Elias, C. Ducker-Benfer and R. Van Eldik, *Eur. J. Inorg. Chem.* **2001**, 1361.

[129] D. Carmona, M. P. Lamata and L. A. Oro, *Eur. J. Inorg. Chem.* **2002**, 2239.

[130] K. Yamanari, R. Ito, S. Yamamoto, T. Konno, A. Fuyuhiro, M. Kobayashi and R. Arakawa, *Dalton Trans.* **2003**, 380.

[131] K. Yamanari, R. Ito, S. Yamamoto, T. Konno, A. Fuyuhiro, K. Fujioka and R. Arakawa, *Inorg. Chem.* **2002**, *41*, 6824.

[132] D. Carmona, F. J. Lahoz, R. Atencio, L. A. Oro, M. P. Lamata, F. Viguri, E. San Jose, C. Vega, J. Reyes, F. Joo and A. Katho, *Chem. Eur. J.* **1999**, *5*, 1544.

[133] Z. Grote, R. Scopelliti and K. Severin, *J. Am. Chem. Soc.* **2004**, *126*, 16959.

[134] L. Mimassi, C. Guyard-Duhayon, M. N. Rager and H. Amouri, *Inorg. Chem.* **2004**, *43*, 6644.

[135] L. Mimassi, C. Cordier, C. Guyard-Duhayon, B. E. Mann and H. Amouri, *Organometallics* **2007**, *26*, 860.

[136] J. Lacour, C. Ginglinger, C. Grivet and G. Bernardinelli, *Angew. Chem. Int. Ed.* **1997**, *36*, 608.

[137] J. Lacour, *Chimia* **2003**, *57*, 168.

[138] J. Lacour and R. Frantz, *Org. Biomol. Chem.* **2005**, *3*, 15.

[139] E. Fischer, *Ber. Dtsh. Chem. Ges* **1894**, *27*, 3189.

[140] A. P. Smirnoff, *Helv. Chim. Acta* **1920**, *3*, 177.

[141] A. von Zelewsky, *Coord. Chem. Rev.* **1999**, *190-192*, 811.

[142] O. Mamula and A. von Zelewsky, *Coord. Chem. Rev.* **2003**, *242*, 87.

[143] P. Hayoz and A. von Zelewsky, *Tetrahedron Letts.* **1992**, *33*, 5165.

[144] D. Sandrini, M. Maestri, M. Ciano, U. Maeder and A. von Zelewsky, *Helv. Chim. Acta* **1990**, *73*, 1306.

[145] L. Yang, A. von Zelewsky and H. Stoeckli-Evans, *Chem. Commun.* **2005**, 4155.

[146] H. Murner, A. von Zelewsky and G. Hopfgartner, *Inorg. Chim. Acta* **1998**, *271*, 36.

[147] O. Mamula, A. von Zelewsky and G. Bernardinelli, *Angew. Chem. Int. Ed.* **1998**, *37*, 290.

[148] O. Mamula, A. von Zelewsky, P. Brodard, C. Wilhelm Schlaepfer, G. Bernardinelli and H. Stoeckli-Evans, *Chem. Eur. J.* **2005**, *11*, 3049.

[149] O. Mamula, A. Von Zelewsky, T. Bark and G. Bernardinelli, *Angew. Chem. Int. Ed.* **1999**, *38*, 2945.

[150] H. Jiang and W. Lin, *J. Am. Chem. Soc.* **2003**, *125*, 8084.

[151] H. Jiang and W. Lin, *J. Organomet. Chem.* **2005**, *690*, 5159.

[152] S. J. Lee and W. Lin, *J. Am. Chem. Soc.* **2002**, *124*, 4554.

[153] S. J. Lee, A. Hu and W. Lin, *J. Am. Chem. Soc.* **2002**, *124*, 12948.

[154] S. J. Lee, J. S. Kim and W. Lin, *Inorg. Chem.* **2004**, *43*, 6579.

[155] H. Jiang and W. Lin, *J. Am. Chem. Soc.* **2006**, *128*, 11286.

[156] B. Kesanli and W. Lin, *Coord. Chem. Rev.* **2003**, *246*, 305.

[157] M. Fujita, J. Yazaki and K. Ogura, *J. Am. Chem. Soc.* **1990**, *112*, 5645.

[158] K. S. Jeong, S. Y. Kim, U.-S. Shin, M. Kogej, N. T. M. Hai, P. Broekmann, N. Jeong, B. Kirchner, M. Reiher and C. A. Schalley, *J. Am. Chem. Soc.* **2005**, *127*, 17672.

[159] P. J. Stang, B. Olenyuk, D. C. Muddiman, D. S. Wunschel and R. D. Smith, *Organometallics* **1997**, *16*, 3094.

[160] P. J. Stang and B. Olenyuk, *Acc. Chem. Res.* **1997**, *30*, 502.

[161] P. J. Stang, *Chem. Eur. J.* **1998**, *4*, 19.

[162] B. Olenyuk, J. A. Whiteford and P. J. Stang, *J. Am. Chem. Soc.* **1996**, *118*, 8221.

[163] P. J. Stang and B. Olenyuk, *Angew. Chem. Int. Ed.* **1996**, *35*, 732.

[164] W. D. Hounshell and K. Mislow, *Tetrahedon Lett.* **1979**, 1205.

[165] M. Nakazaki, K. Naemura and Y. Hokura, *J. Chem. Soc., Chem. Commun.* **1982**, 1245.

[166] M. Fujita, D. Oguro, M. Miyazawa, H. Oka, K. Yamaguchi and K. Ogura, *Nature* **1995**, *378*, 469.

[167] B. M. Goldstein, J. K. Barton and H. M. Berman, *Inorg. Chem.* **1986**, *25*, 842.

[168] K. E. Erkkila, D. T. Odom and J. K. Barton, *Chem. Rev.* **1999**, *99*, 2777.

[169] A. Juris, V. Balzani, F. Barigelletti, S. Campagna, P. Belser and A. von Zelewsky, *Coord. Chem. Rev.* **1988**, *84*, 85.

[170] V. Balzani and A. Juris, *Coord. Chem. Rev.* **2001**, *211*, 97.

[171] X. Hua and A. von Zelewsky, *Inorg. Chem.* **1991**, *30*, 3796.

[172] X. Hua and A. von Zelewsky, *Inorg. Chem.* **1995**, *34*, 5791.

[173] S. Bodige, A. S. Torres, D. J. Maloney, D. Tate, G. R. Kinsel, A. K. Walker and F. M. MacDonnell, *J. Am. Chem. Soc.* **1997**, *119*, 10364.

[174] O. Morgan, S. Wang, S.-A. Bae, R. J. Morgan, A. D. Baker, T. C. Strekas and R. Engel, *J. Chem. Soc., Dalton Trans.* **1997**, 3773.

[175] R. Caspar, H. Amouri, M. Gruselle, C. Cordier, B. Malezieux, R. Duval and H. Leveque, *Eur. J. Inorg. Chem.* **2003**, 499.

[176] B. Bosnich and F. P. Dwyer, *Aust. J. Chem.* **1966**, *19*, 2229.

[177] B. Bosnich, *Acc. Chem. Res.* **1969**, *2*, 266.

[178] B. Bosnich, *Inorg. Chem.* **1968**, *7*, 2379.

[179] B. Bosnich, *Inorg. Chem.* **1968**, *7*, 178.

[180] T. J. Rutherford, M. G. Quagliotto and F. R. Keene, *Inorg. Chem.* **1995**, *34*, 3857.

[181] T. J. Rutherford, P. A. Pelligrini, J. Aldrich-Wright, P. C. Junk and F. R. Keene, *Eur. J. Inorg. Chem.* **1998**, 1677.

[182] D. Tzalis and Y. Tor, *J. Am. Chem. Soc.* **1997**, *119*, 852.

[183] K. Wärnmark, P. N. W. Baxter and J. M. Lehn, *Chem. Commun.* **1998**, 993.

[184] K. Wärnmark, J. A. Thomas, O. Heyke and J. M. Lehn, *Chem. Commun.* **1996**, 701.

[185] K. Wärnmark, O. Heyke, J. A. Thomas and J. M. Lehn, *Chem. Commun.* **1996**, 2603.

[186] F. M. MacDonnell, M.-J. Kim and S. Bodige, *Coord. Chem. Rev.* **1999**, *185-186*, 535.

[187] F. M. MacDonnell and S. Bodige, *Inorg. Chem.* **1996**, *35*, 5758.

[188] T. J. Rutherford and F. R. Keene, *Inorg. Chem.* **1997**, *36*, 3580.

[189] T. J. Rutherford, O. Van Gijte, A. Kircsch-De Mesmaeker and F. R. Keene, *Inorg. Chem.* **1997**, *36*, 4465.

[190] T. J. Rutherford and F. R. Keene, *J. Chem. Soc., Dalton Trans.* **1998**, 1155.

[191] J. Yin and R. L. Elsenbaumer, *Inorg. Chem.* **2007**, *46*, 6891.

6 Chiral Enantiopure Molecular Materials

General Considerations

In this chapter, we will be dealing with inorganic and/or coordination molecular materials. These materials are constructed from elementary 'bricks', and assembled, with or without the aid of another molecular element sometimes referred to as the 'mortar', to form an extended structure. In these superstructures, which may be designated inorganic or coordination polymers, chemical bonds and/or intermolecular interactions join the bricks to each other. The former may be covalent, ionic or hydrogen bonds, while the latter are most often π–π, electrostatic or van der Waals interactions.

6.1.1 Types of Organization

According to their spatial organization, we can distinguish one-, two- and three-dimensional supramolecular networks that we will refer to as 1D, 2D and 3D networks (Figure 6.1).

(i) *1D Networks*

1D networks are characterized in the solid state by a linear, zigzag or helical structure. Each chain is independent of its immediate neighbours and the properties of the network arise from the interactions that develop between the constituent bricks along a given axis. Note, however, that weak interactions between the chains ensure the cohesion of the crystalline structure.

(ii) *2D Networks*

In 2D networks, also known as lamellar structures, there are bonds or interactions between the elementary bricks that lead to the formation of layers; these layers are independent of each other and therefore the properties of these materials are governed by the interactions between the bricks within a layer. The cohesion of the whole material is due, in general, to weak interactions between the layers such as: hydrogen bonds, van der Waals and/or electrostatic interactions.

(iii) *3D Networks*

These networks are characterized by a three-dimensional organization whose geometry depends on the nature of the bricks as well as their assembly in three directions in space. Figure 6.1 represents schematically the base structures for the 1D, 2D and 3D networks.

Chirality in Transition Metal Chemistry: Molecules, Supramolecular Assemblies and Materials   H. Amouri and M. Gruselle
© 2008 John Wiley & Sons, Ltd

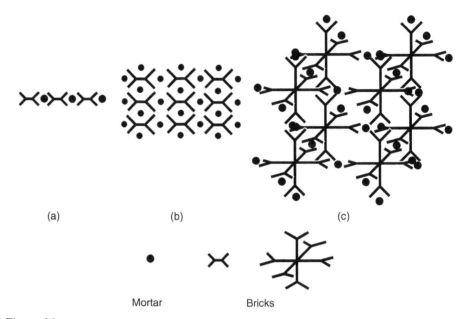

(a)   (b)   (c)

●   ✕   ✖

Mortar   Bricks

Figure 6.1
Schematic representation of (a) 1D, (b) 2D and (c) 3D networks obtained by a combination of 'bricks' and 'mortar'.

6.1.2   Properties

The properties of these molecular superstructures are not, *a priori*, those arising from the simple addition of the properties of their constituent elements. As a result of specific interactions in the network, new properties can appear, for example conductivity, magnetic or optical properties, and so on.

The network structure itself can give rise to particular geometric forms. For example, cavities of suitable size, shape and polarity are the origin of specific recognition, including diastereomeric interactions, phenomena involving a molecular 'host', which is the basis of catalysis and chromatographic separation.

The concept of materials is closely linked to the properties that characterize them. Taking account of this 'technical intentionality' which is attached to materials, and faced with an extensive literature dedicated to 'molecular materials', we will limit our subject matter to those enantiopure chiral materials that are identifiable by specific properties.

6.1.3   Chiral and Enantiopure Materials

Here our interest is in enantiopure chiral molecular materials. The term 'chiral' is often used well beyond its proper meaning, and such use can lead to confusion.[1] *Strictly speaking, a molecule is chiral because it does not contain an improper axis (S_n). The same definition can be used for a molecular material containing many chiral elements. In the case of a solid state organization this means that the space group of the crystalline structure belongs to a chiral space group.* However, this says nothing about the racemic or enantiomeric nature of the material itself.

Clearly 'racemic' and 'enantiopure' do not signify the same thing, each being adjectives that describe a 'chiral material'. Neither is it sufficient just to consider the space group of a material obtained as a single crystal. Even in the case of the space group being chiral, it is possible for the crystal under examination to have been extracted from a racemic collection of crystals. Thus the crystal batch has no effect on polarized light. For a chiral space group, it may also be the case that the crystal under study exists as an indissociable racemic conglomerate.[2,3]

Therefore, to be 'racemic' or 'enantiopure' does not designate the same chiral material. This is a key point, since in its racemic form a chiral material cannot lead to the breaking of spatial symmetry that is responsible for chiroptic properties.

To speak of a chiral material therefore only has any real sense if its resolution and/or enantioselective synthesis can be accomplished. Accordingly, we will develop in particular those strategies that are used to obtain enantiopure chiral materials. Note that, at present, beyond a certain level of complexity of the molecular organization it is not possible to carry out a resolution once the synthesis is complete, and therefore enantioselective syntheses must be used. It is therefore necessary to include in the synthetic strategy one or more molecular elements capable of orienting the overall configuration. These elements are, in general, ligands and/or ready-made molecular bricks that carry the configurationnal information. Starting from racemic or achiral bricks, crystallogenesis can sometimes occur with spontaneous resolution. However, it is not possible to use this route as a synthetic method as it leads to conglomerates and, even in the best cases, to a second-order spontaneous resolution the orientation of which cannot be controlled. The concept of a molecular material is often closely associated with its physical properties. We speak of a material being conductive, magnetic, porous, and so on. In simple multifunctional materials, these properties can coexist with the optical activity without any interaction.[4] In some circumstances, the multiple properties of a material can lead to cross effects that are more complex. This is the case, for example, for magnetochiral anisotropy responsible for the magnetochiral dichroic (MChD) effect[5] (the simultaneous existence of a Cotton and Faraday effect) or electrical magnetochiral anisotropy[6] (the simultaneous existence of electrical conduction and spatial dissymmetry).

The existence of these multiple effects strengthens the study of multifunctional materials and, from this point of view, the availability of enantiopure materials offers an interesting field of research. In effect, the solid state organization of the racemate differs from that of the enantiomers and thus leads to the specific arrangements that determine the particular properties.

We will now move on to the synthesis, organization and properties of chiral enantiopure materials in the fields of conduction, catalysis and chromatographic separation, optics and magnetism.

6.2 Conductors

6.2.1 General Considerations

Organic conductors and superconductors constitute an important class of molecular materials. They are based on π-type organic donors, usually tetrathiofulvalene derivatives (**6.1–6.2**), which form a two-dimensional network with an acceptor, which may be an

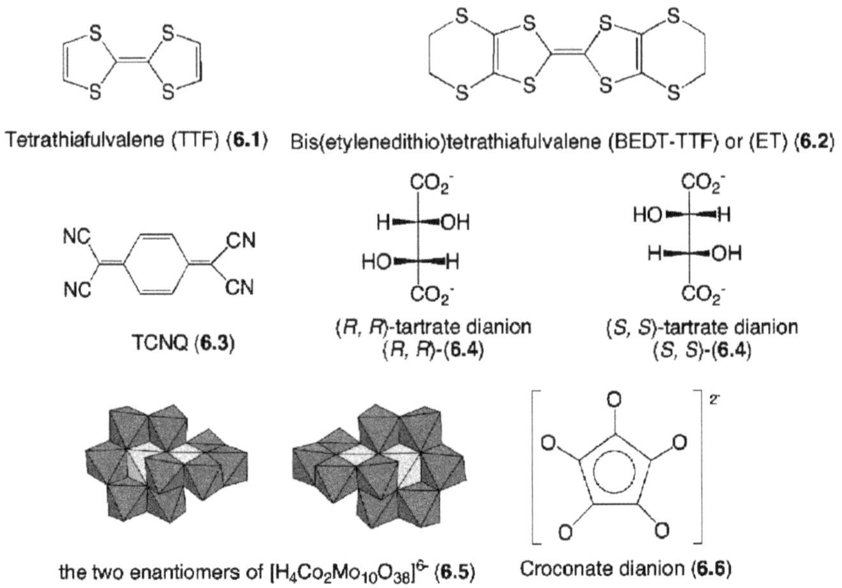

Tetrathiafulvalene (TTF) (**6.1**)   Bis(etylenedithio)tetrathiafulvalene (BEDT-TTF) or (ET) (**6.2**)

TCNQ (**6.3**)

(R, R)-tartrate dianion
(R, R)-(**6.4**)

(S, S)-tartrate dianion
(S, S)-(**6.4**)

the two enantiomers of [H₄Co₂Mo₁₀O₃₈]⁶⁻ (**6.5**)   Croconate dianion (**6.6**)

Figure 6.2
View of some π-type organic donors and anions used in the synthesis of organic conductors.

organic molecule such as tetracyanoquinodimethane (TCNQ) (**6.3**)[7] or an anion such as [PF$_6$]⁻, [AsF$_6$]⁻, [C$_4$H$_4$O$_4$]$^{2-}$ (tartrate anion) (**6.4**)[8], [H$_4$Co$_2$Mo$_{10}$O$_{38}$]$^{6-}$ (polyoxometalate) (**6.5**)[9] or [Fe(C$_5$O$_5$)$_3$]$^{3-}$ (tris(croconate)ferrate(III) anion) (**6.6**)[10] (Figure 6.2).

The properties of these salts depend on the nature of their constituent parts, specifically on their capacity to act as electron donors and acceptors, but also on the crystalline organization of the donor–acceptor system. The crystal packing, as well as the disorder of the crystal structure, plays an important role in the electrical conduction and/or superconductive properties of these compounds.

6.2.2   Why Enantiopure Molecular Conductors?

There are two reasons for studying enantiomerically pure conductors. The first is connected with questions of structure. In effect, it is possible, starting from enantiopure bricks, to prepare crystal structures that are noncentrosymmetric, and which allow the disorder to be limited compared to the racemic derivatives. The second reason is linked to the fact that, for chiral conductors in an enantiopure form, theory predicts the existence of an effect called 'electrical magnetochiral anisotropy' (EMCA) due to the simultaneous breaking of space and time symmetry in the presence of an external magnetic field.[6,11] Thus, the resistivity of a chiral conductor depends on its absolute configuration according to the formula:

$$\mathbf{R}^{D/L}(\mathbf{I}, \mathbf{B}) = \mathbf{R} \circ (1 + \chi^{D/L} \mathbf{I} \mathbf{B} + \beta \mathbf{B}^2)$$

where $\mathbf{R}^{D/L}$ is the resistivity of the left or right form, $\mathbf{I}$ is the intensity of the current through the conductor, $\mathbf{B}$ is the external magnetic field parallel to the current, χ is opposite for the two enantiomers ($\chi^D = -\chi^L$) and β is the longitudinal magneto-resistance.

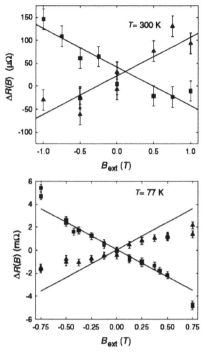

Figure 6.3

Two-terminal magnetochiral resistance anisotropy $\Delta R(I, B_{ext}) = R(I, B_{ext}) - R(-I, B_{ext})$ of D (squares) and L (triangles) bismuth helices (seven turns, 8 mm diameter and 0.8 mm pitch) with $I = 0.2$ A, as a function of the external magnetic field B_{ext}, at 300 K (top, $R_0 = 0.9 \, \Omega$) and 77 K (bottom, $R_0 = 0.2 \, \Omega$) (reproduced with permission from reference[6], copyright 2001, The American Physical Society).

Rikken postulated that the electrical resistance of a chiral conductor (in the sense of being enantiopure) depends linearly on the external magnetic field and the current carried by the conductor, as well as its configuration. The existence of this effect is the direct consequence of the breaking of the time symmetry by the magnetic field, and of the parity by the chirality. Rikken used right- and left-handed macroscopic helices of bismuth, obtained by injecting molten bismuth into a helical mould, and he showed the existence of a magneto-resistance that varies linearly with the applied field and with a slope of the two sign for the opposite configuration of the two helices (Figure 6.3).

Rikken proposed that the EMCA effect could also result from the simultaneous application of a magnetic field and a current to a crystal with an enantiomorphous space group, and that it is a universal property. He showed the existence of this effect in the case of chiral single-walled carbon nanotubes.[12] For most of the investigated tubes, a dependence of the resistance is observed that is odd in both the magnetic field and the current. These observations confirm the existence of EMCA not only for a macroscopic chiral conductor but also for a molecular conductor with chirality on the microscopic level.

6.2.3 Strategies to Obtain Enantiopure Conductors

Since these conductive materials are built from elementary bricks in the form of donors and acceptors, it is possible to introduce chirality by modifying either one or the other. In

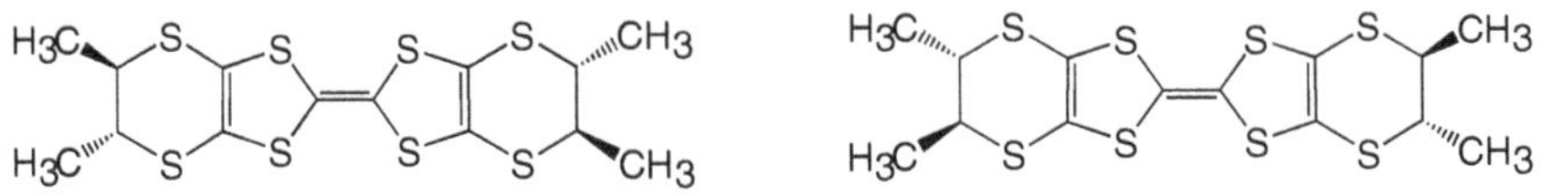

(R, R, R, R)-TMET (R, R, R, R)-(**6.7**)

(S, S, S, S)-TMET (S, S, S, S)-(**6.7**)

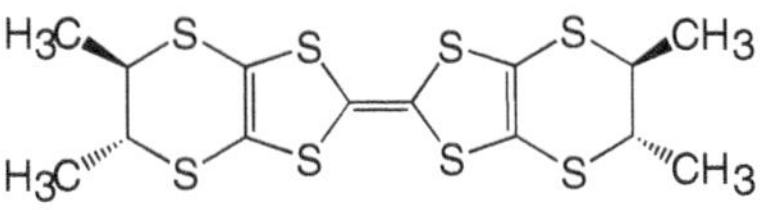

meso-TMET (**6.8**)

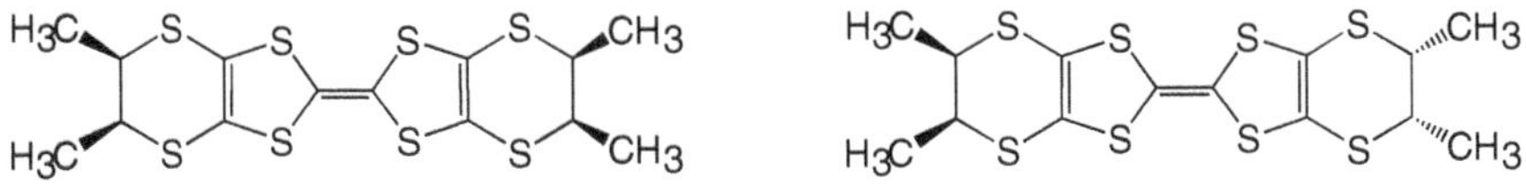

TMET diastereomeric compounds (**6.9**), (**6.10**)

Figure 6.4
View of some tetra-methyl (TMET) substituted ET: enantiomeric TMET (R, R, R, R)-(**6.7**) and (S, S, S, S)-(**6.7**), meso- TMET (**6.8**) and diastereomeric TMET (**6.9–6.10**).

his review, Wallis[13] suggested a strategy based on substituted chiral derivatives of BEDT–TTF (**6.7–6.10**). In general, these derivatives are substituted in the lateral rings of the BEDT–TTF (Figure 6.4).

The substituents may be simple, for example methyl groups, or contain functionality, such as alcohols,[14]–[16] terpenoid substituents,[13] esters,[13] hydroxyamides or oxazolidines.[17] In these last cases, the substituents can be expected to be involved in the supramolecular architecture of the conductor, due to the polar nature of the functions involved and hence their potential interactions (π–π type or hydrogen bonds). The first enantiopure organosulfur donor, reported by Dunitz,[18] was (S,S,S,S)-TMET (S, S, S, S)-(**6.7**), several examples of which are shown in Figure 6.5.

Fourmigué and Avarvari[19] studied racemic and enantiopure mixed valence compounds based on tetrathiofulvalenes incorporating a chiral methyl-oxazoline heterocycle, which they obtained by electrocrystallization in the presence of (n-Bu)$_4$NAsF$_6$. These salts have the formulae: (*rac*)-(TTF-oxa)$_2$AsF$_6$ (*rac*)-(**6.13**), (R)-(TTF-oxa)$_2$AsF$_6$ (R)-(**6.13**) and (S)-(TTF-oxa)$_2$AsF$_6$ (S)-(**6.13**). The racemic compound crystallizes in the triclinic system (space group P-1) with the (AsF$_6$)$^-$ anion on the inversion centre, one donor molecule in a general position in the unit cell and the oxazoline ring almost coplanar with the TTF core but disordered on two positions, corresponding to both (R) and (S) enantiomers on the same site, one as s-cis and the other as s-trans conformers.

The crystalline disorder is due to the presence of the two possible (*cis*) and (*trans*) conformers (Figure 6.6). The enantiopure compounds (R)-(TTF-oxa)$_2$AsF$_6$ (R)-(**6.13**) and (S)-(TTF-oxa)$_2$AsF$_6$ (S)-(**6.13**) crystallize in the P-1 space group with one anion and two independent donor molecules with a (*cis*) and (*trans*) conformation. Overall, the structure of the enantiopure compounds is less disordered, even if the structure of the β type is fairly close to the racemic compound, with a head-to-tail arrangement of the cations.

According to the authors, this difference in the crystal disorder accounts for the observed difference in conductivity between the racemic and enantiopure materials,

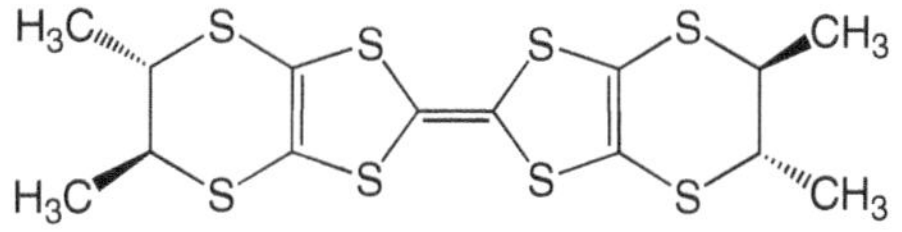

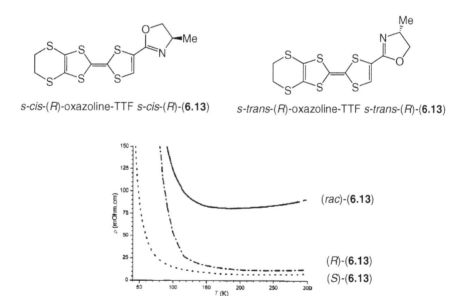

(*S*,*S*,*S*,*S*)-(TMET) (*S*,*S*,*S*,*S*)-(**6.7**)

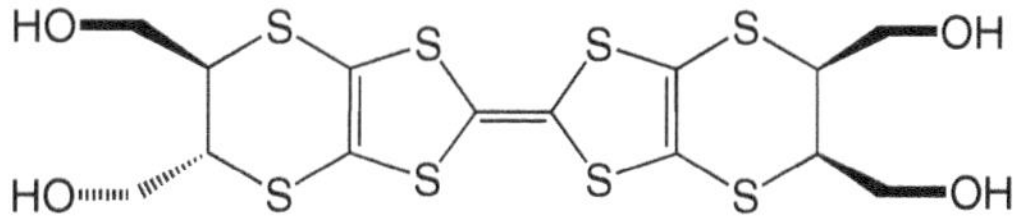

(*R*,*R*,*S*,*R*)-Tetrol-ET (*R*,*R*,*S*,*R*)-(**6.11**)

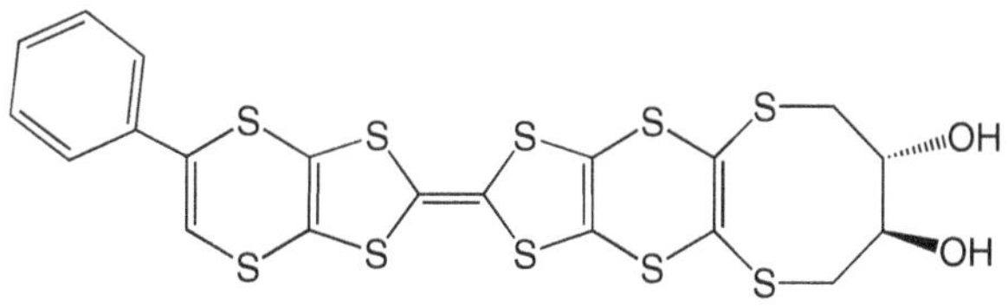

(*R*,*S*)-diol-modified-ET (*R*,*S*)-(**6.12**)

Figure 6.5
Some examples of enantiopure modified ET organosulphur donors.

s-cis-(*R*)-oxazoline-TTF *s-cis-*(*R*)-(**6.13**)   *s-trans-*(*R*)-oxazoline-TTF *s-trans-*(*R*)-(**6.13**)

(*rac*)-(**6.13**)

(*R*)-(**6.13**)

(*S*)-(**6.13**)

Figure 6.6
Temperature dependence of the resistivity for (R)-(TTF-oxa)$_2$AsF$_6$ (R)-(**6.13**), (S)-(TTF-oxa)$_2$AsF$_6$ (S)-(**6.13**) and (rac)-(TTF-oxa)$_2$AsF$_6$ (rac)-(**6.13**) (reproduced with permission from reference[19], copyright 2005, American Chemical Society).

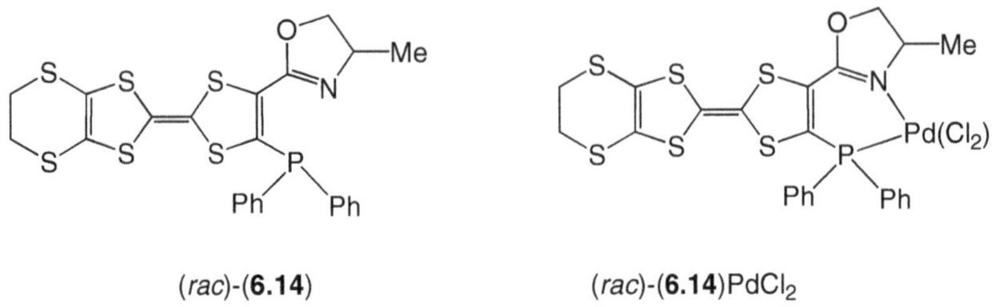

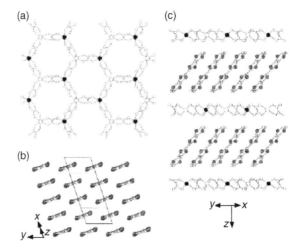

(*rac*)-(**6.14**)   (*rac*)-(**6.14**)PdCl₂

Figure 6.7
Palladium (II) complex (**6.14**)-PdCl₂ based on the EDF-TTF-OX (**6.14**) ligand.

$\sigma_{RT} \approx 10\,S.cm^{-1}$ and $\sigma_{RT} \approx 100\,S.cm^{-1}$, respectively.[19] Recently, they were also able[20] to synthesize the enantiopure EDF-TTF-OX (**6.14**) compounds. The authors showed that complexes of palladium (**6.14**)-PdCl₂ could be formed from these ligands, but these complexes have not yet been obtained in enantiopure form (Figure 6.7).

Coronado approached the construction of conductive chiral materials by using the anionic part to carry the chiral information in the supramolecular architecture and, in this way, obtained the compound [BEDT-TTF]₃[MnCr(C₂O₄)₃] (**6.15**) through electrocrystallization[21]. This compound combines the ferromagnetic properties of the [MnCr(C₂O₄)₃]⁻ anionic subnetwork, below a Curie temperature (T_c) ($T_c = 5.5\,K$), with the room temperature conductivity ($R = 250\,S.cm^{-1}$) of the [BEDT-TTF]₃⁺ cationic sub-network. The anionic network is made up of chiral layers, for example [(Δ)–Mn(Λ)–Cr(C₂O₄)₃]⁻. In the structure obtained, adjacent layers have an opposite configuration (layer n: [(Δ)–Mn(Λ)–Cr(C₂O₄)₃]⁻, layer $n + 1$: [(Λ)–Mn(Δ)–Cr(C₂O₄)₃]⁻). Thus overall the material is achiral, and crystallizes in the P-1 space group (Figure 6.8).

Figure 6.8
Structure of the hybrid material (**6.15**) and the two sublattices. (a) View of the [(Δ)–Mn(Λ)–Cr(C₂O₄)₃]⁻ bimetallic layers (filled and open circles in the vertices of the hexagons represent the two types of metals). (b) Structure of the organic layer, showing the β packing of the BEDT-TTF molecules. (c) Representation of the hybrid structure along the z-axis, showing the alternating organic/inorganic layers (reproduced with permission from reference[21], copyright 2000, Macmillan Publishers Ltd).

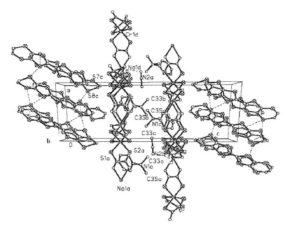

Figure 6.9
View of the alternating cationic and anionic layers in the $\beta''(\text{BEDT-TTF})_2\{[\text{NMe}_3\text{Ph}]\text{NaCR}$ $(\text{C}_2\text{S}_2\text{O}_3)_3(\text{MeCN})\}$ salt (**6.16**) (reproduced with permission from reference[22], copyright 2005, Elsevier).

Other groups have synthesized multifunctional conductive materials (exhibiting both conductivity and magnetism) using BEDT-TTF as the donor and $[\text{Cr}(\text{C}_2\text{S}_2\text{O}_2)_3]^{3-}$ as the chiral anion with D_3 symmetry.[22,23] The structure of these compounds, formed from anionic and cationic layers, is close to that obtained for $[\text{BEDT-TTF}]_3[\text{MnCr}(\text{C}_2\text{O}_4)_3]$ (**6.15**). However, here too, the use of a racemic chiral anion leads to an achiral superstructure $\beta''(\text{BEDT-TTF})_2\{[\text{NMe}_3\text{Ph}]\text{NaCR}(\text{C}_2\text{S}_2\text{O}_3)_3(\text{MeCN})\}$ (**6.16**) (space group P-1) (Figure 6.9).

In another attempt to introduce chirality into a conductive network, Coronado used a complex of Fe(III) with the croconate dianion (**6.6**).[10] This tris-anion is chiral, has D_3 symmetry and is paramagnetic (Figure 6.10). Two salts were obtained, α-BEDT-TTF$_5$[Fe$(\text{C}_5\text{O}_5)_3$].5H$_2$O α-(**6.17**).5H$_2$O and β-BEDT-TTF$_5$[Fe(C$_5$O$_5$)$_3$].C$_6$H$_5$CN β-(**6.17**).C$_6$H$_5$CN. These exhibit metallic and paramagnetic properties, but once again the crystalline compounds formed are achiral with space groups $C2/c$ and P-1 respectively.

Coronado obtained an enantiopure molecular conductor by using Sb$_2$((2R, 3R)-(+)-Tartrate)$_2$ Sb$_2$-(**6.4**)$_2$ as the chiral inductor anion in the self-assembly.[25] Electrocrystallization in the presence of (BEDT–TTF) yields a conductive compound with the formula [BEDT-TTF]$_3$[Sb$_2$((2R, 3R)-(+)-Tartrate)$_2$].CH$_3$CN (**6.18**).CH$_3$CN which crystallizes in the chiral space group $P2_12_12_1$. It is composed of two subnetworks, one anionic and the other cationic, in alternate layers, with the BEDT–TTF perpendicular to the anionic layers and the cationic layers forming a helical structure (Figure 6.11).

The CD spectrum shows intense Cotton effects at 290 and 570 nm (the band being positive at 290 nm and negative at 570 nm for the derivative [BEDT-TTF]$_3$[Sb$_2$((2R, 3R)-(+)-Tartrate)$_2$].CH$_3$CN) (2R, 3R)-(**6.18**).CH$_3$CN. The enantiomer [BEDT-TTF]$_3$ [Sb$_2$((2S, 3S)-(+)-Tartrate)$_2$].CH$_3$CN (2S, 3S)-(**6.18**).CH$_3$CN has a CD spectrum that is symmetrical with the first. The observed Cotton effects are not related to those of the antimonyl tartrate, which is found below 240 nm, but to the BEDT–TTF cationic network, confirming that the whole structure is chiral. Physical measurements show that the material is an enantiopure semiconductor although, except for the CD, it is not possible to attach any particular property to the material's enantiopure chiral character. Following the same synthetic strategy, a salt of BEDT–TTF with the polyoxometalate

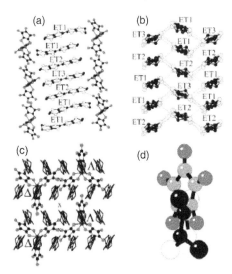

Figure 6.10
Structure of α-BEDT-TTF$_5$[Fe(C$_5$O$_5$)$_3$].5H$_2$O α-(**6.17**).5H$_2$O. (a) View of the alternating cationic and anionic layers. (b) View of the ET layer showing the α packing. (c) View of the anionic layer with the ET molecules above it; the crosses indicate the location of the inversion centres in the organic layer. (d) View of a croconate ring and the terminal S and C atoms of the ET molecules above (black atoms) and below (white atoms) (reproduced with permission from reference[24], copyright 2006, The Royal Society of Chemistry).

anion [H$_4$Co$_2$Mo$_{10}$O$_{38}$]$^{6-}$ was prepared.[9] Starting from the racemic polyoxometalate, the derivative BEDT-TTF$_9$[H$_4$Co$_2$Mo$_{10}$O$_{38}$].4H$_2$O (**6.19**).4H$_2$O is obtained as an achiral crystal (space group *P*21/c). In this case, the structure is also composed of two sub-networks, the layers formed by the BEDT–TTF being perpendicular to the anionic layers (Figure 6.12).

However, it has not been possible to carry out a structural study of the compound obtained from the enantiomer (+)$_{589}$-[H$_4$Co$_2$Mo$_{10}$O$_{38}$]$^{6-}$ by the same electrocrystallization method.

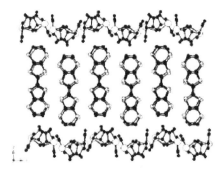

Figure 6.11
View of the alternating layers in the [BEDT-TTF]$_3$[Sb$_2$((2R, 3R)-(+)-Tartrate)$_2$].CH$_3$CN (**6.18**). CH$_3$CN salt (reproduced with permission from reference[8], copyright 2004, Macmillan Publishers Ltd).

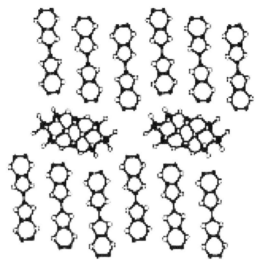

Figure 6.12
Projection of the structure of BEDT-TTF$_9$[H$_4$Co$_2$Mo$_{10}$O$_{38}$].4H$_2$O (**6.19**).4H$_2$O along the b axis, showing the BEDT–TTF perpendicular to the anionic polyoxometalate layers (reproduced with the permission from reference[9], copyright 2005, Elsevier).

6.3 Metallomesogens

6.3.1 General Considerations

A new class of mesogen appeared during the 1990s, that is those incorporating a transition metal, and a very large number of compounds was synthesized leading to new mesophases.[26,27] The introduction of chirality into organic mesophases has long been known, but is much more recent for metallomesogens and can be expected to lead to different effects. These concern not only the organization itself and the characteristics of the mesophases formed, but also their electrical, optical and magnetic properties. As was reported recently by Eelkema and Ferinka[28] *'Coordination complexes owing their chirality either to a chiral ligand or a chiral metal center can be highly effective dopants for induction of cholesteric mesophases'*.

Here we will present the principal types of enantiopure metallomesogens synthesized, as well as their helical twisting power (htp) when used as doping in a nematic phase. We will consider four principal types of enantiopure metallomesogen:

(i) *where the chiral element is attached directly to the metal,*
(ii) *where the chiral elements are carried by the ligands,*
(iii) *where the metal and ligands generate helical chirality, and*
(iv) *those based on phthalocyanines.*

6.3.2 Metallomesogens with the Chiral Element Attached Directly to the Metal

Making use of the chemical property of platinum to bond with a sulphur atom, Fanizzi and Rourke[29] constructed a metallomesogen in which platinum (II) is tetra-coordinated

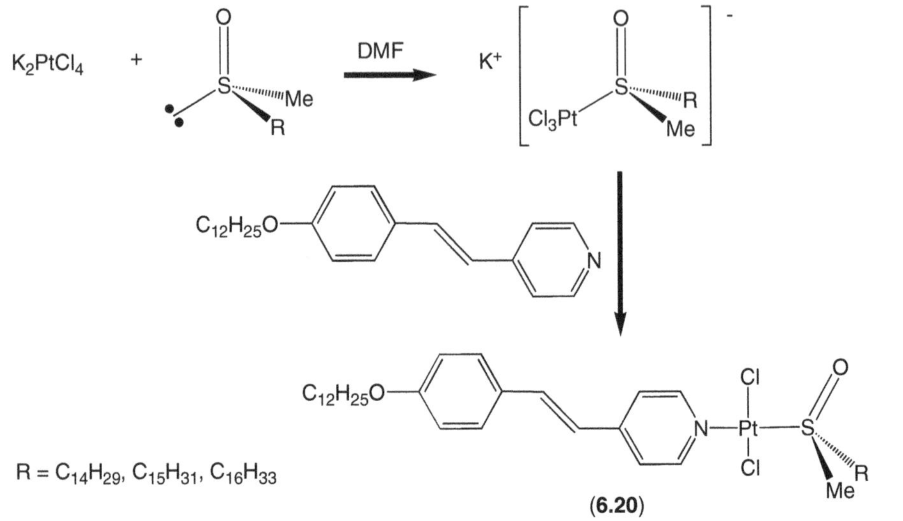

Figure 6.13
Synthetic pathway leading to enantiopure Pt(II)-based metallomesogen.

to the nitrogen of a substituted pyridine, two chlorine atoms and an enantiopure sulphoxide. The synthesis of this compound (**6.20**) is shown in Figure 6.13.

The principal difference observed between the derivative with a racemic sulphoxide and that with the enantiopure form, (*R*)-(**6.20**) or (*S*)-(**6.20**), is in the thermal behaviour. The (*rac*)-(**6.20**) compound exhibits a smectic G phase, followed at higher temperature by smectic F∥, while the enantiopure compounds show only a smectic F. The melting point of the enantiopure compounds is 10 K higher than the corresponding racemic compound.

6.3.3 Metallomesogens with the Chiral Element(s) in the Ligands

Starting from his initial work in 1989 on organometallic compounds with a ferroelectric phase,[30] Espinet developed the chemistry of metallomesogens based on dimeric palladium derivatives of chiral Schiff bases ($Pd_2(\mu\text{-Cl})_2L^*_2$ where L* represents a chiral Schiff base ligand.[31,32] Several examples of these complexes (**6.21-6.23**) are shown in Figure 6.14

All these compounds are mesogenic, whereas the Schiff bases from which they are formed are not. The effect of complexation is therefore the determining factor in the ordering of the imine alkyl chains that leads to an arrangement compatible with the formation of a liquid crystal.

Both the mononuclear palladium complexes, with a chiral alkyl chain in the ether substituent of one of the aromatic rings of the Schiff base, and the dinuclear complexes, where the bridge joining the two palladium atoms is itself chiral, exhibit a blue phase in a glassy state.[33,34] Pure, they form a cholesteric phase, right-handed for the (*S*)-Pd (*S*)-(**6.24**) derivative and left-handed for the (*R*)-Pd_2 (*R*)-(**6.25**) derivative (Figure 6.15).

The (*S*)-(**6.24**) compound shows good solubility in a nematic host (ZLI-1275 Merck) and is an effective doping with a 'molecular twisting power' (*mtp*) of between 1500 and 2400 m^2mol^{-1}, for a doping agent concentration of 17.43 wt% and at a temperature of

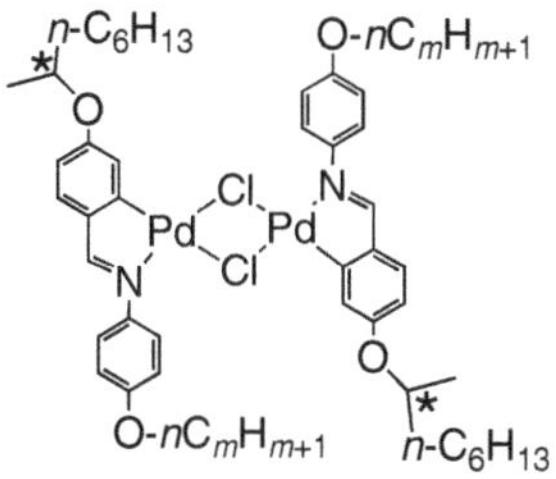

* (S)-(**6.21**) m = 6
(R)-(**6.21**) m = 6
(S)-(**6.21**) m = 8
(R)-(**6.21**) m = 8
(S)-(**6.21**) m = 10
(S)-(**6.21**) m = 14

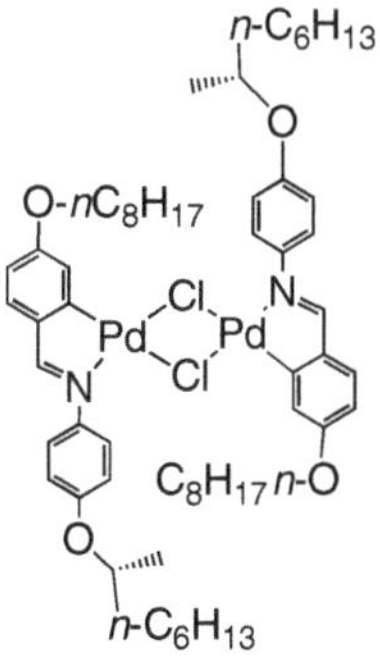

(R)-(**6.22**) 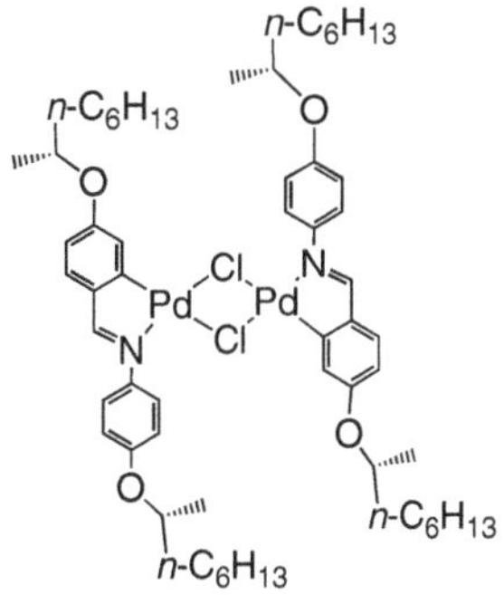   (R)-(**6.23**)

Figure 6.14
Metallomesogens based on dimeric palladium (II) derivatives of enantiopure Schiff bases $Pd_2(\mu\text{-}Cl)_2L^*_2$ (**6.21-6.23**) where L* represents a chiral Schiff base ligand.

30–60°C.[34] In comparison, the most effective calamitic chiral doping agent known – DL21 – has an *mtp* of 28 300 $m^2 mol^{-1}$.[35]

Another way of introducing a chiral element is to synthesize a complex tetracoordinated by a tridentate (pincer) ligand and a monodentate ligand.[36] The chiral element can be present in one (**6.26**) or other of the ligands (**6.27**) (Figure 6.16).

In this way, some nickel and palladium compounds have been synthesized with the intention of obtaining chiral nematic (N*) and smectic (SmC*) phases. The compounds where the chiral element is situated in the chain of the monodentate ligand are not themselves mesogenic. The others are, and two of them have a modest twisting power in Merck MLC-6401 nematic solvent with a right-handed helix.

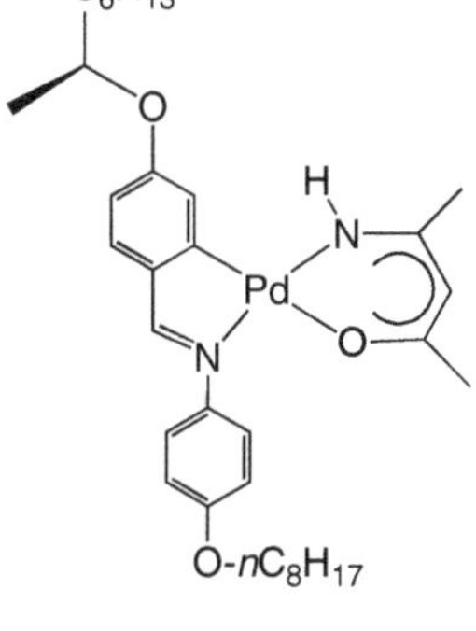

(S)-(**6.24**) 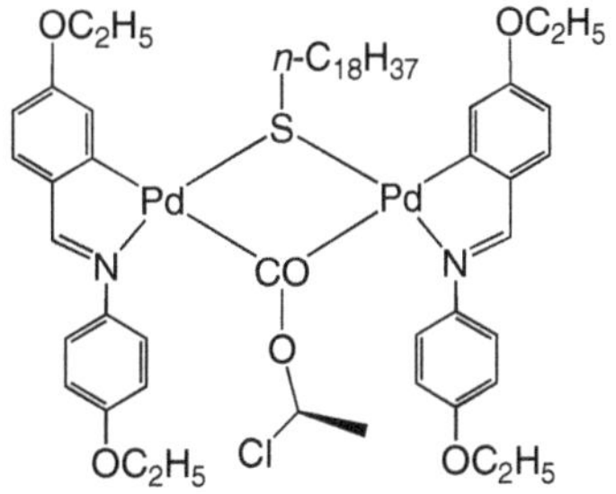   (R)-(**6.25**)

Figure 6.15
(S)-Pd(II) and (R)-Pd(II)$_2$ complexes.

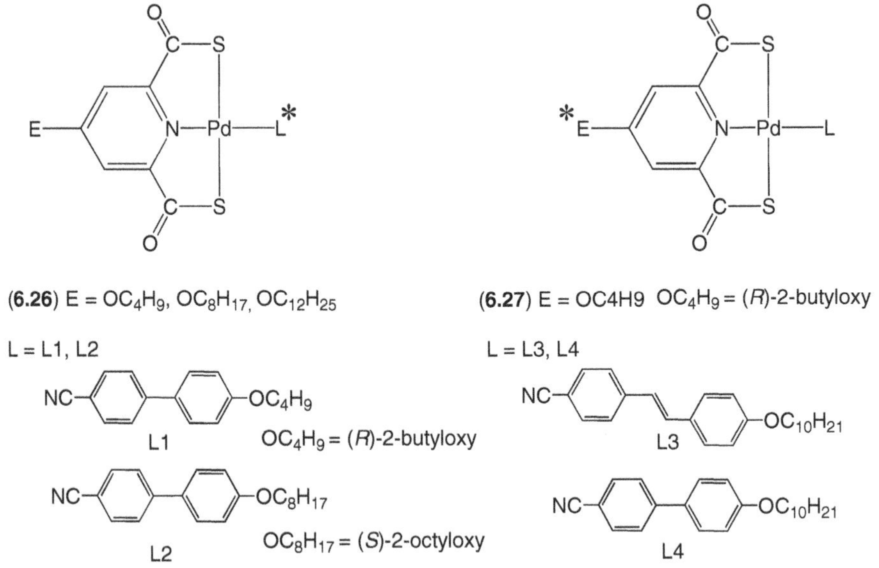

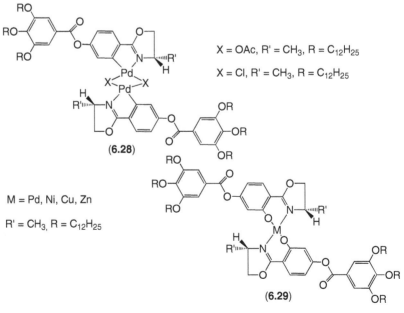

(**6.26**) E = OC$_4$H$_9$, OC$_8$H$_{17}$, OC$_{12}$H$_{25}$ (**6.27**) E = OC4H9 OC$_4$H$_9$ = (*R*)-2-butyloxy

L = L1, L2   L = L3, L4

NC—⬡—⬡—OC$_4$H$_9$ NC—⬡—CH=CH—⬡—OC$_{10}$H$_{21}$

L1   OC$_4$H$_9$ = (*R*)-2-butyloxy   L3

NC—⬡—⬡—OC$_8$H$_{17}$ NC—⬡—⬡—OC$_{10}$H$_{21}$

L2   OC$_8$H$_{17}$ = (*S*)-2-octyloxy   L4

Figure 6.16
Pd(II) complexes tetracoordinated by a tridentate (pincer) ligand and a monodentate ligand.

Metallomesogens formed by 'hexacatenar' chirals have also been synthesized by Serrano, based on tetracoordinated complexes of Pd, Ni, Cu and Zn (**6.28-29**)[37] (Figure 6.17).

The ligands, which are enantiopure oxazolines, have an asymmetric carbon situated close to the metal centre. The presence of the substituents around the metal leads to steric hindrance, which even in the case of the dinuclear palladium derivatives prevents the

X = OAc, R' = CH$_3$, R = C$_{12}$H$_{25}$

X = Cl, R' = CH$_3$, R = C$_{12}$H$_{25}$

(**6.28**)

M = Pd, Ni, Cu, Zn

R' = CH$_3$, R = C$_{12}$H$_{25}$

(**6.29**)

Figure 6.17
Metallomesogens formed by 'hexacatenar' M(II) complexes.

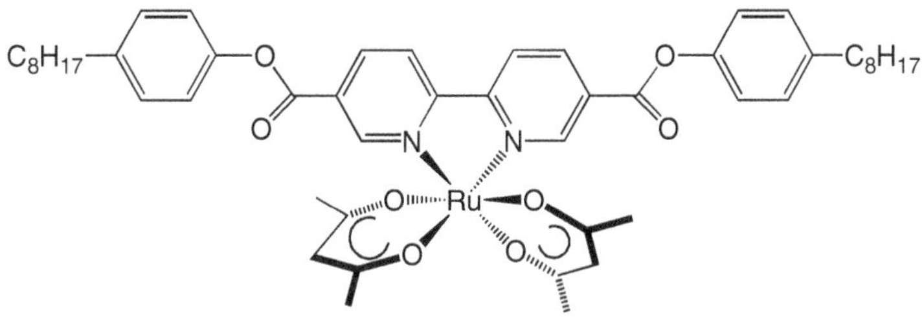

Figure 6.18
View of the (Δ)-(**6.30**) Ru(II) neutral complex.

formation of liquid crystal phases. In certain cases, the use of trinitrofluorenone as an intercalating electron acceptor allows smectic SmA phases to be obtained. The mononuclear derivatives of palladium, nickel and copper (**6.29**) were tested as doping agents in the nematic compound Paliocolor LC242 (BASF).[38] Some of these compounds, where the peripheral chain has six atoms, are capable of inducing a cholesteric phase, although their helical twisting power (*htp*) is only modest.

6.3.4   Metallomesogens Where the Metal and Ligands Generate Helical Chirality

In his work, Hoshino[39,40] mainly considered complexes of ruthenium (II) (**6.30**) coordinated by bidentate ligands [Ru(acac)$_2$L] (acac = acetylacetonato), (L = 5,5′-bis (4-octylphenyloxycarbonyl)-2,2′-bipyridyl) (Figure 6.18).

In these derivatives, the substituted bipyridyl ligand has a mesogenic character. The derivative (*rac*)-(**6.30**) is obtained by the zinc reduction of Ru^{III}(acac)$_3$ followed by addition of the bipyridyl ligand (L). The two enantiomers (Δ)-(**6.30**) and (Λ)-(**6.30**) are separated by chromatography on a column of synthetic laponite doped with [Ru(phen)$_3$]$^{2+}$ (Scheme 6.1).

Comparative study of the racemic derivative (*rac*)-(**6.30**) and that with (Δ) configuration (Δ)-(**6.30**) shows that these complexes do not exhibit mesomorphism. As a monolayer at an aqueous surface they do show a significant difference in that the racemic form exists as aggregates while the derivative with (Δ)-(**6.30**) forms a uniform monolayer in which the molecules are vertically orientated. In this way, their doping properties in nematic phases have been studied.

The technique of 'doping' with enantiopure organic compounds is widely used in the liquid crystal industry. However, the helical potential of hexacoordinated transition metal complexes has been little explored. Nevertheless, the chiral induction power of

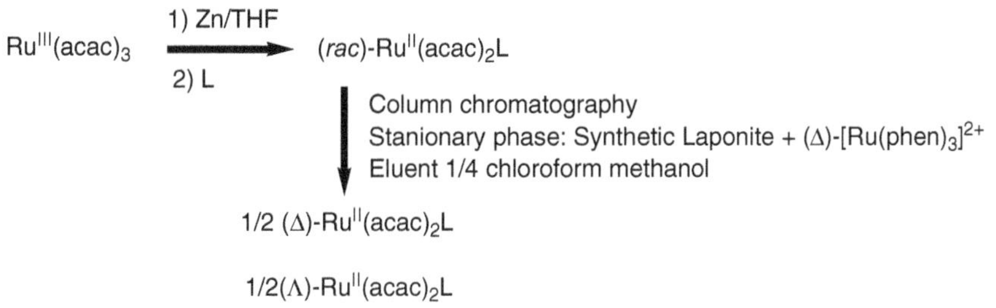

Scheme 6.1

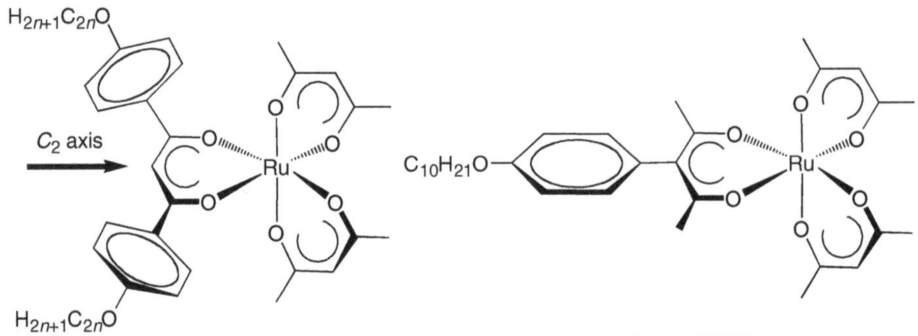

Figure 6.19
View of the L_{per} (Δ)-(**6.31**) and L_{para} (Δ)-(**6.32**) Ru(acac)$_2$L complexes of ruthenium (II).

complexes of the type '[M(L^2)]$_3$' in self-assembly has already been widely shown in the construction of enantiopure 3D networks (see Section 6.5).

The molar $htp(\beta_M = (\delta p^{-1}/\delta x)_{x \to 0}$ (where x is the molar fraction of the doping agent and p is the pitch length (μm)) of the complex (Δ)-(**6.30**) was measured at room temperature in a number of nematic phases, ZLI-1132 Merck-Japan ($\beta_M = -71 \times 10^2 \, \mu m^{-1}$) and MBBA N-(4-methoxybenzylidene)-4-n-butylaniline ($\beta_M = -1.8 \times 10^2 \mu m^{-1}$). These values are close to those obtained with organic compounds of the chiral binaphthyl type. The observations show the formation of chiral nematic phases N* under the influence of the doping agent, with values of pitch (p) close to the visible region of the spectrum. The sense of the induced helix is negative (M). In the case of MBBA, the CD spectrum shows a powerful induced negative effect in the region of the imine chromophores of the MBBA receptor.[40]

Other enantiopure complexes of ruthenium (II) Ru(acac)$_2$L, where L is is a 1,3-diketonate ligand containing phenylene groups, have been synthesized[41] (Figure 6.19).

In these compounds, the alkyl substituents of the aromatic rings of the ligands L are arranged perpendicular L_{per} (**6.31**) or parallel L_{para} (**6.32**) to the C_2 axis of the molecules. A study of these complexes as chiral doping agents in nematic liquid crystals MBBA, EBBA (N-(4-ethoxy-(benzylidene)-4-n-butylaniline), PAA (4, 4'-azoxydianisole) and ZLI-1132 Merck-Japan (ZLI-1132 is a mixture of 4-(-4-alkylcyclohexyl)benzonitrile and 4-(-4-alkylcyclohexyl)-4'-cyanobiphenyl derivatives) showed that, for the doping agent derivatives with (Δ) configuration, in every case the induced helix in the liquid crystal, characterized by its sense (M) or (P), depends on the orientation of the alkyl chains. Thus the compounds L_{per} (Δ)-(**6.31**) result in (M) helices ($htp < 0$), while the compounds L_{para} (Δ)-(**6.32**) induce the formation of (P) helices. The ICD curves confirm that, for the imine transitions of the MBBA, the sense of the helix formed depends not only on the configuration of the doping agent complex, (Δ) ou (Λ), but also on the L_{para} or L_{per} nature of the 1,3-diketonate ligand. Thus, it is possible for the doping agent to be endowed with two structural attributes. One is the ability to orientate the structure of chiral nematic and /or smectic phases according to the direction of the mesogenic axis carried by the metallic complex (Figure 6.20). The other governs the absolute configuration of the helical structure formed by the absolute configuration of the metallic complex with C_2 symmetry.

By using a general synthetic method, many multisubstituted complexes of the type Ru(acac)$_2$L (**6.33**) have been obtained in enantiopure form (Figure 6.21) following the

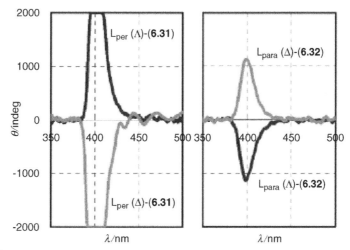

Figure 6.20
ICD spectra for MBBA materials doped with Δ- and Λ-enantiomers of L_{per} (**6.31**) ($n = 12$) (0.1 mol % left) and L_{para} (**6.32**) (0.3 mol % right). Each spectrum was recorded at 30 °C in a glass cell of 25 µm gap at normal incidence (reproduced with permission from reference[41], copyright 2005, American Chemical Society).

reaction scheme shown in Scheme 6.1. The resolution is performed on the neutral compound Ru(acac)$_2$L using as stationary phase a clay (synthetic laponite) doped with (Δ)-[Ru(phen)$_3$]$^{2+}$. All these compounds show an *htp* that is both remarkable and stereospecific for MBBA.[42]

As well as the chiral doping properties of the Ru(acac)$_2$L complexes, Einaga[43] also added the property of photomodulation by introducing a terminal *azo* group to the

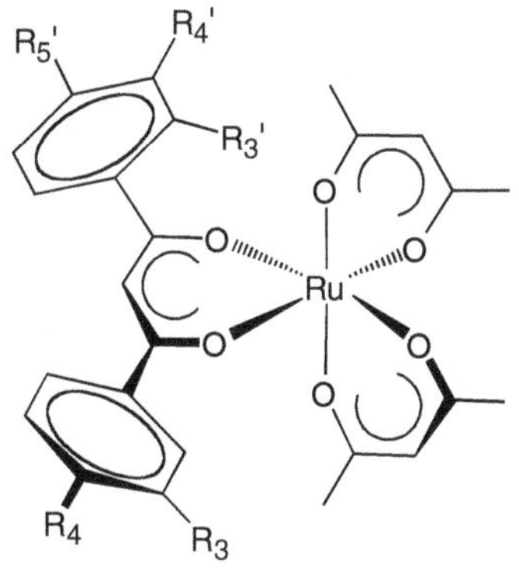

(Δ)-(**6.33**)

R3 = R4= R3' = R4' = R5' = H
R4 = R4' = OC$_8$H$_{17}$; R3 = R3' = R5' = H
R3 = R4 = R4' = OC$_8$H$_{17}$; R3' = R5' = H
R3 = R4 = R3' = R4' = OC$_8$H$_{17}$; R5' = H
R3 = R4 = R3' = R4' = R5' = OC$_8$H$_{17}$

Figure 6.21
Complexes of the type Ru(acac)$_2$L (**6.33**).

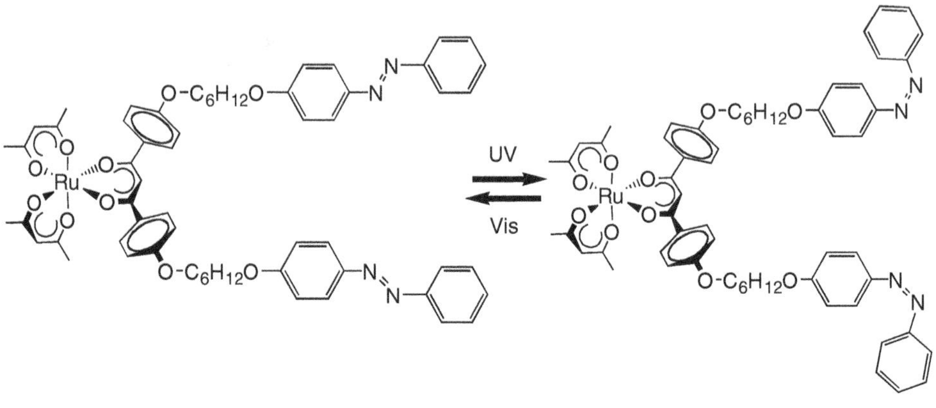

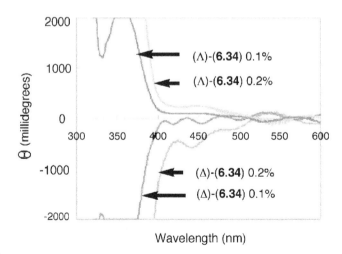

(Λ)-*trans form* (**6.34**)   (Λ)-*cis form* (**6.34**)

Figure 6.22
The functional azo group isomerizes, adopting a *trans* structure under the effect of ultraviolet light, and returning to a *cis* geometry by exposure to visible light.

ligand L. The functional *azo* group is capable of isomerization, adopting a *trans* structure under the effect of ultraviolet light, and returning to a *cis* geometry by exposure to visible light (Figure 6.22).

By introducing this photo-isomerizable group, a particular property is added to the liquid crystal, namely a reversible response to a change in wavelength, and the doping becomes a photoreactive doping agent. An efficient photoreactive doping agent must have a high *htp* for an N* phase, with the value of the *htp* varying significantly under illumination. The compounds (**6.34**), when tested with the nematic ZLI-1132, induce an N* phase where the helical sense is (P) for the *azo* doping agent of (Λ)-(**6.34**) configuration, and (M) for that with (Δ)-(**6.34**) configuration. The helical sense is determined by CD for the π–π* transition of the host molecule at 360 nm, which gives a positive Cotton effect for (P) helicity (Figure 6.23). The intensity of the effect, expressed in θ (mdeg), changes drastically when the sample is illuminated by ultraviolet or visible light.

Figure 6.23
CD spectra of Δ- and Λ-[Ru(acac)₂azo] (**6.34**) in ZLI-1132 at 35.0 °C (reproduced with permission from reference[43], copyright 2006, American Chemical Society).

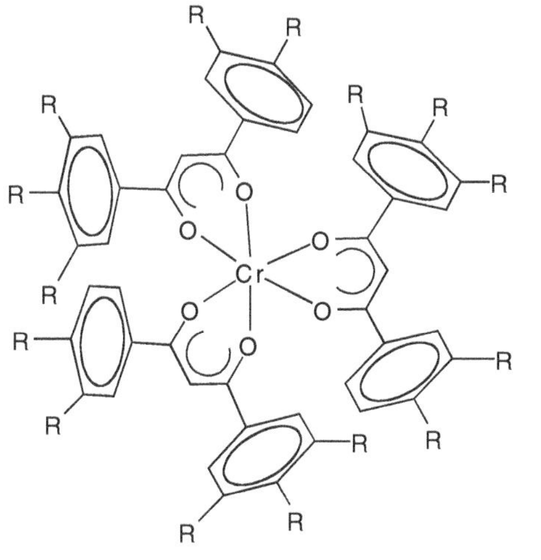

(**6.35**)

Figure 6.24
View of the hexacoordinated chromium (III) complex (rac)-[Cr(L^2)$_3$] (**6.35**).

This change is due to the fact that, according to theory, variation in the intensity of the ICD spectrum is inversely proportional to the pitch of the helix that generates the observed Cotton effect. In this case, the pitch of the induced helix in the nematic host depends on the *cis* or *trans* geometry of the doping agent *azo* group as a result of the different steric requirements of the two isomers.

The racemic hexacoordinated chromium (III) complex (*rac*)-[Cr(L^2)$_3$] with $L = $ 1-3,4,5-trioctyloxyphenyl)-3-(3,4-dioctyloxyphenyl)propane-1,3-dionate (**6.35**) (Figure 6.24) is mesogenic and, at the surface of highly ordered pyrolytic graphite, forms a structure of nanometric order with domains of (Δ) and (Λ) configuration.[44]

With the octahedral compounds of the same type, Swager[45] studied the formation of columnar liquid crystals where the molecules in the mesophase can align themselves along the pseudo-C_3 molecular axis (Figure 6.25).

The compounds studied show two important characteristics.

(i) *The central metal atom determines their configurational stability.*
For the complexes of Fe(III)-(**6.36**) and Mn(III)-(**6.36**) the barrier to interconversion between the (Δ) and (Λ) enantiomers is low, and racemization is rapid, including at room temperature. On the other hand, the complexes of Co(III)-(**6.36**) and Cr(III)-(**6.36**) show configurational stability.

(ii) *Two of the complexes, one of Fe(III)-(**6.36**) and the other of Co(III)-(**6.36**), have ligands substituted by a chiral alkyl chain of defined configuration, that is (R)- or (S)-dihydrocytronellyl.*
Although all of the compounds studied form columnar phases, differences between them become apparent according to the metal involved in the complex. Thus, the fluxional complexes (Fe(III)-(**6.36**) and Mn(III)-(**6.36**)) exhibit a hexagonal columnar phase (Col$_h$) at low temperatures, whereas the nonfluxional complexes show a rectangular columnar phase (Col$_r$). In the case of the Col$_h$ phase, microdomains composed of (Δ) or (Λ) configuration are present. This arises due to the possible

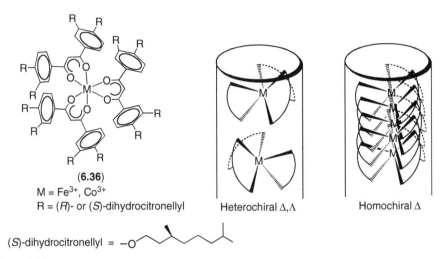

Figure 6.25
View of the columnar heterochiral [(Δ), (Λ)] and homochiral (Δ) arrangements of compound (**6.36**).

arrangement of the ligands when the complex is fluxional. The three possible types of organization are shown in Figure 6.26.

In going from (a) to (b) and then to (c), the structure changes from a hexagonal network with a uniform configuration of all the elements in all the columns, to a hexagonal network where every column has a particular configuration, but with one overall dominant configuration, and finally to a rectangular network where there are equal numbers of columns of the two opposed configurations. This last network is in fact achiral.

A striking demonstration of this hypothesis is shown by the CD study of the Fe(III)-(**6.36**) and Co(III)-(**6.36**) complexes with chiral ligands (Figure 6.27).

In solution, as in the isotropic phase, the Fe(III)-(**6.36**) compound exhibits practically no Cotton effect between 450 and 800 nm. This result is consistent with a rapid

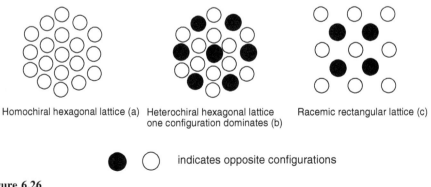

Figure 6.26
Possible organizations in an hexagonal chiral lattice or racemic rectangular lattice for complexes (**6.36**).

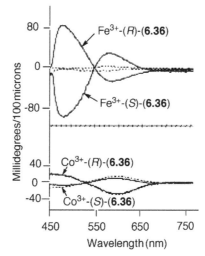

Figure 6.27
CD spectra of Fe^{3+}-(**6.36**) (top) and Co^{3+}-(**6.36**) (bottom) in the Col_h phase at 25 °C (solid line) and in the isotropic phase at 105 °C (dashed line) (reproduced with permission from reference[45], copyright 1999, American Chemical Society).

isomerization of the octahedral centre, while the ligands of (*R*) or (*S*) configuration do not contribute at these wavelengths. For the Co(III)-(**6.36**) compound the CD curves are identical in both the solution and the isotropic phase. The observation of two Cotton effects of opposite sign, according to whether the ligands are of (*R*) or (*S*) configuration, indicates a weak asymmetric induction at the level of the complex under the control of the ligand configuration. The behaviour on cooling is different for the two complexes studied. While there is no change for the Co(III)-(**6.36**) compound, for the Fe(III)-(**6.36**) compound there is a strong increase in the dichroic effect on entering the Col_h phase. The fluxional character of the octahedral complex allows for the possibility of an organization with a higher degree of symmetry. The nature of this organization is controlled by the configuration of the peripheral ligands as the CD curves obtained are perfectly symmetrical for the complexes with (*R*) and (*S*) ligands. In the presence of small quantities of the compound with chiral ligands, the equivalent Fe(III)-(**6.36**) complex with achiral ligands forms a phase that exhibits an intense Cotton effect. The complex with chiral ligands therefore exhibits the phenomenon of chiral induction in the host phase.

6.3.5 Metallomesogens Based on Chiral Phthalocyanines

Phthalocyanines (Pc) constitute an important class of compounds for the formation of liquid crystals. Elements from groups I_A to V_B of the periodic table can coordinate to the four nitrogen atoms of the phthalocyanine macrocycle, leading to more than 70 types of metallic complex. Small divalent ions are accommodated in the centre of the ring forming a planar, tetracoordinated complex, while heavier ions are situated out of the plane[46] (Figure 6.28).

Peripheral substitution of the macrocyclic ring can give rise to chiral phthalocyanine complexes, in particular when the metal is out of the plane, as is shown in Figure 6.29.

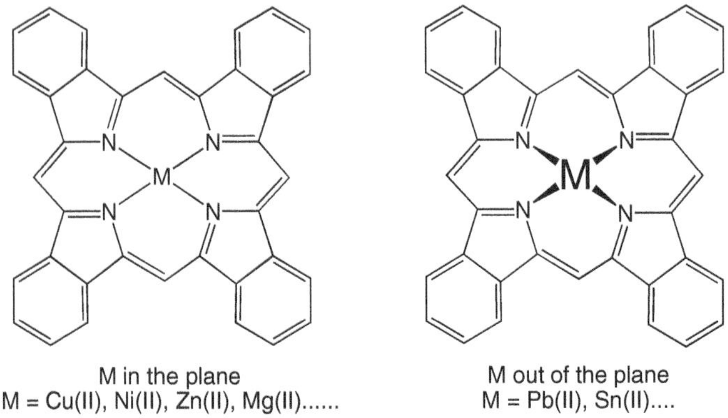

M in the plane
M = Cu(II), Ni(II), Zn(II), Mg(II)......

M out of the plane
M = Pb(II), Sn(II)....

Figure 6.28
View of two types of phthalocyanine forming planar complexes (left) with small divalent ions and
(right) with out-of-plane complexes with larger ions.

When phthalocyanines are substituted by long chains, they give rise to discotic type
columnar mesophases[47] (Figure 6.30).

We will present here some of the principal work that has led to the preparation and
isolation of enantiopure phthalocyanines, as well as some of their properties. Nolte[48]
reported the formation of an optically active chiral Dh* mesophase for octa((*S*)-3,7-
dimethyloctoxy)Pc (**6.37**). The configuration of the stack of substituted phthalocyanines
is a right-handed helix.

Katz[49] surrounded the nucleus of an octaazaphthalocyanine complexed by copper or
nickel with four helicene groups of (M) configuration (**6.38**). These groups give a global
helical structure to the Pc (Figure 6.31). The CD shows that the helices are stacked along

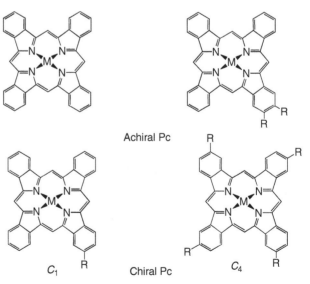

Achiral Pc

C_1 Chiral Pc C_4

Figure 6.29
Macrocycle substitution can afford chiral Pc derivatives.

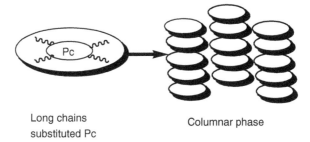

Long chains
substituted Pc

Columnar phase

Figure 6.30
Long chains substitute phthalocyanines to form columnar phases.

an axis perpendicular to the average Pc plane. Although these molecules are highly symmetrical, in the form of Langmuir–Blodgett films on glass they show an important second-order nonlinear optical response.

Kobayashi[50,51] obtained enantiopure phthalocyanines by introducing a helical type chiral element, or a chain containing an asymmetric carbon. The use of two or four chiral binaphthyl substituents [(P) or (S)] or [(M) or (R)], attached to the phthalocyanine via oxo or sulphur bridges, leads to compounds of the type BNpPcM and TNpPcM with M = H$_2$, Zn and Co (Figure 6.32).

CD study in THF solution shows Cotton effects at 700 and 370 nm in addition to the intense effects corresponding to the binaphthyl substituents. These effects, positive for the [(R) or (M)]-binaphthyl and negative for the [(S) or (P)]-ligand, arise from the Q (700 nm) and Soret (370 nm) bands of the conjugated system of the phthalocyanine. These electronic transitions are characteristic of the achiral planar chromophore of

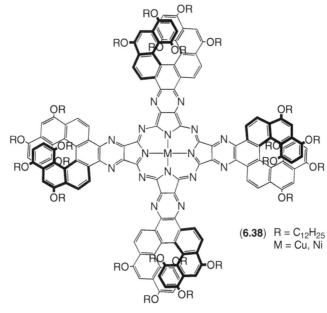

(**6.38**) R = C$_{12}$H$_{25}$
M = Cu, Ni

Figure 6.31
View of a phthalocyanine core surrounded by (M)-substituted helicenes (**6.38**).

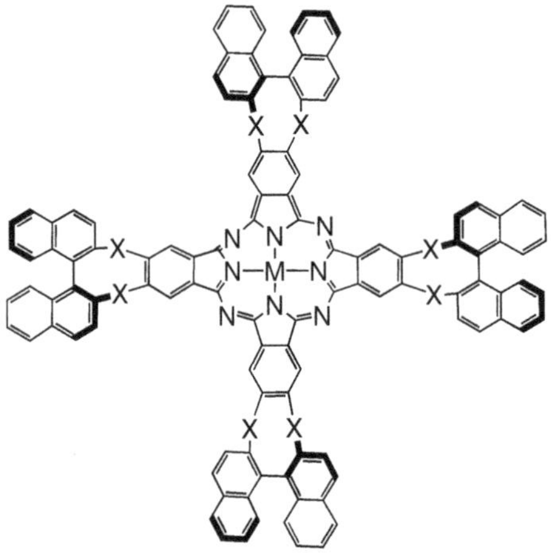

(*S*)-TNpPcM, M = Zn, Cu; X = O, S   (*S*)-BNpPcM, M = Zn, Co

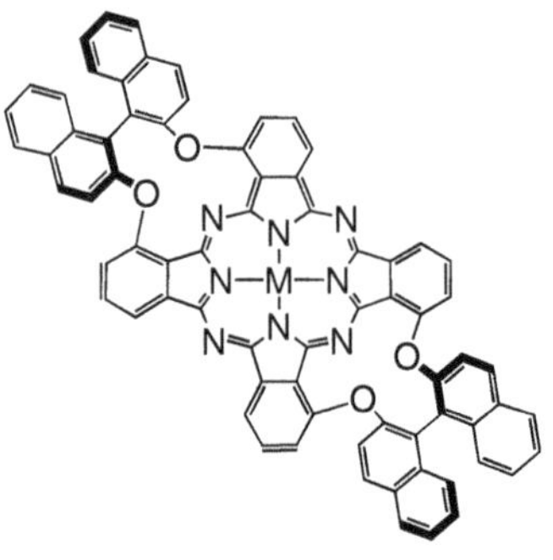

(**6.39**)   (**6.40**)

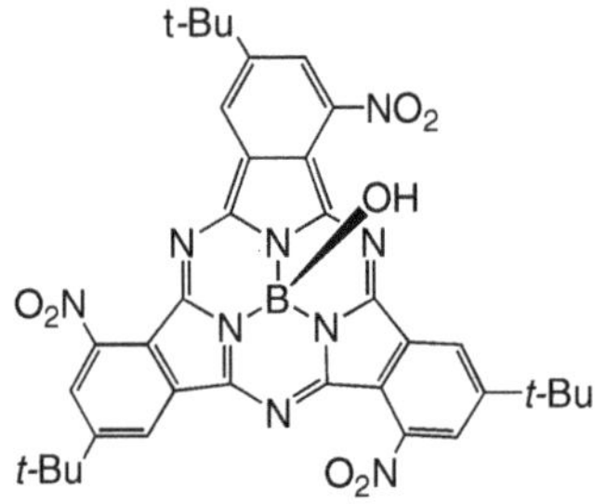

C3 symmetry

(**6.41**)

Figure 6.32
Phathalocyanine complexes **6.39**, **6.40** and **6.41**.

phthalocyanine. Study of this induced dichroism reveals information about the chiral environment of the phthalocyanine chromophore.

A phthalocyanine may also have a chiral structure if the metallic centre coordinated by the nitrogen atoms is out of the plane defined by the Pc ligand. A subphthalocyanine (**6.41**) with C_3 symmetry was resolved by HPLC.[52] The CD showed dichroic effects at 560 and 570 nm (Q_{00}- band) and 280 and 300 nm (Soret band), being either positive or negative according to the enantiomer. Unfortunately, the authors were not able to assign configurations to the enantiomers.

The presence of substituents around the phthalocyanine capable of participating in intermolecular hydrogen bonds is the basis for the formation of chiral superstructures, the configuration of which relates to the configuration of the chiral elements in the peripheral substituents of the Pc. Just as for π–π interactions, these hydrogen bonds are the origin of the forces which allow discotic molecules such as phthalocyanines to organize themselves into superstructures.

(Zn)Pc's have been described containing six chiral chains where the chiral elements (two asymmetric carbons) are racemic and/or of known configuration.[53,54] Their

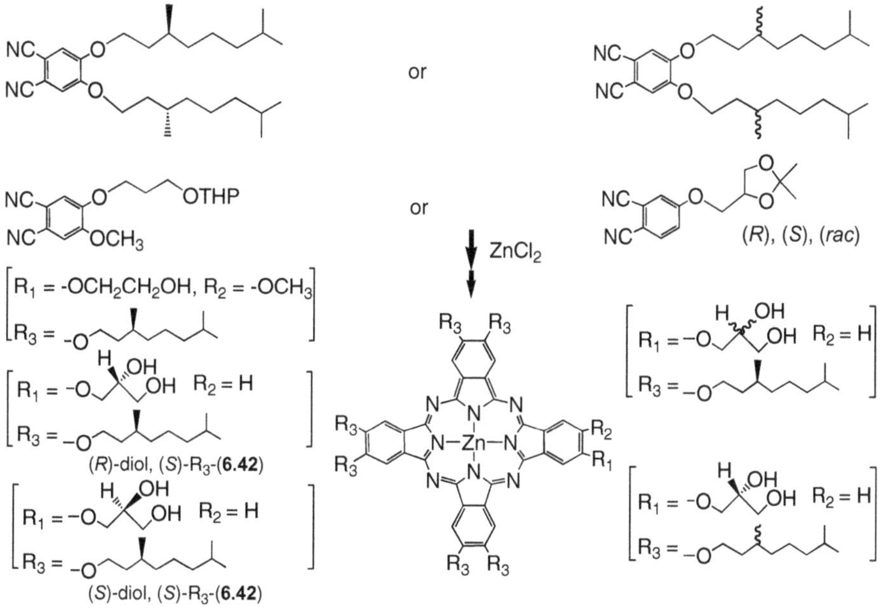

Figure 6.33
Synthesis of unsymmetrical ZnPc.

formulae and syntheses are shown in Figure 6.33. In addition to the six chains that surround them, these Pc's also contain a diol substituent capable of hydrogen bonding.

The study of the absorption spectra of these Pc's is of particular interest. It shows that, in CHCl$_3$ solution, the presence of a diol chain results in the formation of a dimeric superstructure through hydrogen bonding (Figure 6.34). The addition of 0.5% methanol dissociates the dimer by breaking the hydrogen bonds between the two substituted Pc's. In the case of the compounds (S)-diol, (S)-R$_3$-(**6.42**) and (R)-diol, (S)-R$_3$-(**6.42**), the CD curves for CHCl$_3$ are virtually mirror images (Figure 6.35).

For the compound *of* (S)-diol, (S)-R$_3$-(**6.42**), intense Cotton effects are observed for the Q bands of the Pc at 679 (+) and 670 nm (−) as well as at 711 (−) and 698 nm (+). The signs are inverted for (R)-diol, (S)-R$_3$-(**6.42**) and the curves are virtually symmetrical, even though these are diastereomers. These results indicate that a dimer with C_2

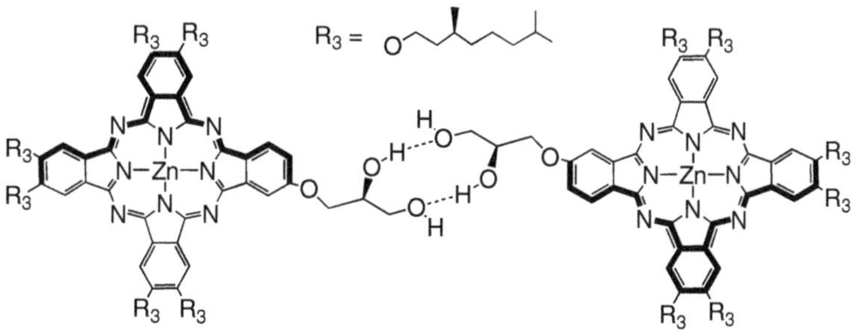

Figure 6.34
Schematic representation of the (S)-diol, (S)-R$_3$-(**6.42**) dimer in CHCl$_3$.

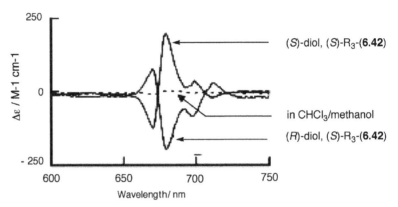

Figure 6.35
CD spectra of (S)-diol, (S)-R$_3$-(**6.42**) and (R)-diol, (S)-R$_3$-(**6.42**) in CHCl$_3$ and a mixture of CHCl$_3$
and MeOH (99.5:0.5) (reproduced with permission from reference[53], copyright 2003, American
Chemical Society).

symmetry is formed. When even a very small amount of methanol is added to the solution
of the Pc the CD spectrum loses all its intensity, clearly showing that it is the dimeric
superstructure that is responsible for the intense Cotton effects observed, since the effects
expected for the diol and the chains are very weak and at lower wavelengths. It is the
configuration of the diols that controls the configuration of the dimeric C_2 helix. The curves
are symmetric as a function of the diol configuration and are unaffected by a change in
configuration of the six lateral chains. The same applies when the chiral elements of the six
chains are racemic, which is not the case for the diol as, in a racemic form, no particular
configuration of the dimer is induced, the dimer itself then being racemic.

In the solid state, the existence of hydrogen bonds depends on the temperature, the
bonds disappearing as the temperature is increased. In the case of (S)-diol, (S)-R$_3$-(**6.42**),
on raising the temperature the structure passes from a lamellar to a hexagonal network,
this transformation being reversible.

6.4   Porous Metalorganic Coordination Networks (MOCN)

6.4.1   General Considerations

The self-assembly of coordination polymers is attracting a growing interest as it allows
the design of materials with large cavities capable of acting as catalysts, in comparison
with well-known catalysts based on zeolites.

Zeolites are a class of aluminosilicates with interconnected cavities and, at present,
there are no chiral zeolites accessible in an enantiopure form. This is because zeolites are
prepared in the presence of a surfactant that must then be removed by heating to 550°C,
conditions under which an enantiopure material cannot be expected to survive. In order to
synthesize porous chiral materials capable of showing, in particular, catalytic properties,
chemists have turned towards the self-assembly of inorganic and/or organometallic bricks
which allows highly selective reactions to take place under relatively mild conditions.
While many porous MOCN type materials exist, those that are chiral and, in particular,
enantiopure, are very rare. This is an area of growing development, and although at

present the catalytic properties of these chiral MOCNs in the field of asymmetric catalysis are generally modest compared with those of homogeneous mononuclear chiral catalysts, they still remain an interesting subject for study.

6.4.2 Main Strategies to Obtain Enantiopure MOCNs

Two types of strategy are used to obtain enantiopure MOCNs. The one employed most often consists of introducing the chiral information into the assembly ligand of the metallic ion network. The molecules most frequently used for this are binaphthol derivatives, tartrate ions, polypyridines substituted with chiral groups and amino acids. It is also possible to prefabricate a chiral building block of known configuration, the self-assembly of which leads to the desired MOCNs.

While it does not constitute, in general, a synthetic method, spontaneous resolution during the course of the cristallogenesis can give rise to seed crystals that are capable of orienting the network configuration through seeding. Thus, Aoyoma[55] observed that self-assembly of the cadmium (II) ion with the ligand 5-(9-anthracenyl)pyrimidine L (**6.43**) leads to a 3D coordination polymer L.Cd-(NO₃)₂.H₂O.EtOH (**6.44**) of helicoidal structure (Figure 6.36). Second-order spontaneous resolution leads to homochiral crystals (space group $P2_1$), where the helix configuration (P) or (M) is random. By using these (P) or (M) crystals to seed the initial reaction solution the compound of desired configuration can be obtained.

We will distinguish between 1D, 2D and 3D networks. Remember that this distinction depends on the fact that, in these networks, the metal–ligand bonds are directed in one, two or three directions in space. This evidently does not exclude the possibility of

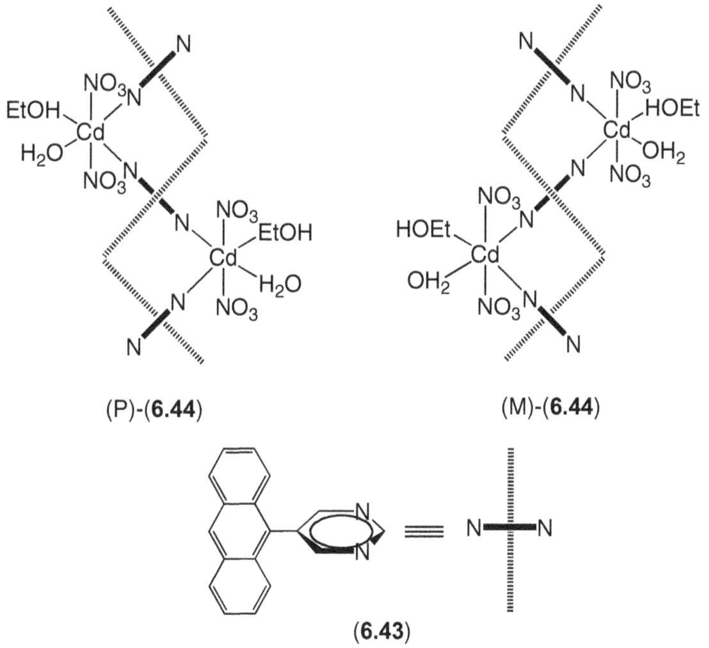

(P)-(**6.44**)   (M)-(**6.44**)

(**6.43**)

Figure 6.36
(P)-and (M)-L.Cd-(NO₃)₂.H₂O.EtOH (**6.44**) helicoidal structure.

three-dimensional cohesion as a result of weak interactions ($\pi-\pi$, electrostatic or van der Waals) between the constituent elements (chains and layers) of the 1D and 2D networks. In reality, in the solid state where the properties of MOCNs are studied, the distinction between these three types of network is rather academic. Attention must be focussed on two recurring aspects of this type of material. The first concerns the very existence of the network other than in the solid and/or crystalline state. Indeed, interaction with a solvent sometimes leads to decoordination and the formation of aggregates. The latter are made up of a greater or lesser number of base units, the properties of which, and in particular the optical properties, are not necessarily the same as those of the network in the solid state. The second aspect lies in the fact that porous crystalline structures often exist only through the presence in their channels of solvent molecules and more or less bulky ions, the removal or exchange of which lead to the collapse of the crystal, these molecules or ions thus assuring its cohesion. Caution must therefore be exercised in the properties expected from these networks based on their formal pore size.

6.4.3 1D MOCNs

In general, the construction of 1D networks proceeds by the coordination of a metallic centre, derived from either a salt or an achiral complex, with a ligand that contains the chiral information and which allows the complexation to progress along a specific axis. The superstructure obtained can acquire its own chirality, as is the case, for example, in the formation of a helix. Thus, Hosseini[56] used the reaction of $CoCl_2$ with the tectonic ligand L (**6.45**), which has C_2 symmetry and one monodentate and one tridentate terminal group, to form infinite chains of $(Co(L)Cl_2.3CH_3CN)$ (**6.46**), arranged in a parallel fashion (Figure 6.37).

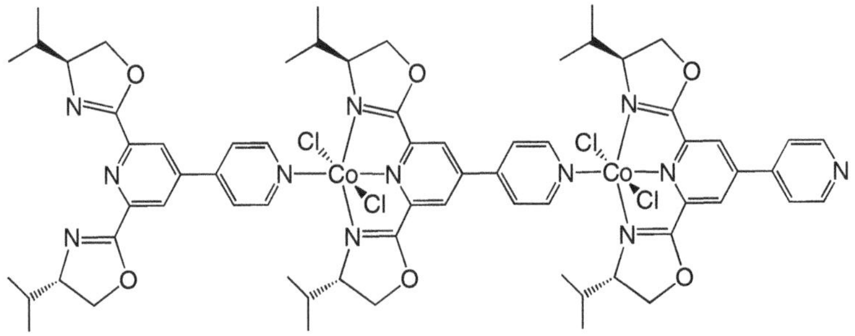

(6.46)

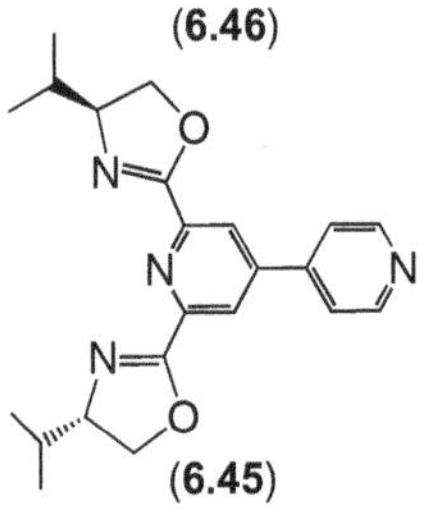

(6.45)

Figure 6.37
Infinite chains of $(Co(L)Cl_2.3CH_3CN)$ (**6.46**) in which the metal centres are coordinated through the enantiopure ligand L (**6.45**).

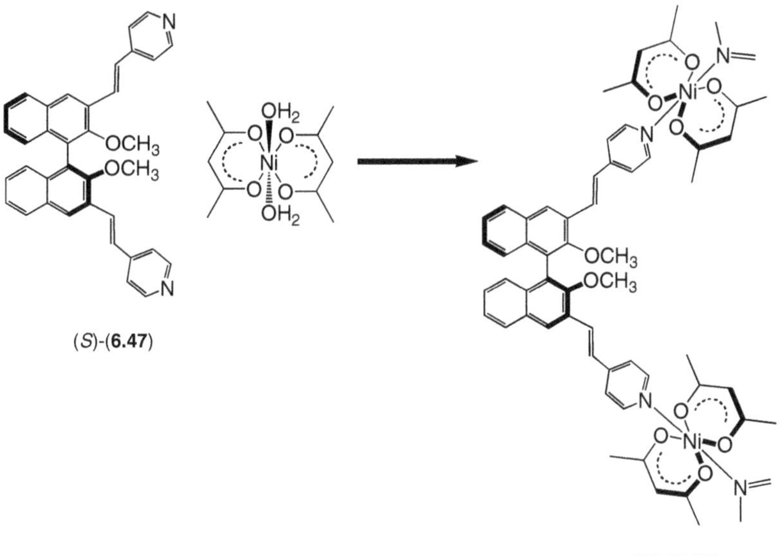

Figure 6.38
Schematic view of one of the helix (**6.48**) formed by reaction of Ni(acac)(H$_2$O)$_2$ with the ligand 2,2'-dimethoxy-1,1'-binaphthyl-3,3'-bis-(4-vinylpyridine) (S)-(**6.47**).

1D compounds of a zigzag or helicoidal type can be obtained by using ligands, which have axial chirality, in particular those that contain the binaphthyl group. In this way, Lin[57] reacted the achiral complex Ni(acac)(H$_2$O)$_2$ with the ligand 2,2'-dimethoxy-1,1'-binaphthyl-3,3'-bis-(4-vinylpyridine) (L) (**6.47**) to obtain a 1D structure (**6.48**) composed of triple helices linked together by van der Waals forces, with π–π interactions leading to stacking of adjacent triplets. The configuration of the homochiral triple helix is controlled by that of the ligand used, being (P) for the ligand with the configuration (*R*)-(**6.47**) and (M) for the (*S*)-(**6.47**) (Figure 6.38).

This is also the case for 1D networks[58] based on silver (Ag$^+$) coordinated to a (*R*)-6,6'-dibromo-1,1'-binaphthyl ligand (**6.49**) substituted by two isonicotinic groups. A helix with (P) configuration (P)-(**6.50**) is formed, its spatial arrangement being as mutually orthogonal columns (Figure 6.39).

Compounds with a zigzag structure along one axis can be obtained by the reaction of 2,2'-dihydroxy-1,1'-binaphthalene-6,6'-dicarboxylic acid (**6.51**) with salts of manganese, copper, cobalt or cadmium. In these networks (**6.52**), hydrogen bonds between the 1D zigzags play a fundamental role in the cohesion of the crystal.[59,60]

Abruña[61–63] obtained enantiopure 1D coordination polymers based on Fe(II) or Ru(II) (Fe-**6.53**) and (Ru-**6.53**) by using a bis-terpyridyl connected by an aromatic ring as the assembly ligand (−)-[ctpy-x-ctpy]-(**6.54**). Such a ligand is chiral due to the terpenyl groups fused to the aromatic rings, and has six N coordination sites, which allows an octahedral geometry around the iron or ruthenium atom. The structures of this ligand and of the resulting polymers are shown in Figure 6.40.

CD study shows that the configuration of the octahedral chiral element is controlled by that of the ligand, being (Δ) for the ligand (−)-[ctpy-x-ctpy] (**6.54**). The chemistry of these coordination polymers is directly related to the work of von Zelewsky[64,65] on the enantioselective formation of chiral mono- and binuclear ruthenium (II) complexes.

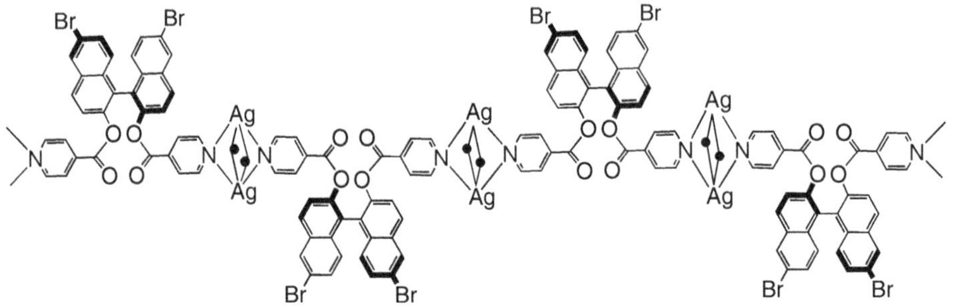

(P)-(**6.50**)

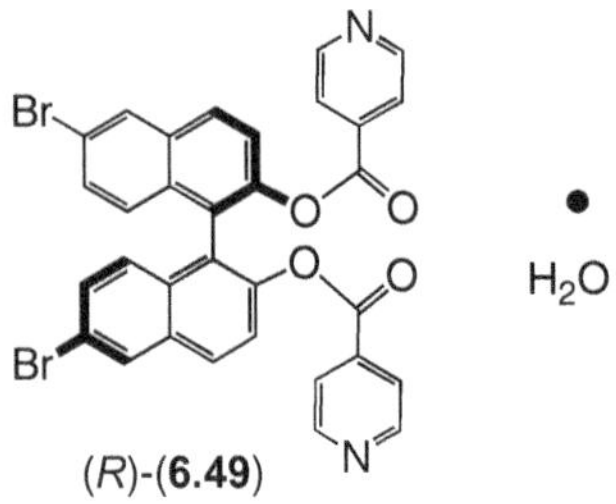

(R)-(**6.49**)

Figure 6.39
1D networks (P)-(**6.50**) based on silver (Ag^+) coordinated to a (R)-6,6′-dibromo-1,1′-binaphthyl ligand substituted by two isonicotinic groups (R)-(**6.49**).

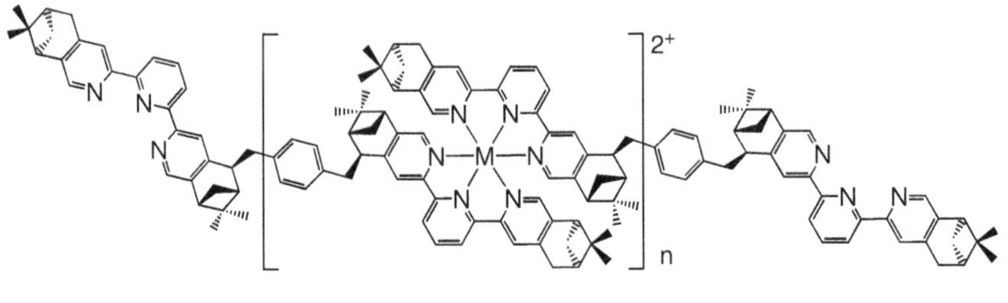

(**6.53**) M = Fe, Ru

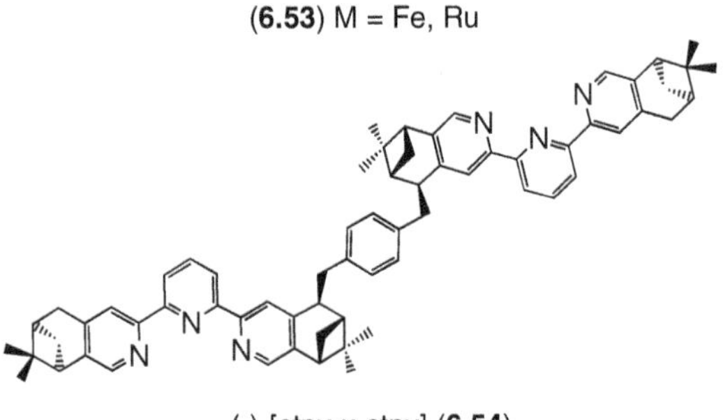

(-)-[ctpy-x-ctpy] (**6.54**)

Figure 6.40
1D (Fe-**6.53**) and (Ru-**6.53**) coordination polymer based on hexacoordinated Fe(II) or Ru(II) metallic ions and the ligand (−)-[ctpy-x-ctpy]-(**6.54**).

6.4.4   2D and 3D MOCNs

Zur Loye[66] observed that many 2D networks of the 'square-grid' type that exhibit channels or pores are interpenetrating, which clearly limits their porous character. He noted that lengthening bifunctional ligands of the *N,N'* type, which a priori should increase the pore size, actually favours the formation of interpenetrating structures. However, the presence of lateral chains helps to reduce the interpenetration and makes possible the formation of networks with infinite, broad channels. Two ligands, one achiral and the other chiral, which give such a result are 9,9-diethyl-2,7-bis(4-pyridylethynyl)fluorene (**6.55**) and 9,9-bis[(*S*)-2-methyl-butyl]-2,7-bis(4-pyridylethynyl)fluorene (**6.56**) (Figure 6.41). In the case of the achiral ligand, the pore size in each layer is 25 × 25 $\mathring{A}^2$, but this is reduced to 16 × 16 $\mathring{A}^2$ if the stacking of the layers is taken into account. The crystal space group is achiral (*P2₁/c*). Using the chiral ligand leads to a layer structure of the same type, but with a chiral space group (*P2₁*). In each layer, the substituents of the fluorene group are uniformly directed towards the interior of the channels. The space delineated by the channels is therefore chiral, although the surface is reduced in size (8 × 8 $\mathring{A}^2$).

Starting from (D)- or (L)-tartaric acid, Kimoon Kim[67] synthesized a chiral ligand (**6.57**) with both carboxylic acid and pyridine functionality (Figure 6.42).

The reaction of this ligand with zinc nitrate leads to a network of the general formula [Zn(μ_3-O)(L-H)₆].2H₃O.12H₂O (POST-1) (**6.58**). In the trinuclear base units, the three zinc atoms and the connecting oxygen atom are in the same plane. Bonds between the zinc ions and the pyridine nitrogens join these units to each other, generating a 2D network structure (Figure 6.43).

These sheets exhibit a hexagonal structure, separated from each other by a distance of 15.5 $\mathring{A}$, and since the sheets stack up on each other, held together by van der Waals bonds,

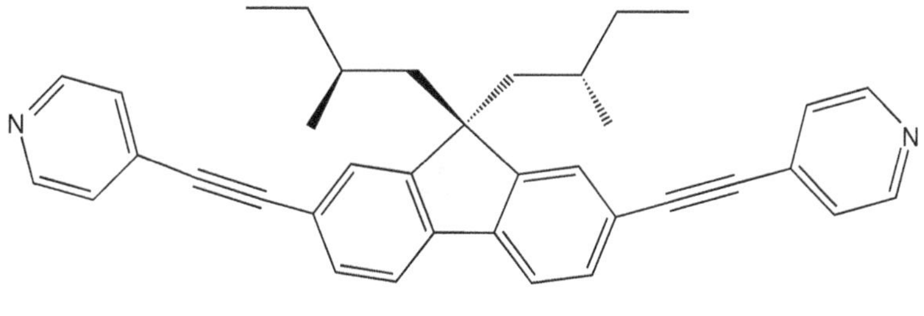

(**6.56**) Chiral ligand

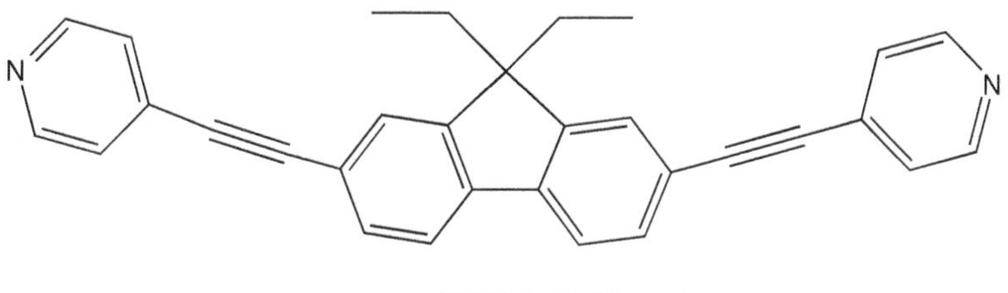

(**6.55**) Achiral ligand

Figure 6.41
9,9-diethyl-2,7-bis(4-pyridylethynyl)fluorene (**6.55**) and 9,9-bis[(S)-2-methyl-butyl]-2,7-bis(4-pyridylethynyl)fluorene (**6.56**).

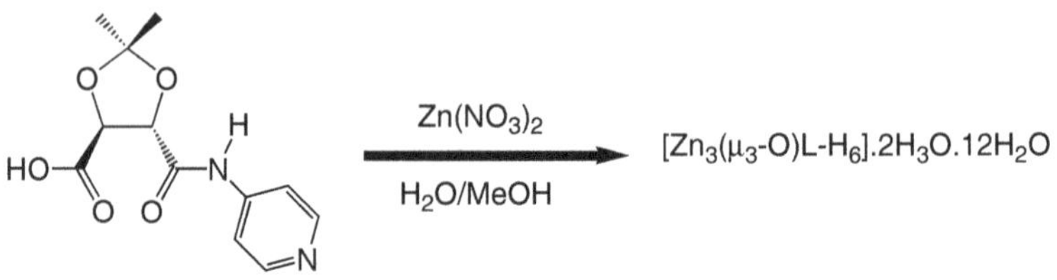

$$[Zn_3(\mu_3\text{-}O)L\text{-}H_6].2H_3O.12H_2O$$

(6.57)   L-(6.58)

Figure 6.42
Chiral organic building block L (**6.57**) obtained from (D)-tartaric acid reacts with Zn(II) leading
to D-POST-1 D-(**6.58**).

the structure possesses wide chiral channels. Each base unit has two protons associated
with two of the six pyridine groups present, and these can be exchanged with Na^+, K^+
and Rb^+ ions. More interestingly, the dication (*rac*)-[Ru(2,2'-bpy)$_3$]$^{2+}$ exchanges with
80% of the possible protons and does so in an enantioselective way since, for the network
produced by the (L)-(**6.57**) ligand, it is the (Δ)-[Ru(2,2'-bpy)$_3$]$^{2+}$ complex, that is
preferentially formed (*ee* = 0.66).

The 2D (**6.58**) network possesses a free pyridine function, which can act as the site for
a catalytic process. Transesterification of the racemic alcohol 1-phenyl-2-propanol with
2,4-dinitrophenyl acetate in the presence of the catalyst L-POST-1 L-(**6.58**) leads to the
(*R*) configuration ester with a modest *ee* of 0.08 (Figure 6.44).

To build homochiral porous metalorganic structures with tuneable pore sizes some
authors[68–70] used camphorate or L-lactic acid as chiral linkers. The strategy consists in
the utilization of two organic ligands, one of them (camphorate or L-lactic acid) bearing
asymmetric centres, while another acts as a rigid spacer to insure the porosity of the
metalorganic framework.

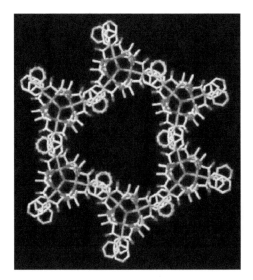

Figure 6.43
The large framework with large pores, which is formed from the trinuclear L-POST-1, L-(**6.58**)
(reproduced with permission from reference[67], copyright 2000, Macmillan Publishers Ltd).

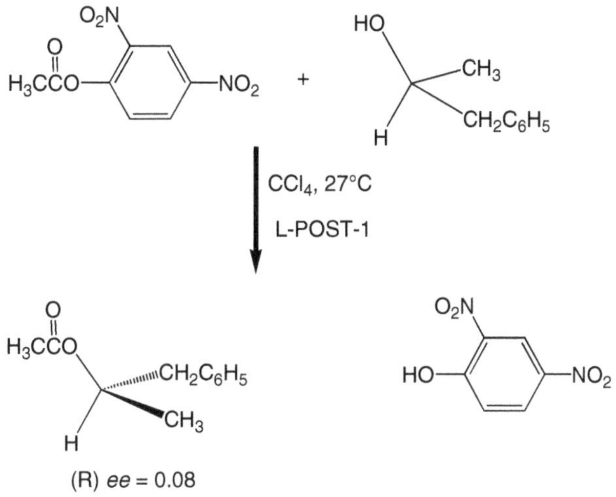

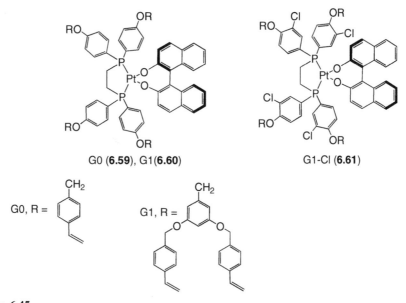

Figure 6.44
Enantioslective transesterification catalysed by L-POST-1 L-(**6.58**).

In order to construct an 'imprinted' chiral cavity around a metal centre with the potential for catalytic activity, Gagné[71] prepared chiral metallodendrimers. He used first and second generation dendrons where the terminating groups were copolymerizable styrenes tetrasubstituted onto a diphosphine ligand chelated to platinum. The tetrahedral coordination sphere of the metal was completed by a (P)-BINOL ligand (**6.59-6.61**) (Figure 6.45).

The copolymerization of metallodendrimers obtained with ethylenedimethacrylate leads to a matrix in which the metal centre is 'protected' by the dendrons to which it

Figure 6.45
First generation dendron (GO (**6.59**), G1 (**6.60**) and G1-Cl (**6.61**)) where the terminating groups were copolymerizable styrenes tetrasubstituted onto a diphosphine ligand chelated to platinum (II).

is coordinated. During the polymerization, some of the (P)-Binol is spontaneously extruded in the case of the dendrons G0 (**6.59**), G1 (**6.60**) and G1-Cl (**6.61**), although this phenomenon is negligible if the aromatic ring attached to the phosphorus is dichlorinated. The materials so obtained have an interesting property. If the (P)-BINOL is subjected to exchange by (*rac*)-6,6′-Bromo-BINOL, between 31 and 47% of the accessible platinum sites participate in the exchange, which occurs with partial retention of configuration. The 6,6′-Bromo-BINOL that replaces the (P)-BINOL has an enantiomeric excess which varies from 0.35 to 0.44 according to the dendrimer used, and the same configuration (P) as the (P)-BINOL initially coordinated to the metal. This shows that the cavity created during the formation of the material is chiral and that its configuration is an imprint of the (P)-BINOL. An attempt to modify the dendrons by substituting an ester for an ether linkage, which would have the effect of making the dendrimer more rigid, did not lead to any significantly more interesting results.[72]

By using the tartrate ion as a bis- or tris-chelating chiral ligand, Williams[73] obtained microporous MOCNs (**6.62**) with lanthanide ions. The originality of this approach is that, due to the thermal stability of the ligand, a hydrothermal route can be used for the synthesis. With erbium(III), the 3D structure $(Er_2(D)-(+)-(Tar)_3(H_2O)_2].3H_2O$ (Er(+)-**6.62**) or $Er_2(L)-(-)-(Tar)_3(H_2O)_2].3H_2O$ (Er(−)-**6.62**), Tar = $[C_2H_4O_6]^{2-}$, ($P1$ space group), possesses hydrophilic channels (5×7 Å^2), due to the presence of the hydroxyl functions which are directed towards the channel interior (Figure 6.46).

By using (L)-(−)-tartrate, Can-Zhong Lu[74] was able to prepare double helicates based on Mo(VI) and Gd(III) (**6.63**). Each interpenetrating double helix possesses a chiral channel, and the connection of the helices to each other through hydrogen bonding forms a 3D network structure (Figure 6.47).

Lin[75] obtained a porous 3D network by reaction of Cd(Cl)$_2$ with the ligand (R)-6,6′-dichloro-2,2′-dihydroxy-1,1′-binaphthyl-4,4′-bipyridine (**6.64**). This network of general formula [Cd$_3$Cl$_3$L$_3$].4DMF.6MeOH.3H$_2$O (**6.65**) crystallizes in the $P1$ space group. The Cd(II) centres, which are octahedrally coordinated and doubly bridged by the chlorine atoms to form a zigzag of [Cd(μ-Cl)$_2$]n chains, serve as secondary building blocks. Each Cd(II) in the chain is coordinated to two pyridyl groups from the ligand. The resulting 3D structure has wide channels with a cross section of 1.6×1.8 nm.

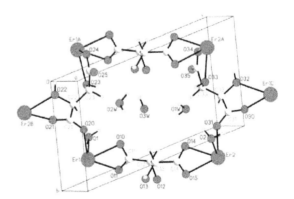

Figure 6.46
Asymmetric channel structure in (Er(−)-**6.62**) from four erbium ions and four L-tartrate ligands (reproduced with permission from reference[73], copyright 2005, The Royal Society of Chemistry).

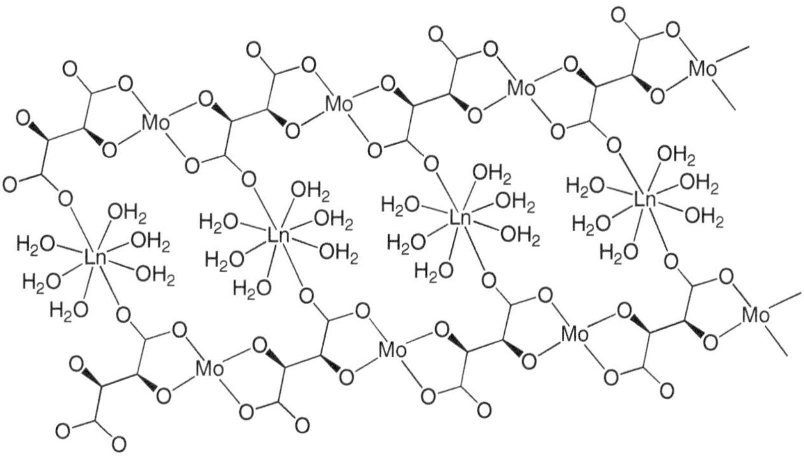

Figure 6.47
Representation of the double left-handed [Mo^{VI}O$_4$Ln^{III}(H$_2$O)$_6$(C$_4$H$_2$O$_6$)$_2$] (**6.63**) helicate.

The 3D network is transformed by the action of Ti(O^{i}Pr)$_4$ to give the compound [Cd$_3$Cl$_3$L'$_3$] (L' = L − 2H$^+$ + Ti(O^{i}Pr)$_2$) (**6.66**) (Figure 6.48) which is found to be an excellent catalyst for the alkylation of aromatic aldehydes by ZnEt$_2$. The enantiomeric excess varies from 0.94 to 0.75, according to the aromatic ring, with an almost quantitative conversion.

Realizing that the introduction of chirality into an extended solid is difficult and unpredictable, Pecoraro[76,77] developed an approach based on chiral metallacrown materials as building blocks. His earlier work in this area was first reported in 1989.[78,79] Metallacrowns can form chiral complexes, the chirality originating from the asymmetric ligands that join the metal ions together (Figure 6.49). The use of chiral β-aminohydroxiamic acids, such as (*R*)- or (*S*)-β-phenylalanine hydroxiamic acid (**6.67**) (H$_2$phenylalanine hydroxiamic acid = pheHA) allows enantiopure 'metallacrowns' carrying chiral information to be synthesized.

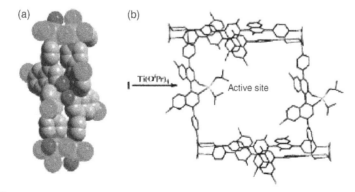

Figure 6.48
(a) Space-filling presentation of the tightly paired **L** (**6.64**) ligands via strong hydrogen bonding and π–π interactions. (b) Schematic representation of the active (BINOLate)Ti(OiPr)$_2$ catalytic sites in the open channels of (**6.66**) (reproduced with permission from reference[75], copyright 2005, American Chemical Society).

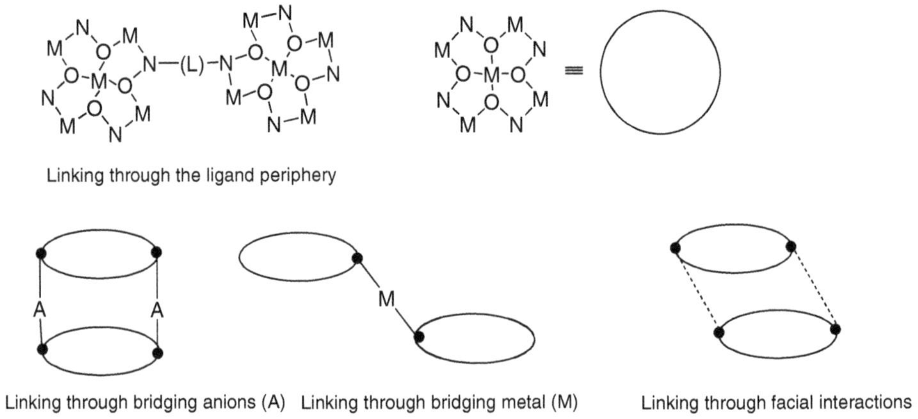

Linking through the ligand periphery

Linking through bridging anions (A)   Linking through bridging metal (M)   Linking through facial interactions

Figure 6.49
Metallacrowns as building blocks for obtaining extended structures.

The association of chiral metallacrowns into networks allows chiral solids to be constructed. An application of this strategy is shown in the synthesis of the enantiomeric copper (II) compounds: $\{Cu(NO_3)_2[12\text{-MC}_{CuN((S)\text{-}\beta\text{-pheHA})}\text{-}4]\}\text{-}\{Cu_2(benzoate)_4\}$ ((S)-**6.68**) or $\{Cu(NO_3)_2[12\text{-MC}_{CuN((R)\text{-}\beta\text{-pheHA})}\text{-}4]\}\text{-}\{Cu_2(benzoate)_4\}$ ((R)-**6.68**). These compounds form crystals (space group $P1$) containing channels (8×9 Å^2).[80]

By pursuing this synthetic strategy, Pecoraro[81] obtained enantiopure amphiphilic helicoidal systems of the formula $[\{Sm(NO_3)\}\{15\text{-MC}_{CuN((R)\text{-}\beta\text{-pheHA})}\text{-}5\}]\text{-}(NO_3)_2$ (**6.69**) that exhibited a fourfold symmetry (crystal space group $P4_1$ for (S)-pheHA and $P4_3$ for (R)-pheHA). Formation of the helix is brought about by the interaction between adjacent 'metallacrowns' which coil up to form a channel in which the exterior, where the benzyl groups are directed, is hydrophobic, while the interior of the channel is hydrophilic. The helical sense is (P) for the ligand (S)-pheHA and (M) for (R)-pheHA (Figure 6.50).

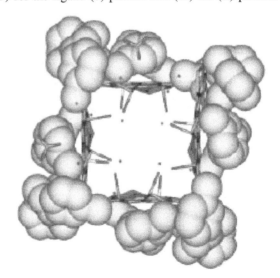

Figure 6.50
View of $[\{Sm(NO_3)\}\{15\text{-MC}_{CuN((R)\text{-}\beta\text{-pheHA})}\text{-}5\}]\text{-}(NO_3)_2$ ((R)-**6.69**) along the z-axis, showing the hydrophilic cavity (reproduced with permission from reference[81], copyright 2002, Wiley-VCH).

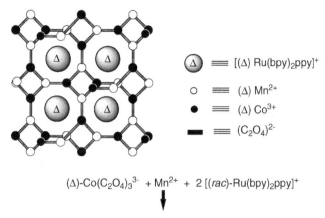

$$(\Delta)\text{-Co}(C_2O_4)_3{}^{3-} + Mn^{2+} + 2\ [(rac)\text{-Ru}(bpy)_2ppy]^+$$

$$\downarrow$$

$$[(\Delta)\text{-Mn}(\Delta)\text{-Co}(C_2O_4)_3(\Delta)\text{-Ru}(bpy)_2ppy]\ \text{(precipitate)} + [(\Lambda)\text{-Ru}(bpy)_2ppy]^+\ \text{(in solution)}$$

Figure 6.51
Resolution of cationic hexacoordinated complexes can be achieved using an asymmetrical matrix effect. The principle of this separation is based on the asymmetry of vacancies in a three-dimensional anionic network.

Gruselle and coworkers constructed the homochiral 3D networks $[(\Delta)–Mn(\Delta)–Co(C_2O_4)_3](\Delta)$-(**6.70**) or $[(\Delta)–Mn(\Delta)–Co(C_2O_4)_3](\Delta)$-(**6.71**)[82,83] to resolve the hexacoordinated Ru(II) complexes (rac)-[Ru(bpy)$_2$ppy]PF$_6$ (rac)-(**6.70**) and [Ru(bpy)$_2$Quo]PF$_6$ (rac)-(**6.71**) with (ppy = phenylpyridine – H$^+$, Quo = 8-quinolate). Although the symmetry of cations (**6.70**) and (**6.71**) is formally C_1, these monocations are capable of occupying the chiral vacancies in the anionic network and with the same configuration as the anionic brick, according to the reaction shown in Figure 6.51. Use of a double excess of the racemic monocation results in resolution, the nonreacted fraction remaining in solution being of the opposite configuration to that consumed in the construction of the network. The enantiomeric excess obtained is of the order of $ee = 0.6$.

6.5 Molecular Magnets

Magnetic molecular materials constitute an important class of compound,[84] and numerous structures have been described in which transition metal ions interact through a ligand, leading to properties resulting from ferro-, ferri- or anti-ferromagnetic interaction. Magnetic molecular materials exist with order 0D (magnet molecules), 1D, 2D and 3D (magnetic networks). Most of them have achiral structures.

6.5.1 Why Enantiopure Molecular Magnets?

The development of research on enantiopure chiral molecular materials has occurred recently.[85] This work has opened up a new field of research into molecular materials possessing both magnetic and optical properties. Such materials appear to be good candidates to present a cross effect between circular dichroism and magnetic circular dichroism, the so-called magnetochiral dichroism (MChD).[86–89] If any chiral system presents MChD, the effect was found to be very weak for diamagnetic[90] and paramagnetic[5,91] compounds due to their low magnetization. The signal presented by

Figure 6.52
Schematic representation of the magnetochiral dichroic (MChD) effect.

an optically active magnet, displaying a large magnetization and important CD and MCD, should be considerably larger. Figure 6.52 shows a schematic representation of the MChD effect: unpolarized light is absorbed in a slightly different manner by an enantiopure molecule (shown schematically in Figure 6.52 as a right hand) possessing a magnetic moment (M), if the light propagates parallel or antiparallel to an external magnetic field (shown schematically as a white vector). This small difference in transmitted light intensity is called the MChD effect.

Rikken and Raupach[92] used the MChD effect to favour the production of one enantiomer over the other in the photochemical racemization of the complex [Cr(C$_2$O$_4$)$_3$]K$_3$. However, the enantiomeric excess obtained as a function of the wavelength of the incident light was very low (*ee* from -1.10^{-4} to $0.5.10^{-4}$), even for a magnetic field as strong as 7.5 T. This phenomenon was predicted by Barron[93] in a provocative article entitled: 'Can a magnetic field induce asymmetric synthesis?'.

6.5.2 Strategies and Synthesis

Enantiopure magnets can sometimes be obtained by spontaneous resolution during crystal growth. Such a process cannot be considered as a preparative method because the control over absolute configuration is uncertain. In many cases, the single crystals obtained are isomorphous and therefore not distinguishable and ultimately the two configurations in the conglomerate cannot be separated. Nevertheless, conglomerates obtained by spontaneous resolution will be considered, because the understanding of their synthesis provides an important source of information about the diastereomeric aspects of self-assembly processes.

From a synthetic point of view, three general strategies to obtain enantiopure molecular magnets have been developed:

(i) *enantioselective synthesis or resolution of chiral ligands with transfer of chiral information to the metal ion assemblies through coordination bonds;*
(ii) *the chiral inductive effect of resolved building blocks in the formation of supramolecular structures; and*
(iii) *the chiral inductive effect of resolved templates.*

6.5.3 Enantioselective Synthesis or Resolution of Chiral Ligands

6.5.3.1 Single-molecule Magnets and Chains (1D Molecular Magnets)

In a cluster, the large number of magnetic centres located in a small volume produces magnetic properties. Metallocrowns (MCs) can provide clusters that possess transition

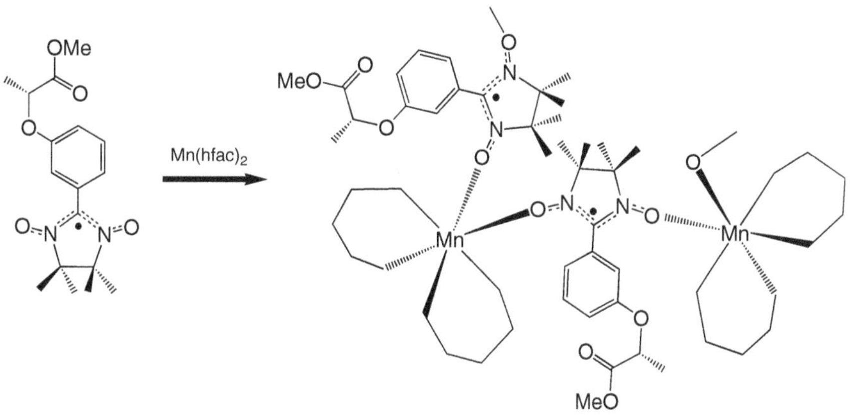

(*R*)-(**6.74**)   [(Λ)-Mn(hfac)$_2$(*R*)-3MNLNN]n (**6.73**)

Figure 6.53
1D (**6.73**) homochiral (Λ) chain based on Mn(II) complex.

metal ions interacting through the ligands. Also, as demonstrated by Pecoraro,[94] chiral MCs can be obtained using (*R*)- or (*S*)-phenylalaninehydroxamic acid (phHA) (**6.67**). [Ln^{III}[15-MCCuII(N)(S)-pheHa-5]](NO$_3$)$_3$ (**6.72**) complex crystallizes as a dimer or a helix, depending on the solvent used. The solid is homochiral and the helix configuration is determined by the configuration of the starting (*R*)- or (*S*)-(phHA), the use of (*S*)-(phHA) leading to a (P)-helix (space group *P*4$_1$), while the use of (*R*)-(phHA) leads to the (M)-helix (space group *P*4$_3$). These complexes do not show an MChD effect, probably due to nonidentity of the chiral and magnetic centres.

Veciana, Amabilino and Luneau[95,96] have reported the synthesis of an enantiopure 1D molecular ferromagnet (ferromagnetic transition below 3K). The complex (**6.73**) formed between (*R*)-methyl[3-(4,5,5,5-tetramethyl-4,5-dihydro-1H-imidazoyl-1-oxy-3-oxide) phenoxyl]-2-propionate ((*R*)-3MLNN) (*R*)-(**6.74**) and [Mn(hfac)$_2$] (hfac = hexafluoroacetylacetonate) crystallizes in the chiral *P*2$_1$2$_1$2$_1$ space group. In the polymeric chain all the hexacoordinated Mn centres have the same (Λ) configuration (Figure 6.53).

Using the ligand 4MLNN (**6.75**) it was observed that reaction of the racemic compound with [Mn(hfac)$_2$] leads to a cyclic dimer (**6.76**) in which the two enantiomers are connected head to tail, while the (*R*)- 4MLNN (*R*)-(**6.76**) crystallizes as a hydrogen-bonded chain molecule (**6.77**), emphasizing that a racemic compound and a pure enantiomer do not necessarily act in the same way[97] (Figure 6.54).

6.5.3.2 2D and 3D Molecular Magnets

Combining hexacyanoferrate (III) anions with Ni(II) complexes derived from the enantiopure ligands *trans*-(*1S, 2S*)- or trans-(*1R, 2R*)-cyclohexane-1,2-diamine (**6.78**) (chxn), Coronado[98] obtained enantiopure layered ferromagnets (*T*$_c$ = 13.8K). The compounds [Ni(*trans*-(1S,2S)-chxn)$_2$]$_3$[Fe(CN)$_6$].2H$_2$0 (1S,2S)-(**6.79**) and [Ni(*trans*-(1R, 2R)-chxn)$_2$]$_3$[Fe(CN)$_6$].2H$_2$O (1R, 2R)-(**6.79**) crystallize in the *P*1 space group.[99] CD measurements of the (*1S, 2S*)-(**6.79**) isomer exhibit positive Cotton effects at 450 nm and 725 nm and a negative one at 585 nm, the (*1R, 2R*)-(**6.79**) isomer exhibiting effects of opposite sign. In contrast, the racemic diamine ligand leads to a centrosymmetric structure belonging to the *P*-1 space group.

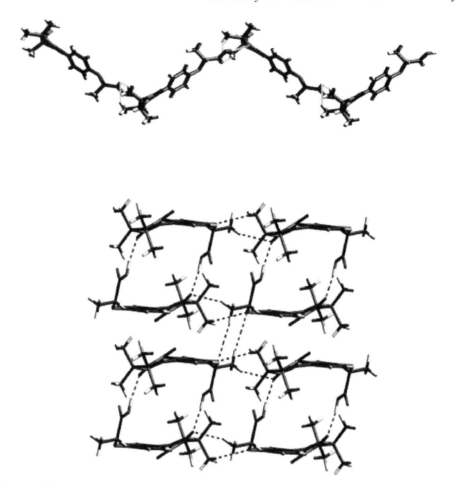

Figure 6.54
Crystal packing of (R)-(**6.77**) (above) and (rac)-(**6.76**) (below) (reproduced with permission from reference[97], copyright 2003, Elsevier).

Enantiopure diamines ((*S*)- or (*R*)-1,2-diamino propane (**6.80**),[100–103], *N*,*N'*-(1,2-cyclohexanediethylethylene)bis(salicylideneiminato) dianion) (**6.81**)[104] are often used to prepare optically active molecular magnets. Enantiopure organic acids were also used as chiral inductive ligands in the design of molecular magnets (tartaric acid,[105] malic acid,[106] camphoric acid,[107] and α-hydroxyacids[108]). Up to now none of these molecular magnets has shown an MChD effect.

6.5.4   Chiral Inductive Effect of Resolved Building Blocks in the Formation of Supramolecular Structures

The oxalate ion $[(C_2O_4)_3]^{2-}$ is a widely used ligand in molecular magnetism. In 1992, Okawa[109,110] described the synthesis of a series of 2D ferromagnets of general formula $NBu_4[M^{II}Cr^{III}(C_2O_4)_3]$ NBu_4-(**6.82**) (Bu = n-C_4H_9); M = $Mn^{2+}, Fe^{2+}, Ni^{2+}, Cu^{2+}$) in which transition metal ions interact through the oxalate ligand. Decurtins[111,112] has

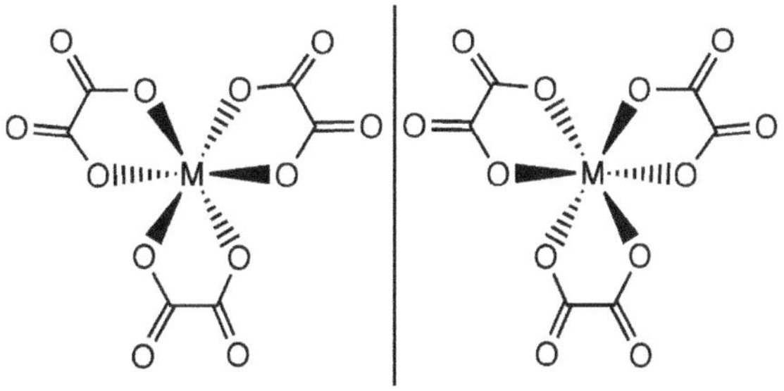

Figure 6.55
Schematic views of the (Δ) and (Λ) enantiomers of the $[M(ox)_3]^{n-}$ hexacoordinated tris-anion.

described the 3D helical networks $Ru(bpy)_3[Mn^{II}Mn^{II}(C_2O_4)_3]$ $Ru(bpy)_3$-(**6.83**) or $Ru(bpy)_3[Li^ICr^{III}(C_2O_4)_3]$ $Ru(bpy)_3$-(**6.84**) with 3D helical structure.

These bimetallic networks are composed of an anionic sublattice $[M1M2(C_2O_4)_3]^{x-}$ and a cationic counterpart $[A]^{x+}$. The charge of each subunit of the anionic sublattice is one or two according to the oxidation state of each metal centre ($[Mn^{II}Cr^{III}(C_2O_4)_3]^{2-}$, $[Li^ICr^{III}(C_2O_4)_3]^-$).

Moreover, in these structures, the metal ions surrounded by three bidentate oxalate ligands in an octahedral manner exhibit a propeller-like chirality. Therefore each chiral element exists as a (Δ) or (Λ) configuration, as shown in Figure 6.55.

Furthermore, the relative configuration of the connected hexacoordinated centres determines the 2D or 3D architecture of the coordination polymers. A heterochiral arrangement $[(\Delta)$-$M1(\Lambda)$-$M2)]$* leads to a 2D network. In this situation the anionic sublattice displays a honeycomb structure while the cationic moiety, which is in general a tetra-alkyl ammonium $(NR_4)^+$ or phosphonium $(PR_4)^+$ ion, is located between the anionic layers. In contrast, a homochiral arrangement $[(\Delta)$-$M1(\Delta)$-$M2)]$* leads to a helical organization of the connected metallic ions (Figure 6.56). The structure of 3D oxalatebased compounds consists of a 10-gon three-connected (10,3) anionic network in which the '$M(C_2O_4)_3$' moiety behaves as a three-connector and the metal ions are located at the 10 summits of a decagon. This topological description can be viewed as three sets of interconnected helices of the same configuration. The anionic structure defines cavities fitted by the cationic counterpart and the cationic sublattice is also organized in a helical manner having the same configuration as the anionic one.

To build such 2D or 3D networks as enantiopure, two geometrical elements must be controlled.[113]

(i) *The relative configuration of the adjacent metal centres.*
    The nature of the cation is the determining factor to orient the reaction towards 2D or 3D structures, in general ammonium or phosphonium ions lead to 2D whereas cations belonging to D_3 symmetry $(Ru(bpy)_3)^{2+}$ or having octahedral structures $(Ru(bpy)_2ppy)^+$ lead to 3D networks.
(ii) *The absolute (Δ) or (Λ) configuration of each hexacoordinated metal centre.*

6.5.4.1 3D Networks

Decurtins[111,112,114] soon recognized the homochirality of all the coordinated metal centres as the driving force of the spontaneous resolution process occurring in the

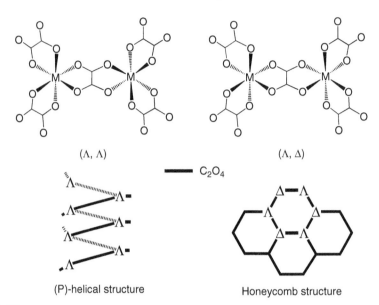

Figure 6.56
Homochiral arrangement of the neighbouring tris(chelated) metal centres leads to a 3D (10, 3) anionic network. Heterochiral arrangement of the neighbouring tris(chelated) metal centres leads to a 2D (6, 3) anionic network.

formation of these molecular materials when starting from (rac)-$[Cr^{III}(C_2O_4)_3]^{3-}$ anionic bricks in the presence of Li^+ and (rac)-$[Ru(bpy)_3]^{2+}$ according to the reaction:

$$(rac)\text{-}[Cr^{III}(C_2O_4)_3]^{3-} + Li^+ + 2(rac)\text{-}[Ru(bpy)_3]^{2+} = [(\Delta)\text{-}Ru(bpy)_3]$$
$$[(\Delta)\text{-}Li^I(\Delta)\text{-}Cr^{III}(C_2O_4)_3] + [(\Lambda)\text{-}Ru(bpy)_3][(\Lambda)\text{-}Li^I(\Lambda)\text{-}Cr^{III}(C_2O_4)_3].$$

Nevertheless, the crystals obtained starting from racemic materials (space group $P2_13$) are often twinned. In pioneering work Gruselle and coworkers[85] demonstrated that using enantio-enriched (Δ)- or (Λ)-$[Cr^{III}(C_2O_4)_3]^{3-}$ allows the possibility of obtaining optically-active 3D networks:

$$(\Delta)\text{-}[Cr^{III}(C_2O_4)_3]^{3-} + Li^+ + 2(rac)\text{-}[Ru(bpy)_3]^{2+} = [(\Delta)\text{-}Ru(bpy)_3][(\Delta)\text{-}Li^I(\Delta)\text{-}Cr^{III}$$
$$(C_2O_4)_3] + [(\Lambda)\text{-}Ru(bpy)_3]^{2+}$$

$$(\Lambda)\text{-}[Cr^{III}(C_2O_4)_3]^{3-} + Li^+ + 2(rac)\text{-}[Ru(bpy)_3]^{2+} = [(\Lambda)\text{-}Ru(bpy)_3][(\Lambda)\text{-}Li^I(\Lambda)\text{-}Cr^{III}$$
$$(C_2O_4)_3] + [(\Delta)\text{-}Ru(bpy)_3]^{2+}.$$

In the same way $[(\Delta)\text{-}Ru(bpy)_2ppy][(\Delta)\text{-}Mn^{II}(\Delta)\text{-}Cr^{III}(C_2O_4)_3]$, (Δ)-$Ru(bpy)_2ppy$ -(**6.82**) and $[(\Lambda)\text{-}Ru(bpy)_2ppy][(\Lambda)\text{-}Mn^{II}(\Lambda)\text{-}Cr^{III}(C_2O_4)_3]$ (Λ) $Ru(bpy)_2ppy$-(**6.82**) ferromagnets were obtained using the $[Ru(bpy)_2ppy]^+$ monocation:[115]

$$(\Delta)\text{-}[Cr^{III}(C_2O_4)_3]^{3-} + Mn^{2+} + 2(rac)\text{-}[Ru(bpy)_2ppy]^+ = [(\Delta)\text{-}Ru(bpy)_2ppy]$$
$$[(\Delta)\text{-}Mn^{II}(\Delta)\text{-}Cr^{III}(C_2O_4)_3] + [(\Lambda)\text{-}Ru(bpy)_2ppy]^+$$

$$(\Lambda)\text{-}[Cr^{III}(C_2O_4)_3]^{3-} + Mn^{2+} + 2(rac)\text{-}[Ru(bpy)_2ppy]^+ = [(\Lambda)\text{-}Ru(bpy)_2ppy]$$
$$[(\Lambda)\text{-}Mn^{II}(\Lambda)\text{-}CrIII(C_2O_4)_3] + [(\Delta)\text{-}Ru(bpy)_2ppy]^+.$$

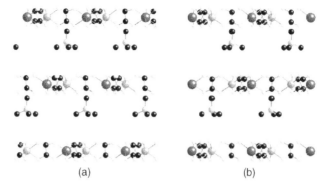

Figure 6.57
Honeycomb layer stacking for 2D compounds crystallizing in (a) the R3c achiral space group, and (b) in the P6$_3$ chiral space group. Only the first atoms of the alkyl chains around the nitrogen atom of the [N(n-C$_4$H$_9$)$_4$]$^+$ cation are shown.

6.5.4.2 2D Networks

2D networks result from a heterochiral arrangement in which adjacent propeller-like chiral elements have opposite configurations. This simple statement has many consequences. Let us consider a single [(Δ)-M1^{II}(Λ)-M2^{III}(C$_2$O$_4$)$_3$]$^-$ anionic layer; this layer is only chiral if the M1, M2 metal centres are different and consequently the d(M1-O) and d(M2-O) distances are significantly different from one another.

A difficulty is also associated with the layer stacking. NnBu$_4$[MnCr(C$_2$O$_4$)$_3$] NnBu$_4$-(**6.82**) obtained from (*rac*)-[Cr^{III}(C$_2$O$_4$)$_3$]$^{3-}$ crystallizes in the *R*c3 space group.[116] The unit cell contains six anionic layers. Adjacent layers correspond to one another through a gliding mirror, which inverts the absolute configuration of two adjacent layers. For example, the layer *n*: [(Δ)-M1^{II}(Λ)-M2^{III}(C$_2$O$_4$)$_3$]$^-$ leads to a [(Λ)-M1^{II}(Δ)-M2^{III}(C$_2$O$_4$)$_3$]$^-$ neighbouring *n* + 1 layer rendering the overall structure achiral (Figure 6.57(a)). Conversely, NBu$_4$[Fe^{II}Cr^{III}(C$_2$O$_4$)$_3$] NBu$_4$-[FeCr](**6.82**) crystallizes in the chiral *P*6$_3$ space group.[117] The unit cell contains two layers of the same configuration (Figure 6.57(b)).

Spontaneous resolution in the crystal growth of NBu$_4$-[FeCr](**6.82**) is not unique[3] and proves the existence of chiral 2D networks when starting from an achiral template cation and a racemic tris(oxalato)metallate. Starting from resolved enantio-enriched (Δ) or (Λ)-[Cr^{III}(C$_2$O$_4$)$_3$]$^{3-}$ crystalline powders were obtained in the reaction with M^{2+} (M = Mn, Fe) and (NR$_4$)$^+$ (R = nBu$_4$; Et, *n*-Pr, *n*-Bu) ions[117]. The *P*6$_3$ space group found by X-ray powder diffraction, as well as the CD spectra, confirm the chiral enantiopure character of these 2D networks (Figure 6.58).

The limitation of the strategy based on the resolution of the anionic tris(oxalato) metallate brick is the relatively fast racemization rate of these complexes when they exist. Therefore, taking into consideration the existence of enantio-specific interactions between the anionic and the cationic part in the 2D and 3D networks, a strategy based on the chiral inductive effect of the template cation was developed. A significant example is the formation of [PPh$_4$MnFe(C$_2$O$_4$)$_3$] (PPh$_4$)[MnFe]-(**6.82**). This compound crystallizes in the *R*3c space group. The tetraphenylphosphonium cation lies between the anionic layers. The P–C bond perpendicular to the anionic layer is located on the threefold axis. The PPh$_4$ cation itself adopts a propeller-like chirality and its point group

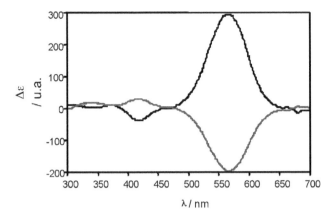

Figure 6.58
Natural circular dichroism curves $NBu_4[(\Lambda)\text{-}Mn(\Delta)\text{-}Cr(C_2O_4)_3]$ NBu_4-(**6.82**) $(-)$-560 nm and $NBu_4[(\Delta)\text{-}Mn(\Lambda)\text{-}Cr(C_2O_4)_3]$ NBu_4-(**6.82**) $(+)$-560 nm.

is D_3. The 'PPh$_4$' cation and the adjacent 'Fe(C$_2$O$_4$)$_3$' fragment of neighbouring layers have the (P, Δ)* relative configurations[118,119] (Figure 6.59).

6.5.5 Chiral Inductive Effect of Resolved Templates

For 3D networks, the backbone of the second strategy uses resolved tris(diimine)metal (II) complexes. The advantage of this method becomes obvious when Ru(II) complexes are used. The inertness of these d^6 low-spin complexes implies a configurational stability of their resolved complexes. They can thus be used to obtain enantiopure 3D bimetallic oxalate-based networks with a total control of the absolute configuration. Optically active

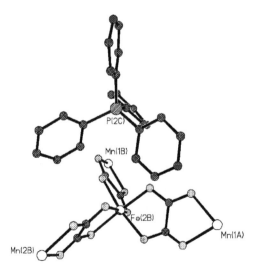

Figure 6.59
View of the relative configurations of $\ll$(P) P(Ph)$_4$ $\gg$ and (Δ) $\ll$ Fe(C$_2$O$_4$)$_3$$\gg$ fragments in the 2D (PPh$_4$)[MnFe]-(**6.82**) network.

$[(\Delta)\text{-Ru(bpy)}_3][(\Delta)\text{-Mn}^{II}(\Delta)\text{-Mn}^{II}(C_2O_4)_3]$ $[(\Delta)\text{-Ru(bpy)}_3\text{-}(\mathbf{6.83})]$ or $[(\Lambda)\text{-Ru(bpy)}_3]$ $[(\Lambda)\text{-Li}^I(\Lambda)\text{-Cr}^{III}(C_2O_4)_3]$ $[(\Lambda)\text{-Ru(bpy)}_3\text{-}(\mathbf{6.84})]$ 3D compounds have been obtained starting from resolved $[\text{Ru(bpy)}^3]^{2+}$ combined with free oxalate and Mn^{2+}[120] (or Cu^{2+})[121] metal ions demonstrating the powerful chiral inductive effect of this cationic template:

$$2\text{Mn}^{2+} + 3(C_2O_4)^{2-} + (\Delta)\text{-}[\text{Ru(bpy)}_3]^{2+} = [(\Delta)\text{-Ru(bpy)}_3][(\Delta)\text{-Mn}^{II}(\Delta)\text{-Mn}^{II}(C_2O_4)_3]$$
$$2\text{Mn}^{2+} + 3(C_2O_4)^{2-} + (\Lambda)\text{-}[\text{Ru(bpy)}_3]^{2+} = [(\Lambda)\text{-Ru(bpy)}_3][(\Lambda)\text{-Mn}^{II}(\Lambda)\text{-Mn}^{II}(C_2O_4)_3].$$

This strategy has been pursued to obtain enantiopure crystals of $[(\Delta)\text{-Ru(bpy)}_2\text{ppy}]$ $[(\Delta)\text{-Mn}^{II}(\Delta)\text{-Cr}^{III}(C_2O_4)_3]$ $(\Delta)\text{-Ru(bpy)}_2\text{ppy-}(\mathbf{6.82})$ and $[(\Lambda)\text{-Ru(bpy)}_2\text{ppy}][(\Lambda)\text{-Mn}^{II}(\Lambda)\text{-}$ $\text{Cr}^{III}(C_2O_4)_3]$ $(\Lambda)\text{-Ru(bpy)}_2\text{ppy-}(\mathbf{6.82})$.[115]

Ferrocenyl substituted ammonium ions have been shown to be efficient template cations of 2D oxalate-based networks.[122] The planar chiral $(pR)\text{-}$ and $(pS)\text{-}1\text{-}(N,N,N\text{-tri}$ $(n\text{-propyl})\text{aminomethyl})\text{-}2\text{-methylferrocene}$ iodide $(\mathbf{6.85})$ (Figure 6.60) lead to the formation of enantiopure $(pR)\text{-}[1\text{-CH}_2\text{N}(n\text{-Pr}_3)\text{-}2\text{-CH}_3\text{-C}_5\text{H}_3\text{FeC}_5\text{H}_5][(\Delta)\text{-Mn}^{II}(\Lambda)\text{-Cr}^{III}$ $(C_2O_4)_3]$ $(pR)\text{-}(\mathbf{6.82})$ and $(pS)\text{-}[1\text{-CH}_2\text{N}(n\text{-Pr}_3)\text{-}2\text{-CH}_3\text{-C}_5\text{H}_3\text{FeC}_5\text{H}_5][(\Delta)\text{-Cr}^{III}(\Lambda)\text{-Mn}^{II}$ $(C_2O_4)_3]$ $(pS)\text{-}(\mathbf{6.82})$.[123]

Recently enantiopure single crystals of $(R)\text{-A}[(\Delta)\text{-Mn}^{II}(\Lambda)\text{-Cr}^{III}(C_2O_4)_3]$ and $(S)\text{-A}$ $[(\Lambda)\text{-Mn}^{II}(\Delta)\text{-Cr}^{III}(C_2O_4)_3]$ were obtained using the methyl$((R)\text{-1-methylpropyl})$di (propyl) ammonium iodide $[(R)\text{-AI}]$ $(R)\text{-}(\mathbf{6.86})\text{I}$ and methyl$((S)\text{-1-methylpropyl})$di (propyl) ammonium iodide $[(S)\text{-AI}]$ $(S)\text{-}(\mathbf{6.86})\text{I}$ salts as templates (Figure 6.61).

MChD studies on the enantiopure $(R)\text{-A}[(\Delta)\text{-Mn}^{II}(\Lambda)\text{-Cr}^{III}(C_2O_4)_3]$ $(R)\text{-A-}(\mathbf{6.82})$ and $(S)\text{-A}[(\Lambda)\text{-Mn}^{II}(\Delta)\text{-Cr}^{III}(C_2O_4)_3]$ $(S)\text{-A-}(\mathbf{6.82})$ single crystals were carried out and confirmed this kind of second-order effect.

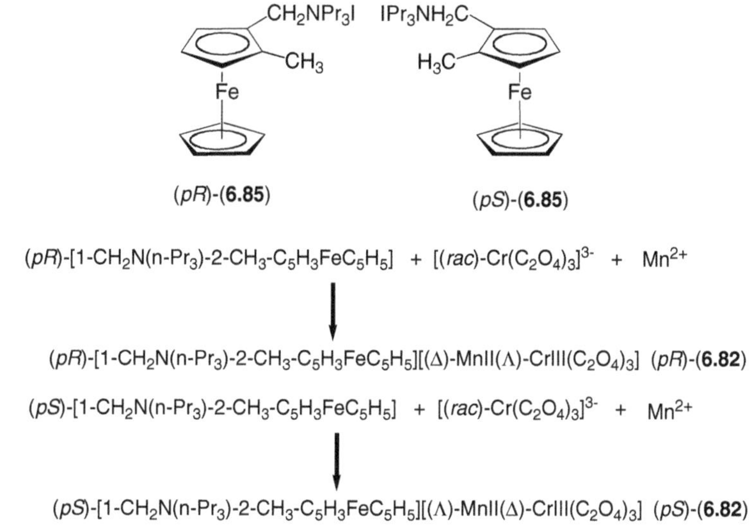

$(pR)\text{-}(\mathbf{6.85})$ $\quad\quad$ $(pS)\text{-}(\mathbf{6.85})$

$(pR)\text{-}[1\text{-CH}_2\text{N}(n\text{-Pr}_3)\text{-}2\text{-CH}_3\text{-C}_5\text{H}_3\text{FeC}_5\text{H}_5]$ + $[(rac)\text{-Cr}(C_2O_4)_3]^{3-}$ + Mn^{2+}

$\downarrow$

$(pR)\text{-}[1\text{-CH}_2\text{N}(n\text{-Pr}_3)\text{-}2\text{-CH}_3\text{-C}_5\text{H}_3\text{FeC}_5\text{H}_5][(\Delta)\text{-MnII}(\Lambda)\text{-CrIII}(C_2O_4)_3]$ $(pR)\text{-}(\mathbf{6.82})$

$(pS)\text{-}[1\text{-CH}_2\text{N}(n\text{-Pr}_3)\text{-}2\text{-CH}_3\text{-C}_5\text{H}_3\text{FeC}_5\text{H}_5]$ + $[(rac)\text{-Cr}(C_2O_4)_3]^{3-}$ + Mn^{2+}

$\downarrow$

$(pS)\text{-}[1\text{-CH}_2\text{N}(n\text{-Pr}_3)\text{-}2\text{-CH}_3\text{-C}_5\text{H}_3\text{FeC}_5\text{H}_5][(\Lambda)\text{-MnII}(\Delta)\text{-CrIII}(C_2O_4)_3]$ $(pS)\text{-}(\mathbf{6.82})$

Figure 6.60
Enantiopure ferrocenyl ammonium salts (pR)-(**6.85**) and (pS)-(**6.85**) are efficient templates to build optically-active 2D oxalate-based networks.

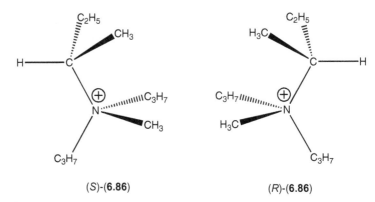

(S)-(**6.86**)   (R)-(**6.86**)

Figure 6.61
Methyl((R)-1-methylpropyl)di(propyl) ammonium (R)-(**6.86**) and methyl((S)-1-methylpropyl) di(propyl) ammonium (S)-(**6.86**).

6.6 Chiral Surfaces

6.6.1 General Considerations

When one talks of chiral surfaces, the first thing that comes to mind is the grafting of chiral molecules onto a support. This aspect is well known and has been used in many applications in the fields of asymmetric synthesis and chiral chromatography. We will not deal any further with this topic, which has been the subject of numerous studies.

We will, on the other hand, explore the principles and, in greater detail, some examples of surface chirality; that is the creation of chiral surfaces as a result of the specific arrangement of adsorbates, or the induction of such surfaces by chiral templates. The chiral organization of molecules at metal surfaces is described and discussed by some authors in fascinating and documented reviews.[124–129] The arrangement of surface atoms plays a decisive role in the very superstructure of adsorbates. Furthermore, the definition of the surface under consideration is of the utmost importance in order to understand the nature of the interactions between the surface and the adsorbed molecules. In general, for the low index planes (100), (110) and (111), the geometries of the last atomic layer are quadratic, rectangular and hexagonal respectively (Figure 6.62).

In Chapter 2 we saw that, while in three dimensions a chiral object is defined by the fact of being nonsuperimposable with its mirror image, in two-dimensional space the

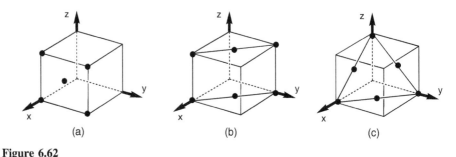

(a)   (b)   (c)

Figure 6.62
The low-index planes (a) (100), (b) (110) and (c) (111) of the fcc crystal Ni, Cu, Ag, Au, Pt, Pd are examples of fcc metals. The crystal planes are denoted by the Miller indices (hkl).

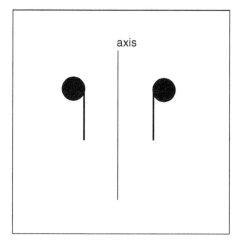

Figure 6.63
In flat land a 2D chiral object can exist; the mirror is reduced to an axis.

same definition applies, but with the exception that the object image is created with respect to an axis in the plane and is nonsuperimposable by geometric translation operations within that plane (Figure 6.63).

Bidimensional molecules, such as 1-nitronaphthalene (**6.87**), acquire a chiral character when adsorbed on a metal surface as their faces (*Si*) and (*Re*) are no longer equivalent: in effect, one face is turned towards, and the other away from, the metal. We will now consider the great richness of these chiral phenomena produced at metal surfaces, in particular.

> (i) *Spontaneous resolution of chiral molecules at a metal surface in 2D space. This phenomenon is often accompanied by self-assembly of the molecules under consideration, creating a chiral bidimensional order of higher-order enantiomerically pure phases in defined domains of the surface. This phenomenon of spontaneous resolution leading to conglomerates, which is rare for 3D chiral molecules, is much less so in 2D space.*
> (ii) *Induction of chirality by enantiopure chiral molecules in 3D, resulting in enantiopure structures at metal surfaces.*
> (iii) *Formation of chiral metal surfaces by electrodeposition in the presence of a chiral ionic medium.*
> (iv) *Formation of chiral metallic nanoparticles.*

Thus, in many cases, surfaces which are by nature achiral can achieve a chiral structure. This is a fundamental point, as numerous catalytic processes occur at a metal surface. The design of such surfaces is therefore of considerable importance in the understanding of metal-catalysed asymmetric synthesis and in the resulting applications.

6.6.2 Spontaneous Resolution of Chiral Molecules at a Metal Surface in 2D Space

6.6.2.1 4-[trans-2-(pyrid-4-vinyl)] Benzoic Acid (PVBA) (6.88) on a Surface of Ag (111)

PVBA is a planar conjugated molecule which exists in the form of two enantiomers (δ) and (λ) in the plane in which it lies. On a surface of Ag (111) PVBA is adsorbed parallel

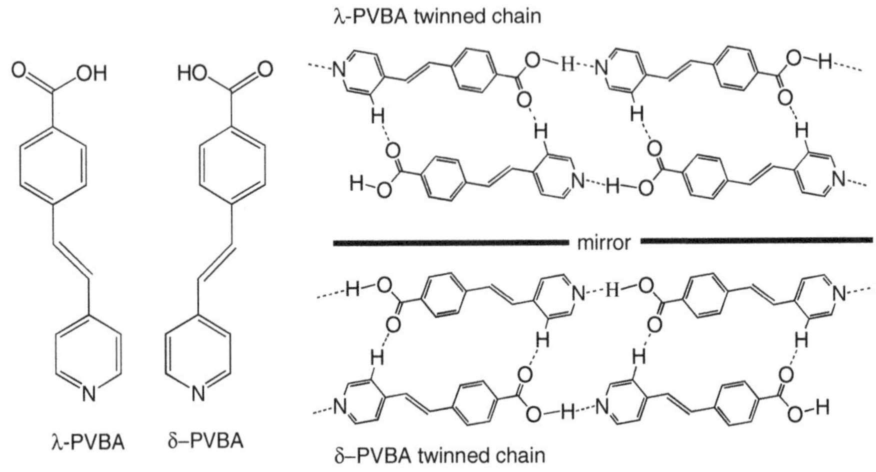

Figure 6.64

1D-supramolecular PVBA (**6.88**) twinned chains on the Ag (111) surface. The spontaneous resolution leads to enantiomorphic domains. The chains run along the [11-2]-direction of the Ag-lattice.

to the plane of the aromatic rings forming infinite chains. The cohesion of these double chains is provided by hydrogen bonding between the PVBA molecules in a head-to-tail arrangement. These double chains are homochiral, and at 300 K form distinct domains of configurations (δ) and (λ) along the [11$\bar{2}$] axis of the metal surface (Figure 6.64).[130,131]

Conducting the same experiment on a Pd (110) surface shows the presence of isolated chiral molecular motifs, instead of the 1D structure obtained on Ag (111). This difference in behaviour reveals an essential point in supramolecular surface chemistry. The final state of self-organization results from an equilibrium between the forces of attraction and repulsion (electrostatic, van der Waals, hydrogen bonds, etc.) that are brought into play at the metal surface. Thus with Pd (110) it is the metal-PVBA interactions that dominate to the detriment of the inter-PVBA hydrogen bonds.

6.6.2.2 1-nitronaphthalene (6.87) on a Surface of Au (111)

In the case of this chiral molecule in 2D, adsorption on the surface of Au (111) occurs parallel to the fused aromatic rings. On lowering the temperature to 70 K, the 1-NP (**6.87**) molecules become organized into decamers, with either the (*Re*)- or (*Si*)-(**6.87**) form predominating in each decamer (Figure 6.65).[132,133]

Here there is also a chiral organization at two levels. Expression of the 2D chirality of (**6.87**) and chiral supramolecular organization defining pure enantiopure domains, images with respect to a given axis of the metal plane under consideration.

6.6.2.3 1,3,5-tricarboxylic Benzoic Acid (1,3,5-TBA) (6.89) and 1,3,4-tricarboxylic Benzoic Acid (1,3,4-TBA) (6.90) on a Surface of Cu (100)

The acids (**6.89**) and (**6.90**) are adsorbed on a surface of Cu (100), the presence of three carboxylic functions evidently being very favourable to the formation of intermolecular hydrogen bonds as well as to the attachment to the metal surface.

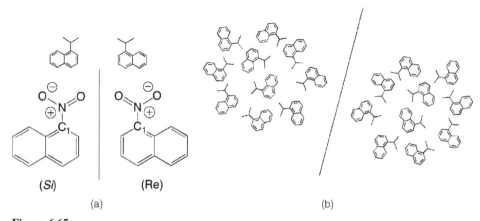

Figure 6.65
(a) View of the (Si) and (Re) configurations of (**6.87**) relative to an axis perpendicular to the picture and pointing to C1. (b) Schematic two chiral decamers adsorption motifs at 50 K on a reconstructed Au (111) surface.

In the case of (1,3,5-TBA) (**6.89**), codeposition of Fe atoms on Cu (100) leads to the formation of a complex Fe(1,3,5-TBA)$_4$ (**6.91**). This 2D complex is chiral (propeller-like) and self-organizes with those of the same configuration leading to 2D phases of opposed (P) and (M) configurations at the surface[134–136] (Figure 6.66).

The acid (1,3,4-TBA) (**6.90**) is attached to the metal by a single carboxylic function, the plane of the aromatic ring making an angle of 25° with respect to the perpendicular to the surface. The two other carboxylic functions are involved in the formation of a hydrogen-bonded network at the metal surface. This 2D network forms two enantiomeric domains for which the [011] axis is the mirror (Figure 6.67).

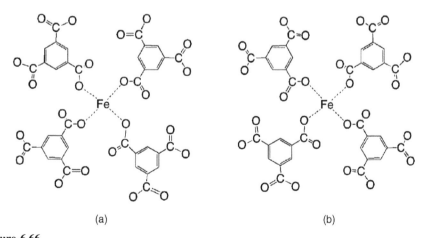

Figure 6.66
Fe(TBA)$_4$ (**6.91**) chiral 2D clusters labelled (a) (P) and (b) (M) on a Cu (111) surface, oriented +75° or −75° with respect to the [001] direction.

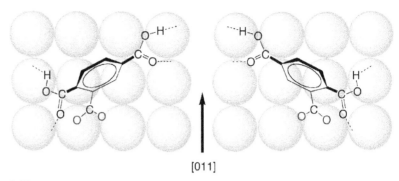

[011]

Figure 6.67
Enantiomorphic domains generated by the adsorption of (1,3,4-TBA) (**6.90**) on Cu (100) surface.
The Cu [001] azimuth represents the mirror.

6.6.3 Induction of Chirality by Enantiopure Chiral Molecules in 3D, Resulting in Enantiopure Structures at Metal Surfaces

So far we have examined situations in which the adsorbed molecule has chirality in 2D, but we will now look at the organization of chiral molecules in 3D at a metal surface. We will examine three cases concerned with axial chirality or chirality centred on carbon.

Ohtani[137] showed that, as a result of the aurophilic character of the thiolate (RS⁻) function, (*R*)- and (*S*)-1,1′-binaphthalene-2,2′-dithiol (BNSH) (**6.92**) (Figure 6.68) attach themselves to the surface of Au (111) in a quasicovalent manner leading to the formation of a 2D phase with a rhombohedral structure. The configuration of this supramolecular arrangement of enantiopure molecules depends on the absolute configuration of the isolated molecule, being (C) for (*S*)-(BNSH) (*S*)-(**6.92**) and (AC) for (*R*)-(BNSH) (*R*)-(**6.92**).

Raval[138–140] and Ernst[141–143] have particularly studied the adsorption of tartaric acid at the surface of Cu (110) and Ni (110). Knowing the role that tartaric acid (**6.4**)-2H⁺ plays in the enantioselective catalytic hydrogenation of β-ketoesters by Ni,[144] as well as its effect in the enantioselective version of the Michael reaction for La(III) catalysts supported on apatites,[145] the understanding of [(**6.4**)-2H⁺]-metal surface interactions has great importance.

On a Ni (110) surface at ambient temperature, double deprotonation of the (*R*, *R*)- [(**6.4**)-2H⁺] occurs to form the ditartrate anion (**6.4**), which is strongly adsorbed at the metal surface.[146] The results of scanning tunnelling microscopy (STM) reveal the formation of a

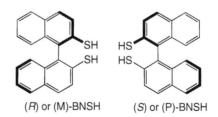

(*R*) or (M)-BNSH   (*S*) or (P)-BNSH

Figure 6.68
Schematic view of (S)-(**6.92**) and (R)-(**6.92**).

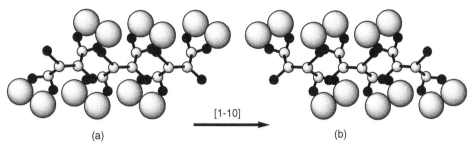

[1-10]

(a)   (b)

Figure 6.69

Adsorption of (a) (R,R)- and (b) (S,S)-(**6.4**) on the Ni (110) surface at room temperature. The picture shows the chiral bitartrate units [(**6.4**)Ni₄] along the [1-10] direction.

short 1D supramolecular organization that extends along the [1$\bar{1}$0] crystallographic axis, the same direction in which the antipode (*S, S*)-(**6.4**) dianion organizes itself.[147] This adsorption is accompanied by the formation of chiral ≪ bitartrate-Ni₄ ≫ units that contribute to the reconstruction of the Ni surface in a chiral form, the geometry of which correlates with the configuration of the starting tartaric acid (Figure 6.69).

On a Cu (110) surface, the behaviour of tartaric acid is different.[138,148,149] The interactions of the TA dianion with the surface, which are dominant in the case of Ni (110), are now weaker, so allowing intermolecular hydrogen bonding to lead to the formation of a supramolecular 2D organization. Molecules of (*R, R*)-(**6.4**) organize themselves into rows of three which are aligned in long parallel chains along the [1$\bar{1}$4] axis ([$\bar{1}$14] if (*S, S*)- [(**6.4**)-2H⁺] is used). These axes do not correspond to any symmetry of the Cu (110) surface, and this destruction of the symmetry elements leads to the formation of enantiomeric chiral surfaces, that is nonsuperimposable images of each other. In order to picture the situation, we can say that along the parallels where the adsorbed molecules are situated there are empty chiral channels (Figure 6.70).

There are other examples in the literature of the adsorption of chiral molecules (cysteine,[150] alanine,[151,152] and alcohols[153]) on metal surfaces. We will not consider these here, but the reader can refer to them in several excellent reviews published on the subject.

6.6.4 Formation of Chiral Metal Surfaces by Electrodeposition in the Presence of a Chiral Ionic Medium

Wildmer[154] showed that a homochiral CuO surface can be formed by electrodeposition at the surface of an electrode by using a solution of tartaric acid to induce the configuration. In this way the use of (*R, R*)-[(**6.4**)-2H⁺] leads to a surface of CuO (1$\bar{1}\bar{1}$), while CuO (11$\bar{1}$) is formed when the crystal growth is carried out in the presence of (*S, S*)-[(**6.4**)-2H⁺]. Conducting the same experiment in the presence of (*rac*)-[(**6.4**)-2H⁺] or (*meso*)-[(**6.4**)-2H⁺] leads to the formation of an achiral surface resulting from the deposition of equal quantities of the two possible surface orientations.

6.6.5 Formation of Chiral Nanoparticles

Chaudret[155] obtained platinum nanoparticles by reducing Pt₂(dba)₃ (**6.93**) (dba = dibenzylideneacetone) with CO in toluene. The addition of enantiopure aminoalcohols and

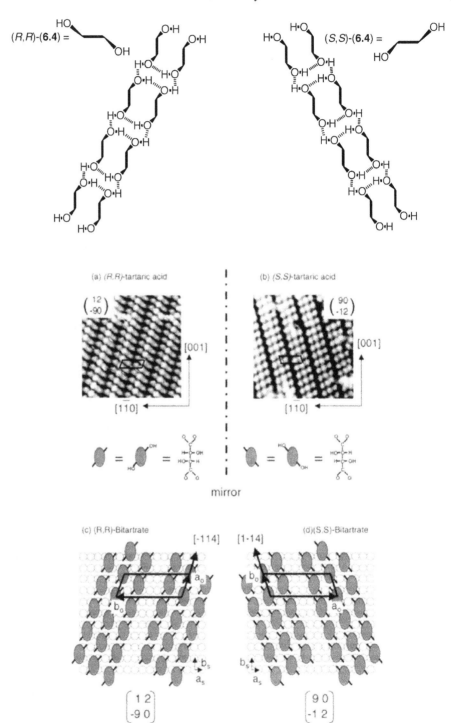

Figure 6.70

Chiral 2D organization of tartrate dianions (**6.4**) on a Cu (110) surface (reproduced with permission from reference[149], copyright 2005, Elsevier).

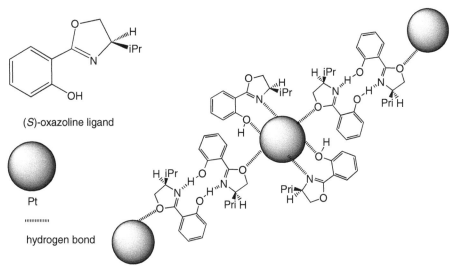

Figure 6.71
Suggested organization of the Pt colloid stabilized by (S)-oxazoline ligand.

oxazolines, which act as bidentate ligands, stabilizes these nanoparticles organized in superstructures (Figure 6.71).

The formation of palladium nanoparticles by reduction with H_2 in THF in the presence of a chiral diphosphine based on xylose (**6.94**) allowed small sized, monodispersed particles to be obtained[156]. These Pd nanoparticles have proven to be good asymmetric catalysts for the CC coupling reaction in the alkylation of (rac)-3-acetoxy-1,3-diphenyl-1-propene by dimethylmalonate under basic conditions (Figure 6.72).

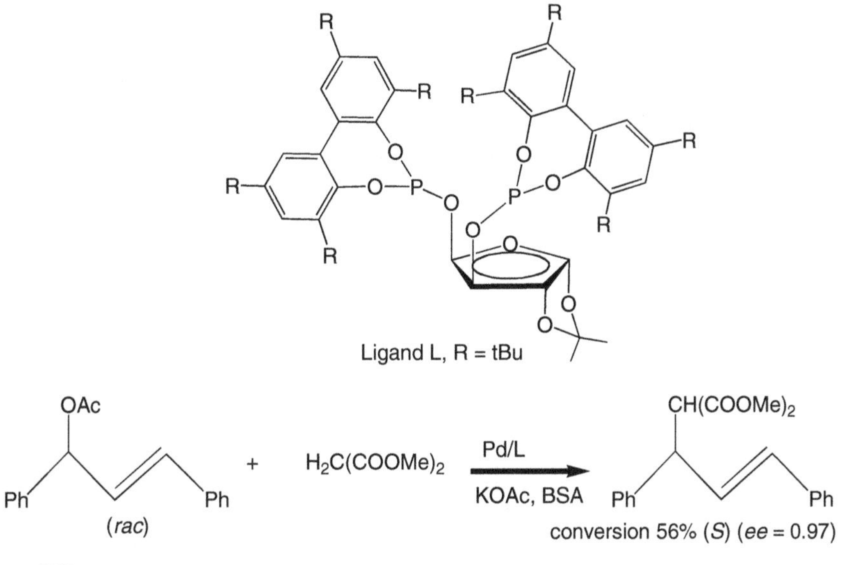

Ligand L, R = tBu

Figure 6.72
Coupling reaction in the alkylation of (rac)-3-acetoxy-1,3-diphenyl-1-propene by dimethylmalonate under basic conditions.

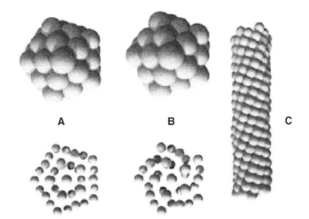

Figure 6.73

Achiral and chiral structural motifs for nanoscale metal structures. A represents an achiral 39-atom structure. B represents the chiral 39-atom structure of D5 symmetry. C illustrates a compact helical nanowire (reproduced with permission from reference[157], copyright 2000, American Chemical Society).

Decomposition of the organometallic precursor [Ru(cod)(cot)] (**6.95**) (cod = 1,5-cyclopentadiene, cot = 1,3,5-cyclooctatriene) by H_2 in the presence of (*R*)-2-aminobutanol or optically-active oxazolines leads to ruthenium nanoparticles. Their properties as a catalyst for the hydrogenation of unsaturated systems have been studied, and show a weak asymmetric induction.

Schaaff and Whetten[157] showed that gold clusters produced by the reduction of Au(I) SG polymers (HSG) = glutathione) and separated by electrophoresis, lead to compounds whose chiroptic properties are intrinsic to the metal cluster. They attributed the Cotton effects observed for energies of 1–6 eV to the structure of the gold cluster itself, suggesting a D_5 or helicoidal symmetry (Figure 6.73).

Silver nanoclusters showing Cotton effects have also been prepared by reduction of Ag(I) salts complexed by DNA.[158] It seems that the formation of optically-active nanoparticles and/or nanoclusters under the influence of chiral ligands depends strongly on the nature and size of the objects under consideration. So that a chiral organization particular to the metal atoms can exist, it is above all necessary that the number of atoms is not so high that the forces that assure the cohesion of the crystal dominate and reorganize the cluster into an achiral structure. This approach to surface chirality is rapidly evolving, and we have only given here a few examples that show the richness and new possibilities offered by this chemistry.[159]

6.7 Summary

Throughout this chapter we have presented some examples of chiral enantiopure molecular materials. We have shown that chirality in molecular materials is not only a particular property governing the rotation sense of polarized light, but it also influences the solid state organization of materials and consequently their physical properties. Multifunctional enantiopure materials can also be at the forefront of such new phenomena as magnetochiral anisotropy or electric magnetochiral anisotropy.

References

[1] P. Cintas, *Angew. Chem., Int. Ed.* **2007**, *46*, 4016.

[2] J. Jacques, A. Collet and S. H. Wilen, *Enantiomers, Racemates and Resolution*, John Wiley and Sons, Inc., New York, **1981**.

[3] G. V. Shilov, N. S. Ovanesyan, N. A. Sanina, L. O. Atovmyan and M. Grusiel, *Russ. J. Coord. Chem.* **2001**, *27*, 567.

[4] R. Clement, S. Decurtins, M. Gruselle and C. Train, *Monash. Chem.* **2003**, *134*, 117.

[5] G. L. J. A. Rikken and E. Raupach, *Nature* **1997**, *390*, 493.

[6] G. L. Rikken, J. Folling and P. Wyder, *Phys. Rev. Lett.* **2001**, *87*, 236602.

[7] A. Dolbecq, M. Fourmigue, P. Batail and C. Coulon, *Chem. Mater.* **1994**, *6*, 1413.

[8] E. Coronado, J. R. Galan-Mascaros, C. J. Gomez-Garcia, A. Murcia-Martinez and E. Canadell, *Inorg. Chem.* **2004**, *43*, 8072.

[9] E. Coronado, S. Curreli, C. Gimenez-Saiz, C. J. Gomez-Garcia and J. Roth, *Synth. Met.* **2005**, *154*, 241.

[10] J. Gomez-Garcia Carlos, E. Coronado, S. Curreli, C. Gimenez-Saiz, P. Deplano, L. Mercuri Maria, L. Pilia, A. Serpe, C. Faulmann, and E. Canadell, *Chem. Commun.* **2006**, 4931.

[11] G. L. J. A. Rikken, C. Strohm and P. Wyder, *Phys. Rev. Lett.* **2002**, *89*, 133005.

[12] V. Krstic, S. Roth, M. Burghard, K. Kern and G. L. J. A. Rikken, *J. Chem. Phys.* **2002**, *117*, 11315.

[13] J. D. Wallis and J.-P. Griffiths, *J. Mater. Chem.* **2005**, *15*, 347.

[14] T. Ozturk, C. R. Rice and J. D. Wallis, *J. Mater. Chem.* **1995**, *5*, 1553.

[15] F. Leurquin, T. Ozturk, M. Pilkington and J. D. Wallis, *J. Chem. Soc., Perkin Trans. 1* **1997**, 3173.

[16] G. A. Horley, T. Ozturk, F. Turksoy and J. D. Wallis, *J. Chem. Soc., Perkin Trans. 1* **1998**, 3225.

[17] C. Rethore, M. Fourmigue and N. Avarvari, *Tetrahedron* **2005**, *61*, 10935.

[18] J. D. Wallis, A. Karrer and J. D. Dunitz, *Helv. Chim. Acta* **1986**, *69*, 69.

[19] C. Rethore, N. Avarvari, E. Canadell, P. Auban-Senzier and M. Fourmigue, *J. Am. Chem. Soc.* **2005**, *127*, 5748.

[20] C. Rethore, M. Fourmigue and N. Avarvari, *Chem. Commun.* **2004**, 1384.

[21] E. Coronado, J. R. Galan-Mascaros, C. J. Gomez-Garcia and V. Laukhin, *Nature* **2000**, *408*, 447.

[22] E. I. Zhilyaeva, G. V. Shilov, O. A. Bogdanova, R. N. Lyubovskaya, R. B. Lyubovskii, N. S. Ovanesyan, S. M. Aldoshin, C. Train, M. Gruselle, *Synth. Met.* **2005**, *148*, 251.

[23] E. I. Zhilyaeva, G. V. Shilov, O. A. Bogdanova, R. N. Lyubovskaya, R. B. Lyubovskii, N. S. Ovanesyan, S. M. Aldoshin, C. Train and M. Gruselle, *Mater. Sci.* **2005**, *22*, 565.

[24] C. J. Gomez-Garcia, E. Coronado, S. Curreli, C. Gimenez-Saiz, P. Deplano, M. L. Mercuri, L. Pilia, A. Serpe, C. Faulmann and E. Canadell, *Chem. Commun.* **2006**, 4931.

[25] E. Coronado, R. Galan-Mascaros Jose, J. Gomez-Garcia Carlos, A. Murcia-Martinez and E. Canadell, *Inorg. Chem.* **2004**, *43*, 8072.

[26] A. M. Giroud-Godquin and P. Maitlis, *Angew. Chem., Int. Ed.* **1991**, 370.

[27] P. Espinet, M. A. Esteruelas, L. A. Oro, J. L. Serrano and E. Sola, *Coord. Chem. Rev.* **1992**, 215.

[28] R. Eelkema and B. L. Feringa, *Org. Biomol. Chem.* **2006**, *4*, 3729.

[29] F. P. Fanizzi, V. Alicino, C. Cardellicchio, P. Tortorella and J. P. Rourke, *Chem. Commun.* **2000**, 673.

[30] P. Espinet, J. Etxebarria, M. Marcos, J. Perez, A. Remon and J. L. Serrano, *Angew. Chem., Int. Ed.* **1989**, *101*, 1076.

[31] M. J. Baena, J. Barbera, P. Espinet, A. Ezcurra, M. B. Ros and J. L. Serrano, *J. Am. Chem. Soc.* **1994**, *116*, 1899.

[32] M. J. Baena, P. Espinet, M. B. Ros and J. L. Serrano, *J. Mater. Chem.* **1996**, *6*, 1291.

[33] J. Buey and P. Espinet, *J. Organomet. Chem.* **1996**, *507*, 137.

[34] J. Buey, P. Espinet, H.-S. Kitzerow and J. Strauss, *Chem. Commun.* **1999**, 441.

[35] G. Heppke, D. Lötzsch and F. Oestreicher, *Z. Naturforsch, Teil A* **1986**, 1214.

[36] P. Espinet, E. Garcia-Orodea and J. A. Miguel, *Chem. Mater.* **2004**, *16*, 551.

[37] M. Lehmann, T. Sierra, J. Barbera, J. L. Serrano and R. Parker, *J. Mater. Chem.* **2002**, *12*, 1342.

[38] K. Meyer, K. Siemensmeyer, K.-H. Etzbach and P. Schuhmacher, *Vol. WO97/00600*, **1995**.

[39] K. Okamoto, Y. Matsuoka, N. Wakabayashi, A. Yamagishi and N. Hoshino, *Chem. Commun.* **2002**, 282.

[40] N. Hoshino, Y. Matsuoka, K. Okamoto and A. Yamagishi, *J. Am. Chem. Soc.* **2003**, *125*, 1718.

[41] J. Yoshida, H. Sato, A. Yamagishi and N. Hoshino, *J. Am. Chem. Soc.* **2005**, *127*, 8453.

[42] Y. Matsuoka, H. Sato, A. Yamagishi, K. Okamoto and N. Hoshino, *Chem. Mater.* **2005**, *17*, 4910.

[43] T. Mitsuoka, H. Sato, J. Yoshida, A. Yamagishi and Y. Einaga, *Chem. Mater.* **2006**, *18*, 3442.

[44] N. Kakegawa, N. Hoshino, Y. Matsuoka, N. Wakabayashi, S.-i. Nishimura and A. Yamagishi, *Chem. Commun.* **2005**, 2375.

[45] S. T. Trzaska, H.-F. Hsu and T. Swager, *J. Am. Chem. Soc.* **1999**, *121*, 4544.

[46] J. Simon and P. Bassoul, *Design of Molecular Materials*, John Wiley & Sons, Ltd, Chichester, **2000**.

[47] K. Ohta, R. Higashi, M. Ikejima, I. Yamamoto and N. Kobayashi, *J. Mater. Chem.* **1998**, *8*, 1979.

[48] C. F. van Nostrum, A. W. Bosman, G. H. Gelinck, P. G. Schouten, J. M. Warman, A. P. M. Kentgens, M. A. C. Devillers, A. Meijerink, S. J. Picken, *et al.*, *Chem. Eur. J.* **1995**, *1*, 171.

[49] J. M. Fox, T. J. Katz, S. Van Elshocht, T. Verbiest, M. Kauranen, A. Persoons, T. Thongpanchang, T. Krauss and L. Brus, *J. Am. Chem. Soc.* **1999**, *121*, 3453.

[50] N. Kobayashi, *Chem. Commun.* **1998**, 487.

[51] N. Kobayashi, R. Higashi, B. C. Titeca, F. Lamote and A. Ceulemans, *J. Am. Chem. Soc.* **1999**, *121*, 12018.

[52] N. Kobayashi and T. Nonomura, *Tetrahedron Lett.* **2002**, *43*, 4253.

[53] M. Kimura, T. Kuroda, K. Ohta, K. Hanabusa, H. Shirai and N. Kobayashi, *Langmuir* **2003**, *19*, 4825.

[54] M. Kimura, M. Sano, T. Muto, K. Hanabusa, H. Shirai and N. Kobayashi, *Macromolecules* **1999**, *32*, 7951.

[55] T. Ezuhara, K. Endo and Y. Aoyama, *J. Am. Chem. Soc.* **1999**, *121*, 3279.

[56] A. Jouaiti, M. W. Hosseini and N. Kyritsakas, *Chem. Commun.* **2002**, 1898.

[57] Y. Cui, H. L. Ngo and W. Lin, *Chem. Commun.* **2003**, 1388.

[58] A. Jouaiti, M. W. Hosseini, N. Kyritsakas, P. Grosshans and J.-M. Planeix, *Chem. Commun.* **2006**, 3078.

[59] Y. Cui, H. L. Ngo, P. S. White and W. Lin, *Chem. Commun.* **2003**, 994.

[60] Y. Cui, H. L. Ngo, P. S. White and W. Lin, *Inorg. Chem.* **2003**, *42*, 652.

[61] S. Bernhard, K. Takada, D. J. Diaz, H. D. Abruna and H. Murner, *J. Am. Chem. Soc.* **2001**, *123*, 10265.

[62] S. Bernhard, J. I. Goldsmith, K. Takada and H. D. Abruna, *Inorg. Chem.* **2003**, *42*, 4389.

[63] J. A. Barron, S. Glazier, S. Bernhard, K. Takada, P. L. Houston and H. D. Abruna, *Inorg. Chem.* **2003**, *42*, 1448.

[64] P. Hayoz, A. Von Zelewsky and H. Stoeckli-Evans, *J. Am. Chem. Soc.* **1993**, *115*, 5111.

[65] H. Muerner, P. Belser and A. von Zelewsky, *J. Am. Chem. Soc.* **1996**, *118*, 7989.

[66] N. G. Pschirer, D. M. Ciurtin, M. D. Smith, U. H. F. Bunz and H.-C. zur Loye, *Angew. Chem., Int. Ed.* **2002**, *41*, 583.

[67] J. S. Seo, D. Whang, H. Lee, S. I. Jun, J. Oh, Y. J. Jeon and K. Kim, *Nature* **2000**, *404*, 982.

[68] N. Dybtsev Danil, L. Nuzhdin Alexey, H. Chun, P. Bryliakov Konstantin, P. Talsi Evgeniy, P. Fedin Vladimir and K. Kim, *Angew Chem Int Ed Engl.* **2006**, *45*, 916.

[69] D. N. Dybtsev, M. P. Yutkin, E. V. Peresypkina, A. V. Virovets, C. Serre, G. Ferey and V. P. Fedin, *Inorg. Chem.* **2007**, *46*, 6843.

[70] D. Bradshaw, T. J. Prior, E. J. Cussen, J. B. Claridge and M. J. Rosseinsky, *J. Am. Chem. Soc.* **2004**, *126*, 6106.

[71] J. Becker and J., M. Gagné, R., *Organometallics* **2003**, 4984.

[72] S. M. Voshell and M. R. Gagne, *Organometallics* **2005**, *24*, 6338.

[73] S. Thushari, J. A. K. Cha, H. H.-Y. Sung, S. S.-Y. Chui, A. L.-F. Leung, Y. Yu-Fong and I. D. Williams, *Chem. Commun.* **2005**, 5515.

[74] C.-D. Wu, C.-Z. Lu, S.-F. Lu, H.-H. Zhuang and J.-S. Huang, *Dalton Trans.* **2003**, 3192.

[75] C.-D. Wu, A. Hu, L. Zhang and W. Lin, *J. Am. Chem. Soc.* **2005**, *127*, 8940.

[76] V. L. Pecoraro, J. J. Bodwin and A. D. Cutland, *J. Solid State Chem.* **2000**, *152*, 68.

[77] G. Mezei, C. M. Zaleski and V. L. Pecoraro, *Chem. Rev.* **2007**, 4933.

[78] V. L. Pecoraro, *Inorg. Chim. Acta* **1989**, *155*, 171.

[79] M. S. Lah and V. L. Pecoraro, *J. Am. Chem. Soc.* **1989**, *111*, 7258.

[80] J. J. Bodwin and V. L. Pecoraro, *Inorg. Chem.* **2000**, *39*, 3434.

[81] A. D. Cutland-Van Noord, J. W. Kampf and V. L. Pecoraro, *Angew. Chem., Int. Ed.* **2002**, *41*, 4667.

[82] M. Brissard, M. Gruselle, B. Malezieux, R. Thouvenot, C. Guyard-Duhayon and O. Convert, *Eur. J. Inorg. Chem.* **2001**, 1745.

[83] M. Brissard, H. Amouri, M. Gruselle and R. Thouvenot, *CR Chimie* **2002**, *5*, 53.

[84] O. Khan, *Molecular Magnetism*, Wiley-VCH Verlag GmbH, Weinhem, **1993**.

[85] R. Andres, M. Gruselle, B. Malezieux, M. Verdaguer and J. Vaissermann, *Inorg. Chem.* **1999**, *38*, 4637.

[86] V. A. Markelov, M. A. Norikov and A. A. Turkin, *JETP Lett.* **1977**, *25*, 378.

[87] N. B. Baranova and B. Y. Zeldovich, *Mol. Phys.* **1979**, *38*,1085.

[88] G. Wagniere and A. Meier, *Chem. Phys. Lett.* **1982**, *93*, 1085.

[89] L. D. Barron and J. Vrbancich, *J. Mol. Phys.* **1984**, *51*, 715.

[90] P. Kliendienst and G. Wagnière, *Chem. Phys. Lett.* **1998**, *288*, 89.

[91] G. L. Rikken and E. Raupach, *Phys. Rev. E* **1998**, 5081.

[92] E. Raupach, G. L. J. A. Rikken, C. Train and B. Malezieux, *Chem. Phys.* **2000**, *261*, 373.

[93] L. D. Barron, *Science* **1994**, 1491.

[94] C. M. Zaleski, E. C. Depperman, J. W. Kampf, M. L. Kirk and V. L. Pecoraro, *Inorg. Chem.* **2006**, *45*, 10022.

[95] M. Minguet, D. Luneau, E. Lhotel, V. Villar, C. Paulsen, D. B. Amabilino and J. Veciana, *Angew. Chem., Int. Ed.* **2002**, *41*, 586.

[96] J. Vidal-Gancedo, M. Minguet, D. Luneau, D. B. Amabilino and J. Veciana, *J. Phys. Chem. Solids* **2004**, *65*, 723.

[97] M. Minguet, D. Luneau, C. Paulsen, E. Lhotel, A. Gorski, J. Waluk, D. B. Amabilino and J. Veciana, *Polyhedron* **2003**, *22*, 2349.

[98] E. Coronado, C. J. Gomez-Garcia, A. Nuez, F. M. Romero and J. C. Waerenborgh, *Chem. Mater.* **2006**, *18*, 2670.

[99] E. Coronado, J. Gomez-Garcia Carlos, A. Nuez, M. Romero Francisco, E. Rusanov and H. Stoeckli-Evans, *Inorg. Chem.* **2002**, *41*, 4615.

[100] K. Inoue, H. Imai, P. S. Ghalsasi, K. Kikuchi, M. Ohba, H. Okawa and J. V. Yakhmi, *Angew. Chem., Int. Ed.* **2001**, *40*, 4242.

[101] H. Imai, K. Inoue, M. Ohba, M. Okawa and K. Kikuchi, *Synth. Met.* **2003**, 91.

[102] K. Inoue, K. Kikuchi, M. Ohba and H. Okawa, *Angew. Chem., Int. Ed.* **2003**, *42*, 4810.

[103] W. Kaneko, S. Kitagawa, and M. Ohba, *J. Am. Chem. Soc.* **2007**, *129*, 248.

[104] H.-R. Wen, C.-F. Wang, Y.-Z. Li, J.-L. Zuo, Y. Song and X.-Z. You, *Inorg. Chem.* **2006**, *45*, 7032.

[105]   E. Coronado, R. Galan-Mascaros Jose, J. Gomez-Garcia Carlos and A. Murcia-Martinez, *Chem. Eur. J.* **2006**, *12*, 3484.

[106]   A. Beghidja, P. Rabu, G. Rogez and R. Welter, *Chem. Eur. J.* **2006**, *12*, 7627.

[107]   M.-H. Zeng, B. Wang, X.-Y. Wang, W.-X. Zhang, X.-M. Chen and S. Gao, *Inorg. Chem.* **2006**, *45*, 7069.

[108]   A. Beghidja, G. Rogez, P. Rabu, R. Welter, M. Drillon, *J. Mater. Chem.* **2006**, *16*, 2715.

[109]   H. Tamaki, Z. J. Zhong, N. Matsumoto, S. Kida, M. Koikawa, N. Achiwa, Y. Hashimoto and H. Okawa, *J. Am. Chem. Soc.* **1992**, 6974.

[110]   H. Tamaki, M. Mitsumi, K. Nakamura, M. Matsumoto, S. Kida, H. Okawa and S. Lijima, *Mol. Cryst. Liq. Cryst.* **1993**, *233*, 257.

[111]   S. Decurtins, H. W. Schmalle, P. Schneuwly and H. R. Oswald, *Inorg. Chem.* **1993**, *32*, 1888.

[112]   S. Decurtins, H. W. Schmalle, P. Schneuwly, J. Ensling and P. Guetlich, *J. Am. Chem. Soc.* **1994**, *116*, 9521.

[113]   M. Gruselle, C. Train, K. Boubekeur, P. Gredin and N. Ovanesyan, *Coord. Chem. Rev.* **2006**, *250*, 2491.

[114]   S. Decurtins, H. W. Schmalle, P. Schneuwly, R. Pellaux and J. Ensling, *Mol. Cryst. Liq. Cryst.* **1995**, *273*, 167.

[115]   R. Andres, M. Brissard, M. Gruselle, C. Train, J. Vaissermann, B. Malezieux, J.-P. Jamet and M. Verdaguer, *Inorg. Chem.* **2001**, *40*, 4633.

[116]   L. O. Atovmyan, G. E. Shilov, R. N. Lubovskaya, E. I. Zhilyaeva, N. S. Ovanesyan, S. I. Pirumova, I. G. Gusakovskaya and Y. G. Morozov, *JETP Lett.* **1993**, *58*, 766.

[117]   N. S. Ovanesyan, V. D. Makhaev, S. M. Aldoshin, P. Gredin, K. Boubekeur, C. Train and M. Gruselle, *Dalton Trans.* **2005**, 3101.

[118]   S. Decurtins, H. W. Schmalle, H. R. Oswald, A. Linden, J. Ensling and P. Guetlich, *Inorg. Chim. Acta* **1994**, 65.

[119]   G. V. Shilov, L. O. Atovmyan, N. S. Ovanesyan, A. A. Pyalling and L. Bottyan, *Russ. J. Coord. Chem.* **1995**, 288.

[120]   F. Pointillart, C. Train, K. Boubekeur, M. Gruselle and M. Verdaguer, *Tetrahedron: Asymmetry* **2006**, *17*, 1937.

[121]   F. Pointillart, C. Train, M. Gruselle, F. Villain, H. W. Schmalle, D. Talbot, P. Gredin, S. Decurtins and M. Verdaguer, *Chem. Mater.* **2004**, *16*, 832.

[122]   B. Malezieux, R. Andres, M. Brissard, M. Gruselle, C. Train, P. Herson, L. L. Troitskaya, V. I. Sokolov, S. T. Ovseenko, T. V. Demeschik, N. S. Ovanesyan and I. A. Mamed'yarova, *J. Organomet. Chem.* **2001**, *637–639*, 182.

[123]   M. Gruselle, R. Thouvenot, B. Malezieux, C. Train, P. Gredin, T. V. Demeschik, L. L. Troitskaya and V. I. Sokolov, *Chem. Eur. J.* **2004**, *10*, 4763.

[124]   M. Jacoby, *Chem. Eng. News* **2002**, 43.

[125]   S. M. Barlow and R. Raval, *Surf. Sci. Rep.* **2003**, *50*, 201.

[126]   V. Humblot, S. M. Barlow and R. Raval, *Prog. Surf. Sci.* **2004**, *76*, 1.

[127]   S. De Feyter and F. C. De Schryver, *Chem. Soc. Rev.* **2003**, *32*, 139.

[128]   A. Baiker, *Catal. Today* **2005**, *100*, 159.

[129]   K.-H. Ernst, *Top. Curr. Chem.* **2006**, *265*, 209.

[130]   J. V. Barth, J. Weckesser, G. Trimarchi, M. Vladimirova, A. De Vita, C. Cai, H. Brune, P. Guenter and K. Kern, *J. Am. Chem. Soc.* **2002**, *124*, 7991.

[131]   J. Weckesser, A. De Vita, V. Barth Johannes, C. Cai and K. Kern, *Phys. Rev. Lett.* **2001**, 8709.

[132]   K. Bohringer, K. Morgenstern, W. D. Schneider and R. Berndt, *Angew. Chem., Int. Ed.* **1999**, 821.

[133]   K. Bohringer, K. Morgenstern, W. D. Schneider, R. Berndt, F. Mauri, A. De Vita and R. Car, *Phys. Rev. Lett.* **1999**, 324.

[134]   P. Messina, A. Dmitriev, N. Lin, H. Spillmann, M. Abel, V. Barth Johannes and K. Kern, *J. Am. Chem. Soc.* **2002**, *124*, 14000.

[135] A. Dmitriev, H. Spillmann, N. Lin, V. Barth Johannes and K. Kern, *Angew. Chem., Int. Ed.* **2003**, *42*, 2670.

[136] H. Spillmann, A. Dmitriev, N. Lin, P. Messina, V. Barth Johannes and K. Kern, *J. Am. Chem. Soc.* **2003**, *125*, 10725.

[137] B. Ohtani, A. Shintani and K. Uosaki, *J. Am. Chem. Soc.* **1999**, *121*, 6515.

[138] M. O. Lorenzo, C. J. Baddeley, C. Muryn and R. Raval, *Nature* **2000**, *404*, 376.

[139] R. Raval, *Nature* **2003**, *425*, 463.

[140] V. Humblot, O. Lorenzo Maria, J. Baddeley Christopher, S. Haq and R. Raval, *J. Am. Chem. Soc.* **2004**, *126*, 6460.

[141] S. Romer, B. Behzadi, R. Fasel and K.-H. Ernst, *Chem. Eur. J.* **2005**, *11*, 4149.

[142] M. Parschau, B. Behzadi, S. Romer and K.-H. Ernst, *Surf. Interface Anal.* **2006**, *38*, 1607.

[143] M. Parschau, S. Romer and K.-H. Ernst, *J. Am. Chem. Soc.* **2004**, *126*, 15398.

[144] H. U. Blaser, *Tetrahedron: Asymmetry* **1991**, 843.

[145] K. Mori, M. Oshiba, T. Hara, T. Mizugaki, K. Ebitani and K. Kaneda, *New J. Chem.* **2006**, *30*, 44.

[146] V. Humblot, S. Haq, C. Muryn, W. A. Hofer and R. Raval, *J. Am. Chem. Soc.* **2002**, *124*, 503.

[147] T. E. Jones and C. J. Baddeley, *J. Mol. Catal.* **2004**, *216*, 223.

[148] R. Fasel, J. Wider, C. Quitmann, K.-H. Ernst and T. Greber, *Angew. Chem., Int. Ed.* **2004**, *43*, 2853.

[149] V. Humblot and R. Raval, *Appl. Surf. Sci.* **2005**, *241*, 150.

[150] A. Kühnle, L. T. R., B. Hammer and F. Besenbacher, *Nature* **2002**, *415*, 891.

[151] D. I. Sayago, M. Polcik, G. Nisbet, C. L. A. Lamont and D. P. Woodruff, *Surf. Sci.* **2005**, *590*, 76.

[152] S. M. Barlow, S. Louafi, D. Le Roux, J. Williams, C. Muryn, S. Haq and R. Raval, *Surf. Sci.* **2005**, *590*, 243.

[153] C. F. McFadden, P. S. Cremer and A. J. Gellman, *Langmuir* **1996**, *12*, 2483.

[154] R. Widmer, F.-J. Haug, P. Ruffieux, O. Groening, M. Bielmann, P. Groening and R. Fasel, *J. Am. Chem. Soc.* **2006**, *128*, 14103.

[155] M. Gomez, K. Philippot, V. Colliere, P. Lecante, G. Muller and B. Chaudret, *New J. Chem.* **2003**, *27*, 114.

[156] S. Jansat, M. Gomez, K. Philippot, G. Muller, E. Guiu, C. Claver, S. Castillon and B. Chaudret, *J. Am. Chem. Soc.* **2004**, *126*, 1592.

[157] T. G. Schaaff and R. L. Whetten, *J. Phys. Chem. B* **2000**, *104*, 2630.

[158] J. T. Petty, J. Zheng, N. V. Hud and R. M. Dickson, *J. Am. Chem. Soc.* **2004**, *126*, 5207.

[159] H. Behar-Levy, O. Neumann, R. Naaman and D. Avnir, *Adv. Mater.* **2007**, *19*, 1207.

Index

Note: page numbers may include both text and figures, except for those in italics which refer only to figures.